Installing, Troubleshooting, and Repairing Wireless Networks

Installing, Troubleshooting, and Repairing Wireless Networks

Jim Aspinwall

McGraw-Hill
New York Chicago San Francisco Lisbon
London Madrid Mexico City Milan New Delhi
San Juan Seoul Singapore Sydney Toronto

The McGraw-Hill Companies

Cataloging-in-Publication Data is on file with the Library of Congress

1 2 3 4 5 6 7 8 9 0 DOC/DOC 0 9 8 7 6 5 4 3

P/N 141071-6
PART OF
ISBN 0-07-141070-8

The sponsoring editor for this book was Judy Bass and the production supervisor was Sherri Souffrance. It was set in Century Schoolbook by Patricia Wallenburg.

Printed and bound by RR Donnelley.

This book was printed on recycled, acid-free paper containing a minimum of 50% recycled, de-inked fiber.

DEDICATION

Most authors select one or a few people that have inspired them through their work—and for those few special people who have inspired me there are many, many more who have fueled their inspirations and ability to inspire. I cannot limit myself to the select few without calling to mind the many—by name or your inspirational efforts.

Events of the past 2–3 years have provided the truly exceptional opportunity and pleasure of working with so *many* wonderful people applying themselves in many different fields and ways towards worthy causes. Just as I feel strongly about acknowledging the work of the people behind projects such as this, I feel moved to elation and tears by people applying themselves towards the basics of life that make it possible to write, produce and ultimately read the work we produce.

There is so much that touches us one way or the other—and it can affect us and those around us deeply and most importantly—and I feel it deserves some thought and taking advantage of an opportunity for a call to action. To that, my modest words to recognize and apply positive energy to all efforts of awareness, enlightenment, encouragement, education and action. Technology is nothing without the people we share it with.

First and perhaps specifically—to the literally thousands and thousands of people who have given incredible amounts of time and physical effort participating or in support of charitable efforts to raise funds for those baffling diseases we have yet to solve. Almost anyone can fix myriad computer problems—it takes thousands, perhaps millions of us together working towards treatment and cures for the many cancers and disorders that alter our lives or the lives of those we know and love in some way, and too all to many who unfortunately pay the ultimate sacrifice. The monsters must be conquered.

And so to the walkers, crews and volunteers of the Avon breast cancer fundraising efforts—we know the love, the work, the tears—eventually we will know the cure. We do not want any more of our families or friends to know this monster. "Just a little bit farther" my friends!

To the efforts of the participants and teams contributing to leukemia and lymphoma. One of my "fan club" of people frequently asking for help with her computer was recently diagnosed with leukemia—I cannot cure her disease but I hope someone can so that she and others are *able* to continue to experience and accomplish computer challenges, and more importantly the essential qualities a long healthy life has to offer!

To those who work towards detecting and solving diabetes—some of my very best mentors have been affected and I want them and others to be well and mentoring others.

To mentors and teachers—the ability and dedication to share information to enrich our lives through raising interest and improving abilities is so special indeed. Hopefully you are inspired directly or indirectly by the lives you help move forward.

To our audience—those new to my work and those who make up the market and inspiration to produce such works. I would have no reason to do this if it were not for you.

To my wonderful wife Kathy—an all too frequent "author's widow" through the creation of this work and my hobbies, and a two-time Avon walker and crew member who exposed me to the most awesome opportunities for awareness and inspiration I may ever know. I love you.

"Good job. Keep going!"

CONTENTS

FOREWORD

Wi-Fi or 802.11b is being heralded as the "next big thing." It is being used to create "hotspots" or points of wireless access to the Internet and beyond in airports, hotels, coffee shops and other places where the mobile workforce congregates.

Wi-Fi is also being installed in more homes and offices than ever before. The Wi-Fi association states that in 2002 more than thirty million Wi-Fi devices were sold at retail and in 2003 this number could more than double. Notebook computer companies are building Wi-Fi capability into most of their products and the cost of the infrastructure (access points) as well as PC card devices has plummeted.

Wi-Fi proponents believe that Wi-Fi connections will become nearly ubiquitous over the next few years with both "for fee" and free access to the Internet becoming so prevalent that the need for wide-area wireless data systems will be compromised and all of us will be able to wirelessly connect via this technology almost anywhere. New players are emerging almost daily. The latest, Cometa, is a joint venture between AT&T, Intel, and several investment companies. Cometa's goal it is to install and operate 20,000 hotspots in the top fifty cities in the United States. Today there are about 5,000 hotspots in operation—adding 20,000 more is a big undertaking.

But is Wi-Fi really easy to use? Is it secure? Will notebook and PDA users flock to hotspots? At the moment, there are more questions than answers in the Wi-Fi world. A newer, faster version of Wi-Fi is being embraced by the consumer electronics industry to be used for distribution of video and audio content in our homes. Computer makers have decided to hedge their bets by building both versions of Wi-Fi into their products starting in 2003.

Jim Aspinwall is a "hands on" person. We have worked together installing radio systems on mountaintop radio sites and we have spent many hours discussing the issues surrounding the convergence of computing and communications. Jim is particularly well qualified to introduce you to the world of Wi-Fi and draws on his experience in both the computer and communications fields as he explains this complex topic in simple, easy-to-understand terms. I believe that you will find the following pages to be a well written, valuable source of information.

Andrew M. Seybold
Editor-in-Chief
Forbes/Andrew Seybold's Wireless Outlook

ACKNOWLEDGMENTS

No book can be effective without information being shared by some, amplified by others, and absorbed by many. This is my fourth work in 12 years—meager by the standards of those who are able to churn thousands of pages each year for us to enjoy—regardless a labor of enjoyment and enlightenment.

There are many to acknowledge as such an effort is conceived, created, implemented, and distributed. Writing a book may be the easiest part. Producing one that delivers useful information, looks good, and makes sense is the task of many people behind the scenes. In this case there is no question where to start in acknowledging and appreciating those involved.

Judy—my beloved acquisitions editor. She got stuck with me and "IRQ, DMA, & I/O" back in 1995 and encouraged me through a lot of interesting work since that eventually led to the conception of this project. I am forever grateful!

Patty—our typesetter on this project. She makes my words look like a real book.

The production staff at McGraw-Hill. I do not know you all by name or the work you do, but thank you very much for doing it!

Tim Pozar—RF engineering genius. Years of experience and a balanced nature towards real world applications of RF and technology are rare assets. Making human readable sense of the FCC rules is but one of his contributions to this field. Thank you Tim!

Andy Seybold—can insight get any better? If it's radio, Andy's done it. If it's computing, Andy's done it. To put them together and figure out the reasonable from the fantasy in economic or practical value is unique. Thank you Andy for your inspiration and good words!

BAWUG—the Bay Area Wireless User Group. I've snooped and dropped a few little "bombs" on their mailing list and attended a couple of meetings, and must acknowledge that technology is truly driven by some very sharp, wary, and forward thinking people.

Some first-class contributing hardware and software vendors. Sharon at HyperLink Technologies. The folks in marketing and PR at LinkSys, Orinoco Wireless, Amphenol-Connex, Funk Software, WiMetrics, Wild-Packets, AirMagnet, and several others.

CHAPTER 1

Wireless Essentials

This book focuses on what is commonly known as 802.11 and WiFi wireless networking technologies—their implementation and problem-solving. To set perspective, it briefly covers the wireless context, history, benefits, costs, and governmental issues related to the current state of wireless networking.

The term *wireless* is generic, and while it is typically synonymous with radio, it is not limited to radio. Wireless can also be defined as ultrasonic (sound) or infrared (light) wave communication between two devices.

When wireless is used in the context of radio wave (the portion of known spectrum between sound and light waves) communications, dozens of issues come into play—most of them regulatory and technical.

In terms of networking, wireless replaces the patch cables, patch panels, hubs, and network adapters or hard-wiring between a computer, printer, or similar device and another; or replaces larger scale common network equipment with a different style of network adapter—essentially a "data radio"—perhaps an additional external antenna and the airwaves. Of course the connection between the data radio or wireless networking adapter and an antenna involves a wire—but far less wire than dragging a cumbersome cable across the floor around your living room or a meeting room at the office, and much less than trying to get wired to a network connection a few miles away.

When you think of wireless networking, visions of connecting to the Internet, or a home or office network with your personal computer (PC) or personal digital assistant (PDA) as readily as using a cellular phone come to mind. While desirable, this ideal situation is not as easily achievable or reliable as the creation and maintenance of subscriber-based cellular phone services—backed by major corporations with billions of dollars and a recurring revenue stream. Some of these complexities will become obvious as we delve into this chapter.

Wireless networking also brings up security issues—your data are no longer safely tucked within the confines of a set of wires over which you have control—because radio signals have few tangible or controllable boundaries. It is this lack of tangible, controllable boundaries that makes wireless both attractive and complex in many ways.

Wireless Defined

Wireless has been, and will be, a part of everyday life and computing in a couple of different forms. The first is infrared or light-wave communications between specific devices and a virtual port on a PC. Other implementations use radio communications—from National Semiconductor's Airport modules, which essentially create an over-radio null-modem extension between two PCs' serial ports for use with programs such as pcAnywhere or LapLink, through cellular-phone-aware modems, to a variety of 400 MHz, 900 MHz, and 2400 MHz (2.4 GHz) radio connections. All of these have in common the benefit of not having to use a variety of different connectors, adapters, and wires, to transfer data between one device and another.

Wireless Equals Infrared

Infrared (IR) or light-wave communications is most prevalent in remote control devices for televisions, VCRs, and stereos, and has found its way into computer keyboards, pointing devices, printers, and PDAs. IR devices are good miniature models for studying some aspects of radio frequency wave communications—particularly for higher frequency applications such as cellular phones, Bluetooth, and 802.11 wireless networking devices.

Light waves are a strictly line-of-sight means of getting information from one device to another. This means that both devices must have a clear, unobstructed visual path between them. Obstructions may be as obvious as the back of your favorite chair blocking the (invisible) light beam from reaching the TV, or as subtle and baffling as the smoked-glass door on your stereo cabinet. A less obvious form of interference with IR remote control devices is bright artificial or natural sunlight overwhelming the detector in the appliance you are controlling, making it difficult or impossible to determine what command you are sending.

There are some work-arounds to the challenges presented to line-of-sight communications. You can, of course, play tricks with mirrors to "bend" light around corners, which is a valuable example of the characteristics of radio wave behavior. You may also place detectors and retransmitters or repeaters in between the devices to extend or bend the signals around corners too—which is exactly the service network repeaters and back-to-back radio links do for radio-based systems. If you could see IR light waves and play with them like you can laser beams and simple flashlights, you could study these constraints quite easily. Your next best substitute would be to use a theatrical fogger or some dry ice and flashlights or a laser pointer to play with light dispersion, distortion, reflection, and refraction to get an idea of how radio waves work.

IR devices are easy implement. Light waves are not regulated by any institution or government interests, and the applications for them are a lot simpler than radio systems and high-speed networking. We cannot neglect that visible light communications—perhaps that of smoke signals and certainly that of naval intership communications by light beacons—share the line-of-sight, low-cost, ease of implementation and limited benefits of IR communications.

Wireless Equals Radio Frequencies

Moving up in capabilities, benefits, and complexity, but also down in wavelength, we come to the radio spectrum of wireless communications. Using radio in general—and for high-speed data specifically—is more complex for several reasons:

- The radio frequency (RF) spectrum is regulated (in the United States) by the Federal Communications Commission (FCC). Typically in other countries the postal or telecommunications agency is charged with regulating radio use, and most countries also participate in global radio consortiums. What can and cannot be placed on and make use of specific portions of RF spectrum, and resolving interference and jamming disputes, is the domain of these agencies.
- Exposure to RF waves is regulated (in the United States) by the FCC and Occupational Safety and Health Administration (OSHA) guidelines (and is the subject of much environmental debate and cellular telephone safety issues). Although wireless networking products emit very low levels of RF power, they are often placed in the company of other much higher powered RF devices that have known detrimental effects on human tissue. Simply climb-

ing radio towers to install radio equipment is also covered by safety regulations.

- Radio waves require sometimes complex and definitely precise antenna construction to emit and receive the signals.
- Placement of RF antennas external to the devices of interest, for best performance, is contrary to desired discretion and concealment and requires special cable to make a connection between the radio and the antenna.
- Radio waves are subject to myriad and different environmental and atmospheric conditions, both at very low frequencies (such as the AM broadcast band from 530–1750 KHz) and very high frequencies (such as the cellular and microwave bands above 800 MHz) that can affect their range (through absorption and reflection) and alter their distribution (through atmospheric reflection and refraction).
- Radio waves are also subject to path alteration from natural and man-made terrain and structural elements—from tree leaves to metal building frames and office desks and chairs.
- Radio waves require stringent frequency controls to be transmitted and received properly, whereas light waves need not be as precise for their given applications.
- Controlling the path of radio waves—getting a desired signal where you want it and not where you don't—is technically difficult.
- Keeping interfering signals away from, or from affecting the sensitive radio receiver, is a complex technical issue. Since many wireless networking systems are located near much higher powered RF devices, interference or signal swamping is very much an issue.
- Conveying data over radios involves complex modulation or embedding schemes to be received properly.
- Radio engineering, development, and implementation is highly specialized and, thus, by nature expensive.
- In general, good radio systems are not inexpensive to build or package.

There are good reasons to list all of these issues because they are very real—though sometimes intangible—elements to the overall success of commercial (e.g., corporate, service provider) and noncommercial (e.g., home, general public) wireless networking. Responsible networking equipment manufacturers must take into consideration all of these elements and adhere to most of them in order to design, make, sell, and support their products—legally, if not also ethically. A responsible wireless networking service provider—whether a grassroots "Internet everywhere" initiative or a commercial venture selling wireless access to the Internet—must consider most of these elements and adhere to many of them. A responsible and reasonable wireless networking system implementer and end-user should know about all of these elements and comply with many of them.

For all of these parties, being aware of, understanding, and being able to deal with the absolute technical details, as well as the magic of RF transmission and reception, is essential to building, using, and maintaining a successful wireless network.

Federal Regulations—FCC. If you are into technical details, find a friendly RF engineer to accompany you while browsing the FCC's Web site and specifically trying to "grock" Part 15 of Title 47 of the Code of Federal Regulations (CFR)—the part of the Federal regulations the FCC uses to help manage the radio spectrum and those devices used in wireless networking and other services.

When you venture into wireless networking, you are *not* alone—others are using the same spectrum you will be—and not just other wireless networking users, but devices and users of other radio services. Other services that share the wireless networking spectrum are:

- Devices that fall into Part 15 of the ISM band (2400–2483 MHz)
- Devices that fall into the U-NII band
- Industrial, Scientific, and Medical (ISM)—Part 18
- Satellite Communications—Part 25
- Broadcast Auxiliary—Part 74
- Stations in the Maritime Services—Part 80
- Aviation Services—Part 87
- Land Mobile Radio Services—Part 90
- Amateur Radio—Part 97
- Fixed Microwave Services—Part 101
- Federal Usage (NTIA/IRAC)

Most of the following sections are excerpts from Tim Pozar's *Regulations Affecting 802.11 Deployment* paper, which describes the existing rules and regulations as they pertain to the radio spectrum used by wireless networking and other radio services.

INTRODUCTION TO THE TECHNOLOGY. 802.11 is a standard group under the Institute of Electrical and Electronics Engineers, Inc. (IEEE®) that develops standards related to wireless and wired Ethernet transmission. This includes the actual Physical layer, such as 802.11a, and 802.11b modulation schemes.

802.11b is a direct sequence spread spectrum (DSSS) technology that, in the United States, occupies eleven channels that center on frequencies in the ISM band, from 2.412 to 2.462, in 5 MHz steps. The spectrum used by 802.11b is 22 MHz wide. As the channels are smaller than the occupied bandwidth, you really have only three channels (1, 6, and 11) that are usable in a small area, or else you may run into interference.

802.11a does not use direct sequence. Instead, it uses a modulation scheme called orthogonal frequency division multiplexing (OFDM). OFDM uses fifty-two 300 KHz-wide carriers grouped into one channel that is 20 MHz wide. With the slower symbol speed of OFDM and the forward error correction incorporated into 802.11a, it is more resilient to multipath interference. However, because 802.11a is at more than double the frequency of 802.11b, there is greater free space loss. 802.11a will only have about 18 percent of the signal that 802.11b will have with the same gain antennas and transmitter power.

Whereas 802.11b occupies a band known as the ISM band, 802.11a occupies a section of spectrum known as unlicensed national information infrastructure (U-NII) band. This band was approved in 1997 and was promoted by the group WINForum, which was made up of individuals and companies such as Apple Computer.

The band takes up 300 MHz of spectrum and is divided into three 100 MHz sections. The first two are next to each other, and the third is 375 MHz up from the top of the second band. The low band runs from 5.15 GHz to 5.25 GHz, the middle band runs from 5.25 GHz to 5.35 GHz, and the high band runs from 5.725 GHz to 5.825 GHz.

REGULATIONS AND LAWS THAT WILL AFFECT DEPLOYMENT OF 802.11 WIRELESS NETWORKS. The spectrum is managed by a number of different organizations. The most visible to the general public is the FCC. The FCC manages civilian and state and local government usage of the radio spectrum. You will be directly affected by this regulatory organization.

The FCC has a set of rules and regulations that define the use of spectrum, as well as policies and procedures for working with the FCC. You can read these in hard copy by ordering the *Code of Federal Regulations, Title 47* from the Government Printing Office (GPO) at: http://bookstore.gpo.gov.

Companies such as Pike and Fischer (http://www.pf.com) offer subscription services to the updated FCC regulations and other policies and proposed rules. There are also free, although slightly dated, versions of the FCC rules, such as the *Hypertext FCC Rules Project* run by Harold Hallikainen at: http://www.hallikainen.com/FccRules.

Harold's site actually indexes the GPO's on-line version of the Rules. Go directly to the GPO's on-line access of the rules at: http://www.access.gpo.gov/nara/cfr/cfr-table-search.html.

As the GPO's site points to all of the CFR, you want the section known as *Title 47—Telecommunication*.

Enforcement. The Commission has authority to investigate any user of the band. In fact, it can actually come on site and inspect the operation of the equipment:

> 15.29(a)—Any equipment or device subject to the provisions of this part, together with any certificate, notice of registration, or any technical data required to be kept on file by the operator, supplier, or party responsible for compliance of the device shall be made available for inspection by a Commission representative upon reasonable request.

At this point in time, the FCC has very limited resources for enforcement, as the trend for the last couple of decades is deregulation and reduction of staffing in the enforcement bureaus. The FCC will likely only visit you if there is a complaint. There have been rare reports of the FCC going after wireless Internet service providers (WISPs) when they interfered with Part 97 (amateur radio) users. Working with the co-users of these bands is in your best interest, as they will be the ones complaining.

The National Telecommunications and Information Administration (NTIA) works with the Interdepartmental Radio Advisory Committee (IRAC), which manages federal use of the spectrum. You likely will not hear from them unless you do something really wrong.

POWER LIMITS. Ideally, a well engineered path will have just the amount of power required to get from point "A" to point "B" with good reliability. Good engineering will limit the signal to only the area being served. This has the effect of reducing interference and providing a more efficient use of the spectrum. Using too much power will cover more area than is needed and has the potential to wreak havoc on other users of the band. As 802.11 is designed for short-range use, such as in offices and homes, it is limited to very low power.

- 802.11b—FCC 15.247

  Point-to-multipoint: You are allowed up to 30 dBm or 1 watt of transmitter power output (TPO) with a 6 dBi antenna, or 36 dBm or 4 watts of effective radiated power over an isotropic antenna (EIRP). The TPO needs to be reduced 1 dB for every dB of antenna gain over 6 dBi.

  Point-to-point: The FCC encourages directional antennas to minimize interference to other users. The FCC, in fact, is more lenient with point-to-point links by only requiring the TPO to be reduced by 1/3 of a dB instead of a full dB for point-to-multipoint.

  More specifically, for every 3 dB of antenna gain over a 6 dBi antenna, you need to reduce the TPO 1 dB below 1 watt. For example, a 24 dBi antenna is 18 dB over a 6 dBi antenna. You would have to lower a 1-watt (30 dBm) transmitter 18/3 or 6 dB to 24 dBm or 1/4 watt.

- 802.11a—FCC 15.407

  Point-to-multipoint: As described before, the U-NII band is divided into three sections. The low band runs from 5.15 GHz to 5.25 GHz and has a maximum power of 50 mW (TPO). This band is meant to be in-building only, as defined by the FCC's Rules and Regulations Part 15.407 (d) and (e):

  > (d) Any U-NII device that operates in the 5.15–5.25 GHz band shall use a transmitting antenna that is an integral part of the device.
  >
  > (e) Within the 5.15–5.25 GHz band, U-NII devices will be restricted to indoor operations to reduce any potential for harmful interference to co-channel MSS operations.

The middle band runs from 5.25 GHz to 5.35 GHz, with a maximum power limit of 250 mW. Finally, the "high" band runs from 5.725 GHz to 5.825 GHz, with a maximum transmitter power of 1 watt and antenna gain of 6 dBi or 36 dBm, or 4 watts EIRP.

Point-to-point: As with 802.11b, the FCC does give some latitude to point-to-point links in 15.407(a)(3). For the 5.725 GHz to 5.825 GHz band, the FCC allows a TPO of 1 watt and up to a 23 dBi gain antenna without reducing the TPO 1 dB for every 1 dB of gain over 23 dBi.

15.247(b)(3)(ii) does allow the use of any gain antenna for point-to-point operations without having to reduce the TPO for the 5.725 GHz to 5.825 GHz band. You should look at which part your equipment is certified under to see what restrictions you have for EIRP.

EQUIPMENT LIMITATIONS AND CERTIFICATION. Part 15 devices are designed to be installed and used by the general public. With this in mind, the Commission wants them to be as "idiot proof" as possible. It has severe limitations on what you can do with this gear. For instance, the Rules state:

> 15.203—An intentional radiator shall be designed to ensure that no antenna other than that furnished by the responsible party shall be used with the device.

A bit further, the Rules repeat the same sentiment:

> 15.204(c)—Only the antenna with which an intentional radiator is authorized may be used with the intentional radiator.

The basics of certification can be found in FCC 2.901 through 2.1093. The requirement for Part 15 devices can be found in 15.201. Equipment can be certified a couple of ways—as a component or as a "system." In the case of a component, you can have a piece of equipment known as a transmitter, an amplifier, or an antenna. All can be mixed and matched with each other. If you have equipment certified as a system, it cannot be used with other equipment. See 15.203 and 15.204.

> 15.204(b)—A transmission system consisting of an intentional radiator, an external radio frequency power amplifier, and an antenna, may be authorized, marketed, and used under this

part. **However, when a transmission system is authorized as a system, it must always be marketed as a complete system and must always be used in the configuration in which it was authorized. An external radio frequency power amplifier shall be marketed only in the system configuration with which the amplifier is authorized and shall not be marketed as a separate product.** (Author added boldface for emphasis.)

In other words, you cannot take an access point that is certified as a system and attach an antenna that is not a part of its certification. You can, however, recertify equipment. If you purchase gear on the street, there is nothing to stop you from reselling this gear at a profit or loss. In fact, you could recertify this equipment too. There is some question about whether you need approval from the manufacturer. I talked to one communications law attorney and he said approval is not needed.

Certification is an involved process and can be costly. You should contract with many of the consultants in this field for guidance.

INTERFERENCE. The labeling requirement in Part 15.19 states:

This device complies with Part 15 of the FCC Rules. Operation is subject to the following two conditions: (1) This device may not cause **harmful interference**, and (2) this device must accept any interference received, including interference that may cause undesired operation.

Description. Of course, interference is typically the state of the signal in which you are interested, while it is being destructively overpowered by a signal in which you are not interested.

The FCC has a specific definition of "harmful interference":

Part 2.1(c)—**Harmful interference**—Interference which endangers the functioning of a radio-navigation service or of other safety services or seriously degrades, obstructs, or repeatedly interrupts a radio-communication service operating in accordance with these [International Radio] Regulations.

In Part 15, it is repeated as:

> Part 15.3(m)—**Harmful interference**. Any emission, radiation, or induction that endangers the functioning of a radio navigation service or of other safety services or seriously degrades, obstructs, or repeatedly interrupts a radio-communications service operating in accordance with this chapter.

As there are other users of this band, interference will be a factor in your deployment. The 2.4 GHz band is a bit more congested than the 5.8 GHz band, but both have co-users that need to be watched.

Federal regulations—OSHA. This perhaps obscure aspect of wireless networking is very important, as more and more individuals and companies are hoping to expand their wireless operations and benefit from strategically located and high-elevation sites, where radio equipment and antennas already exist. Although the lower RF power levels of most wireless networking equipment dictate that we place more equipment closer to the users, rather than high atop mountains and buildings with vast line-of-sight views, higher elevations are used for some wireless implementations. As such, network engineers or computer technicians who once only had to worry about banging their head on the bottom of a desk to plug in an Ethernet cable, now have to be concerned about falling off ladders, rooftops, and tall pieces of steel structure to ply their trade.

I have some pet peeves about this aspect of wireless networking. One is that the Occupational Safety and Health Administration (OSHA) essentially dictates some of the tools and practices that must be used when installing wireless (or any) equipment on elevated locations—most commonly radio towers, but areas of rooftops are of concern as well. The other is that many are ignorant of or ignore the spirit, intent, and practical aspects of such regulations. We would hope that people climbing ladders, towers, and working near the edges of roofs would embrace some common sense—but then very little is common among any group of people, and sense is an intangible based on experience—and in this case, the realities of gravity and solid geometry.

I have been climbing radio towers and working on rooftops since I was 14 years old (trees and jungle gyms before then). I'm self-motivated enough by a dislike of pain and having to fix what I break to climb safely with safety equipment and a keen sense of being aware of my surroundings—miss that ledge or step, or lose a grip without a safety tether, and gravity takes over. The frailty of the human body is no match for Mother Earth or structural lumber, stee,l or concrete, or falling tools or equipment. Nor are the delicate tissues inside any match for the hundreds, thousands, or even millions of watts of RF energy emitted from commercial radio systems, FM, or TV broadcast stations.

Before 1995 or so, no one thought much about the hazards of working near radio transmitters. Oh, a few hearty souls full of bravado have claimed to "feel a little warm" when working closer to some antennas than others, and many have joked about warming their lunch in front of radar antennas, but we all climbed and worked amidst significant fields of RF radiation with little or no caution until OSHA told us how much RF energy might hurt us. It is not uncommon to have to wait until after hours or nightfall for some tower climbing operations to begin—when stations could reduce power or turn off their transmitter, or schedule less-watched times to get transmitter power reduced without impacting the economic value of a broadcast schedule. Fortunately, I've never, at least knowingly, climbed into the path of severe RF energy exposure, felt any unusual warming (it usually gets colder the farther up you climb), or had any known adverse effects from the RF I have been exposed to (though others may differ as to the state of my mind sometimes).

Until I attended an OSHA certification course, I regularly strapped on a full recreational rescue harness and used two safety lanyards—the type used by rock climbers—when climbing radio towers. While I believe for me they are as safe, more comfortable, and certainly lighter than the OSHA-required variety, to maintain OSHA compliance, I must now wear an American National Standards Institute (ANSI)-standard industrial harness that weighs twice as much, and costs three to four times as much as my gear from my favorite sporting goods outlet, REI. None of this assures that I will not drop a wrench onto one of my friends below, but they may be more assured that I will not fall on top of them, destroying thousands of dollars of someone else's equipment on the way down.

HUMAN EXPOSURE TO RADIO FREQUENCY RADIATION.[1] This book does not cover the pseudo-scientific arguments of human exposure to

[1]Parts of the discussion of FCC and OSHA regulations are excerpts from *Regulations Affecting 802.11 Deployment* by Tim Pozar, of Late Night Software and the Bay Area Wireless Users Group, pozar@lns.com.To obtain a full copy of Tim Pozar's *Regulations Affecting 802.11 Deployment* paper, visit http://www.lns.com/papers/part15/.

RF radiation. Instead, it addresses the current ANSI limits, as related to human exposure to RF fields. However, keep in mind that cellular telephone companies have run into groups that are using this pseudo-science to delay or stop deployment of cell phone installations via city and county governments.

Once 802.11 deployment becomes more popular, these groups may have an impact on your deployment. After all, they know what microwave ovens can do, and 802.11b runs at the same frequency. The FCC's concern is:

> At the present time there is no federally mandated radio frequency (RF) exposure standard. However, several nongovernment organizations, such as the American National Standards Institute (ANSI), the Institute of Electrical and Electronics Engineers Inc. (IEEE), and the National Council on Radiation Protection and Measurements (NCRP) have issued recommendations for human exposure to RF electromagnetic fields....
>
> On August 1, 1996, the Commission adopted the NCRP's recommended Maximum Permissible Exposure limits for field strength and power density for the transmitters operating at frequencies of 300 KHz to 100 GHz. In addition, the Commission adopted the specific absorption rate (SAR) limits for devices operating within close proximity to the body as specified within the ANSI/IEEE C95.1-1992 guidelines. (See Report and Order, FCC 96-326.) The Commission's requirements are detailed in Parts 1 and 2 of the FCC's Rules and Regulations [47 C.F.R. 1.1307(b), 1.1310, 2.1091, 2.1093]—from http://www.fcc.gov/oet/rfsafety.

This bulletin breaks down exposure limits for workers exposed around the equipment and for the general public. At 2.45 GHz, it is 4.08 mW/cm^2 for unlimited time exposures for workers, and 1.63 mW/cm^2 for 30 minutes for the general public. As this is energy absorbed over time, workers can raise or lower the mW/cm^2 for a controlled situation by decreasing or increasing the time exposed. It would be hard to regulate this for the public, so you should not apply this "time versus exposure" calculation for the public.

The Office of Engineering and Technology (OET) Bulletin Number 65 (August 1997), *Evaluating Compliance with FCC Guidelines for Human Exposure to Radiofrequency Electromagnetic Fields*, at http://www.fcc.gov/oet/info/documents/bulletins/#65, shows how to calculate these fields.

As an example, a near-field calculation of a 2-foot aperture dish (24 dBi) with 1/4 watt of power applied (maximum EIRP for point-to-point) has almost a 1 foot area in front of the dish that would be considered "controlled," and 2-foot area in front of the dish with limited exposure for the general public. Simply place your dishes out of the way, at above head height. The FCC has a page that covers many of these issues at: http://www.fcc.gov/oet/rfsafety.

Local regulations.[2] When installing antennas for clients, you may run into local ordinances and homeowner agreements that would prevent installations. Thanks to associations such as the Satellite Broadcasting and Communications Association (SBCA), who lobbied the FCC, the FCC has stepped in and overruled these ordinances and agreements.

For a good introduction to this topic, read Roy Trumbell's paper at: http://www.lns.com/sbe/antenna_mounts.html. This rule should only apply to broadcast signals such as TV, DBS, or MMDS. It could be argued that the provision for MMDS could cover wireless data deployment as…

> 1.4000—Restrictions impairing reception of television broadcast signals, direct broadcast satellite services, or multichannel multipoint distribution services:
>
> 1.4000(a)(1)(i)—An antenna that is: (A) Used to receive direct broadcast satellite service, including direct-to-home satellite service, or to receive or transmit fixed wireless signals via satellite, and (B) One meter or less in diameter or is located in Alaska;…
>
> 1.4000(a)(2)—For purposes of this section, "fixed wireless signals" means any commercial non-broadcast communications signals transmitted via wireless technology to and/or from a fixed customer location. Fixed wireless signals do not include, among other things, AM radio, FM radio, amateur (HAM) radio, Citizen's Band (CB) radio, and Digital Audio Radio Service (DARS) signals.

There are conditions:

> 1.400(c)—In the case of an antenna that is used to transmit fixed wireless signals, the provisions of this section shall apply only if a label is affixed to the antenna that:

---

[2] See footnote 1 on page 13.

(1) Provides adequate notice regarding potential radiofrequency safety hazards, e.g., information regarding the safe minimum separation distance required between users and transceiver antennas; and

(2) References the applicable FCC-adopted limits for radiofrequency exposure specified in 1.1310 of this chapter.

Questions such as, "Can traffic such as Multicast IP fall into these rules?" and "What percentage of traffic must be broadcast?" need to be resolved before you can use this section of the FCC rules.

- Height Limitations
  Local Ordinances: Most, if not all, cities regulate the construction of towers. There will be maximum height (e.g., 300 feet in Oakland, or 10 feet for a mast on a residence in Fremont), zoning of the antenna/tower (residential or commercial), construction (e.g., no antennas 15 feet above the tower in Oakland or 300 feet setback in Fremont), and aesthetic (e.g., what color, how hidden) regulations. Depending on these factors, you will have to jump over various hurdles with each city and installation.

  The Federal Aviation Administration (FAA) and the FCC tower registration:

  > The FAA is very concerned about airplanes bumping into objects. Part 17.7(a) of the FCC Rules and Regulations describes:
  >
  > Any construction or alteration of more than 60.96 meters (200 feet) in height above ground level at its site.

Details can also be found in the U.S. Department of Transportation Advisory Circular AC70/7460-1K. If your tower falls into this category, then it is necessary to register it with the FCC, as per Part 17.4.

Wireless Networking Radio Spectrum

Wireless devices have typically occupied five different portions of radio spectrum depending on their application and the state of technology and regulations, and not exclusively—usually other devices and services share the RF spectrum. Briefly:

- **49 MHz band:** Once used by the Airport wireless serial cable connection manufactured by National Semiconductor—now obsolete. By nature of the size and power of equipment, this band accommodates only short-range communications for small consumer devices.
- **420–450 MHz:** Typically considered the amateur radio UHF spectrum filled with repeaters, intersight links, and amateur television (ATV) signals. Home weather stations and wind sheer radar systems also use this spectrum.

  This UHF spectrum is quite popular as it offers the advantage of small equipment and antennas, reasonable station-to-station range, and easily constructed and maintained repeater systems offering a 10–50 mile range with moderate power levels. The range for low-power (100 mW to 1 W) devices usually does not exceed 1–2 miles.
- **800 and 900 MHz bands:** Mostly occupied by analog and digital cellular phone systems, this spectrum also contains many trunked two-way radio services, Nextel cellular services, high-power paging transmitters, two-way communications, and amateur radio operations (925–935 MHz). Some of this spectrum had been occupied by the now defunct Metricom wireless Internet access service. A variety of remote controls, such as garage door openers and automotive security systems, also use 900 MHz for short-burst data transmissions. (Metricom's Ricochet service has been acquired and may be redeployed in some areas.)

  This spectrum is best known for excellent building penetration at reasonable power levels, although paging transmitters typically pump 250–350 watts into high-gain antennas, making their effective radiated power as much as 3000 watts.

  Typical deployment of these high-UHF systems is more like cellular telephone systems—several lower power stations located in grid-like fashion proximate to the users of the services—which is how you get hundreds of portable cellular telephones to work so well. FCC regulations and allowable technology limit the data throughput using this spectrum to well below 64 kbps. Signal range at 100 mW power levels may be 1–5 miles, with directional antennas at 10–30 miles.

- **2.4 GHz:** The current and most prevalent 802.11b wireless networking spectrum is also occupied by a variety of medical, consumer, amateur radio, Bluetooth, and other services. The bandwidth available and technologies using 2.4 GHz allow for as good or better than wired 10BaseT Ethernet data throughput, but do not be surprised if the microwave oven in your kitchen or favorite coffee shop interrupts your surfing! With 100 mW power levels and built-in antennas, the signal range will be about a mile or so, with external directional antennas and a clear line-of-sight path up to 10 miles.
- **5 GHz:** The spectrum for emerging 802.11a wireless networking is also shared by other services. The range for 802.11a devices will be half or less than that of 802.11b 2.4 GHz devices.

There has been significant evolution of wireless technologies, and there are a lot of unseen neighbors out there, as described in the FCC regulations section. Getting along may be tough, but be assured that someone else is watching. Take this to heart not only when considering problems with wireless networking, but when considering security and reliability as well.

Summary

At this point, you may be in awe that wireless networking exists at all, when government regulations and safety considerations and getting along with everyone else on a tiny speck of radio spectrum you cannot even see or touch. Fortunately, the equipment vendors are the ones saddled with most of the responsibility to adhere to the regulations, until the product gets into your hands—then compliance and safety become your responsibility.

Compliance is easy if you follow the recommended practices the manufacturers provide with their original and add-on equipment. Stay with the defined system of equipment you purchase to be on the safe side. If you modify anything outside of the system or exceed power and radiated signal strength, you could be in violation of the regulations, or worse, cause safety and health problems for yourself or others. The following chapters on basic system components, system design, and example systems will help you create and maintain a safe, legal, and reliable wireless network.

CHAPTER 2

Wireless Network Criteria and Expectations

There are generally three well-known types or deployments of wireless networks:

- The simple local area network (LAN) that you would find at home or in a small office
- A campus or neighborhood LAN that you would find emanating from a home or central location to cover roughly a square mile or less—often called a *hotspot*, where wireless activity may be available
- A metropolitan area network covering several square miles, from which several mobile and portable users benefit

These are typically point-to-multipoint installations where one or many access points together are used to distribute a single network to multiple client systems. Lesser known, but equally useful and beneficial, are point-to-point relay systems to interconnect different networks or facilities.

Each of these types of networks may be associated with one of the following types of services:

- Personal/private use by an individual or family
- Publicly shared use by those known and familiar to the host/provider of the network
- Private network use to serve a business and its employees
- Subscription-based networks or Internet service providers (ISPs) available to anyone paying to obtain the service as you would obtain dial-up, digital subscriber line (DSL), or cable Internet access

Similar to the subscription services that make wide area access available to the general public are several growing efforts to deploy free wireless Internet services to the public in different communities—Seattle and San Francisco being among them.

The U.S. government sees wireless services as a way to solve the "last mile" problems of spreading high-speed Internet access to the general public, especially in areas where cable TV and phone service providers have not or will not deploy cable or DSL services to their subscribers because they will not recover the high costs of these services with relatively few subscribers.

Most of the issues with all of these types of wireless networks are about the same—how much signal can you get how far away, what is in the path of the signal, and how can you make the signal better? What typically differs is the type of equipment used, as well as how it is installed, configured, secured, and maintained. There will also be cost differences in the equipment and type of installation. External antennas and cabling cost extra. Mounting an antenna at home is free, but putting an access point or a wireless relay/bridge system atop a building will usually incur monthly fees.

Performance—What To Expect

The success of any network, any project for that matter, is based on expectations, perceptions, specifications, and factors, and of course actual performance—that is, does it work?

Chances are, a reasonable/feasible, properly designed and implemented wireless networking system will work flawlessly for you. So the first steps are to define and understand reasonable/feasible and properly designed, and implemented in this context.

Reasonable and feasible have both an economic and a practical aspect. The economics of wireless networking are discussed in the next section, but expect a 30–40 percent savings versus conventional wired networks. The practical aspects, including design, implementation, and maintenance have to consider several physical, logistical, and administrative aspects. Consider the following a basic reality check and checklist for your implementation:

- Do you need wireless technology?
  - Is this a permanent or temporary installation?
  - Are you unable to freely or practically access areas to string cables?
  - Are you prevented by lease, contract, or policy from running wires?
  - Will you always have control over the security and access to your cabling?
  - Do you currently have a wired network?
  - Is there an aesthetic reason to go wireless?
  - Do you need a temporary peer-to-peer setup?
  - Do you travel and need or want more than dial-up connection speeds?

- Is the site wireless-friendly?
  - Are there sources of interference that cannot be eliminated?
  - Will a wireless network system interfere with other devices?
  - Do technical or security policies preclude broadcasting your network traffic through a wireless system?
  - Does the structure facilitate wireless technology with little or no metallic obstruction?
- Can you use wireless technology?
  - What distances are you hoping to cover?
  - Do you have a line-of-sight path to all systems?
  - What data throughput speeds do you need?
  - Can you adequately secure your data over a wireless connection? Do you care?
  - Are all of your systems wireless capable—current or recent hardware, operating systems, and applications?
  - Will some of your systems still need to be wired (older technology)?
- Who will design, install, and maintain your wireless system?
  - Do you or your vendor understand and have experience doing wireless?
  - Do you or your vendor have access to analytical equipment or software tools to survey your site as part of the design phase and to troubleshoot implementation problems?
  - Will there be enough skilled resources to administer your network?
- Can you afford wireless?
  - In the simplest forms of wireless implementation, as an alternative or replacement for a wired LAN, wireless networking has significant cost advantages over wires. If you need to cover greater distances or bend around corners to get between systems, you will need intermediate sites and equipment. This topic is covered in "The Cost of Wireless" section in this chapter.

As you can see, creating a wireless network can be more involved than a jaunt to the local computer store or on-line shop, grabbing a few wireless cards and access points, and plugging things in—they just might not work. Many of these issues are covered in depth in the following sections and in subsequent chapters.

Do You Need Wireless Technology?

Those who cannot or will not run wires—apartment dwellers or those restricted by office lease or the physical structure itself from running cables across easements, civil boundaries, etc.—are obvious candidates for using wireless networking.

Shared office facilities, where tenants may share a common telephone/network equipment and cabling room, are also good candidates for wireless—to reduce the risks of bandwidth or data theft, tampering, or encountering old or inadequate wiring.

When using temporary office space, as for a campaign headquarters, charity event/race/marathon, emergency operations center, or field post, or while awaiting the completion of a permanent office, certainly do not waste the time and money involved in deploying a wired LAN infrastructure.

Wireless is ideal for travelers and commuters who need to stay connected to corporate or personal communications and can find a location at many large airports, urban cafés, public libraries, and some college campuses having wireless services. Free and subscription-based wireless services are being deployed more and more. Unfortunately, you may have to maintain subscriptions to many service providers in order to be able to connect, as well as be familiar with the many different wireless network connection parameters and subscription log-on methods to get and stay connected.

Using wireless network adapters is ideal for setting up a quick peer-to-peer network between friends, much as you might use the infrared connection features of personal digital assistants (PDAs) to beam information back and forth.

Is the Site Wireless-Friendly?

The issue of other devices and wireless services interfering with your wireless network can be the biggest barrier to a successful implementation. There are both technical and social engineering means of determining if wireless networking might work.

The first technical method is to simply acquire one access point and one client wireless network card, preferably on a laptop personal computer (PC), and set up a simple wireless connection to an existing network. Walk around with the laptop and try to use the network in as many places of interest as possible. Many of the client-side adapters include signal strength monitoring software so that you can see how strong and reliable your wireless connection will be. If you approach a piece of equipment that interferes with the wireless signal, your received signal strength will probably drop below acceptable levels and you will lose your connection to the network.

Loss of connection may be intermittent, rather than based on a specific location or simply proximity to other equipment, and this may be an indication that another wireless service or an appliance that affects your signal is in use nearby. Pay attention to this when microwave ovens and special equipment may be in use more often than at other times. Of course, interference from the microwave oven in the company cafeteria is a great excuse to stop working, take a break, and get away from the computer.

More technical, often preferred, and hyped by many wireless networking consultants is a complete radio frequency (RF) site survey performed with a spectrum analyzer—a highly technical piece of test equipment that can see details of both large and small portions of RF spectrum—identifying, qualifying, and quantifying the types of signals it receives. In some cases, the analyzer can also tell you what type of signal is being received, if it is not obvious by the visual display and characteristics of the spectrum. Unless the received signal can be demodulated to reveal the information within, and that information contains the identity of who is responsible for the transmission, it may be impossible to tell who is generating that signal. Moving the spectrum analyzer's antenna closer to or farther from different areas, or using a directional antenna, can tell you proximity or locate the transmitting device.

A spectrum analysis may not be conclusive evidence as to whether the site will accommodate wireless networking, because 802.11a and 802.11b use sophisticated modulation and signal processing techniques, a signal may get through 100 percent of the time even in the presence of interference. You will only know by trying it.

Conversely, unless a spectrum analyzer is present and monitoring the right portion of the RF spectrum for several days, a typical 1–2 hour "quick check" of a site may miss very significant interference that could render your network useless for several minutes or hours. Similarly, a clean, interference-free site today could become cluttered with new interference as other networks, appliances, or services come online nearby.

To enhance your confidence in your site's ability to accommodate wireless, do a little walk-around/talk-around investigation, and not just before you install your system. Do so frequently to help determine if nearby building tenants, new occupants, or other sources of interference are about to be introduced into your environment.

Can You Use Wireless Technology?

One of the most common questions about wireless is, How far will it go? As with most answers about technical things, it depends. 802.11b was designed with native, unmodified, unenhanced devices to extend the length of a 10BaseT Ethernet wire by 300 meters. This equals 985 feet, about a city block, or 0.18 miles. Unobstructed, unimpeded with line-of-sight, 802.11b will do just that and probably more. But who is going to hold their laptops above their heads or mount an access point itself on a rooftop to communicate digitally?

In most real-world cases, two native 802.11a devices will do well to clear 100 feet before the signals fade or are reflected too much to make a reliable connection. You may be able to add external antennas to your wireless equipment, overcome obstructions, and generally improve near-field penetration or increase range.

If you simply need to improve straight distance range, look for a directional antenna, or a pair of them, to provide approximately 8, 12, or 16 dB of signal gain. These may provide up to 10–12 miles of range between devices—not bad if you want to walk around a city park with a directional antenna attached to your laptop, attracting the attention of others.

To get this kind of range, one of the devices needs to be mounted high above surrounding terrain and buildings—which means finding space at a commercial radio site or a friend's house atop a hill or high-rise building. (I would be keenly interested to know if anyone successfully builds a solar-powered access or relay point and hides it in a tree someplace just to prove that wireless can be free and everywhere.)

If that meager 100 feet of coverage around your office bothers you, or you cannot seem to stay connected to the LAN during critical presentations in the conference room, then installing an omnidirectional antenna with 3–6 dB of gain will add penetration.

Remember, the primary intent of wireless is to get you off the 10BaseT CAT 5 cable tether. Stretching that invisible nonwire to cover neighborhoods and vast metropolitan areas involves just a little engineering and significant financial investment, which will be covered in later chapters. At this point, keep in mind that you are trying to get what amounts to a beam of light, or a reflection thereof, through an obstructed maze in a fog bank—and you will have a little better understanding of what you are up against with some wireless systems.

When you start trying to use wireless beyond the desktop, the issues of interfering with other devices and wireless services, as well as any security or policy issues that may preclude or prohibit the use of wireless, may or may not be obvious.

As a potentially interfering party, you should be mindful of other services. It would not be a good thing to discover that your wireless equipment interfered with medical diagnostic equipment, aircraft or military systems, or otherwise violated the Federal Communications Commission (FCC) rules by making an amateur radio system unusable. Doctors or medical technicians may not be able to discern, locate, or identify a source of interference with their instruments, but technical people such as amateur radio operators, who generally associate with engineers at various levels, can muster considerable resources to pinpoint interfering equipment.

If interference is not an issue, then certainly where you choose to apply wireless networking may be an issue. Radio signals will reflect off metal surfaces, but will not bend around corners. Unless you can establish a precise reflector, you cannot count on your signal getting around, much less through, metal reinforced walls, metal doors, elevators, dense plumbing, electrical wiring, or similar often hidden obstructions. One of the most common and troublesome hidden obstructions you can encounter is the wire screening used as a support for stucco and concrete construction materials. Another is aluminum siding. These are especially troublesome if you are trying to use your wireless gear between your inside home office and your patio or the neighbor's home. Those who live in wood or vinyl sided structures are better off in this regard. Metal screening and siding, as well as dense metal framing and plumbing or electrical tubing, will block and reflect wireless signals.

Look around you now and consider how many metallic objects are near you. Then walk around and consider how many more objects are between all the places where you would put wireless equipment. Consider everything from your computer monitor and case, file cabinet, recipe box, mini-blinds, window frames and screens, toaster, microwave oven, coffee maker, range vent hood, oven, cooktop, refrigerator, pots and pans, canisters, soup cans, a roll of aluminum foil, door knobs, hinges, faucet handles, VCR, DVD player, TV set, lamp bases, cubicle walls, and towel dispensers, down to your gold pen and favorite metal travel mug. Inside your walls are electrical wires, conduit, gas, water and vent pipes, metal framing pieces, and hundreds of screws or nails. Each of these is a possible point of reflection for a radio signal. The tiniest objects may be the most significant, as a 2.4 GHz wireless signal wave is only a couple of inches long—matching almost perfectly with a common construction nail. Your signal may also be absorbed by natural objects—trees, plants, leaves, and moist earth.

Blocked or absorbed wireless signals simply mean that the received signal will be weaker than desired, making your network unreliable. Reflected wireless signals, even when you have a line-of-sight path between the transmitter and receiver, can cancel out or jumble the desired signal, making it unusable. It is also possible, especially in nonline-of-sight conditions, for the reflected signals to be stronger than the original signal. Think of a blocked wireless signal like dense fog decreasing visibility and light levels. Think of reflected wireless signals like a mirror ball with light dancing in different directions. You do not see the original light source, just the reflections, which may be decorative, but not very useful to light an object.

You may expect out-of-the-box 802.11b wireless equipment to reach a few hundred feet, 100–300 feet being the typical advertised range. Because 802.11a equipment uses higher frequencies, it is typically limited to 50–100 feet without additional antennas.

Distance and overall obstruction/reflection density are significant technical influences on the success of a wireless network. Distance can be overcome with the use of external antennas (if your device provides such a connection), repeating or network bridging stations to extend the network, and additional access points to distribute the wireless network farther or into difficult to reach places. Neighborhood, campus, and metro area networks require the use of higher elevations at one end to overcome obstructions and improve line-of-sight path opportunities, as well as higher gain antennas and transmitter signal amplifiers to extend their range. Obviously, the more equipment you have to deploy to make the network work, the more expensive it will be.

If interference, signal blocking, or reflections are not of concern, you may have other sources of interference keeping you from deploying wireless networking—company or other policy being one of them, as well as the risk of signal and, thus, data theft being the other. Without very tight directional antenna patterns, it is possible to receive almost any wireless signal if you can get close enough to it. Most of the time, highly directional antennas are used only to extend a wireless signal between two fixed points, or a mobile user with a directional antenna and a fixed point with a nondirectional antenna. They are generally too large, inconvenient, and expensive to use for each and every client workstation.

A large retail chain store—a computer store selling wireless equipment no less—experienced someone receiving signals from its checkout systems and intercepting the data, including customer information and credit card numbers. The unknown assailant did not hack into the network, but merely listened to and stored what was heard. Wireless networking enthusiasts entertain themselves by driving and even walking around towns and campuses sniffing out wireless network signals—often finding hundreds of different wireless networks in operation within urban downtown areas. Wireless signals essentially cannot be contained. Like a smoker trying to sneak a puff in the restroom, a tell-tale whiff can be detected.

Knowing that wireless signals can be picked up by anyone, as if they had plugged into your wired LAN systems, means that you should probably provide some form of additional security for your data. Then, if someone does get your data, it will be unreadable or useless to them. While 802.11a and 802.11b do provide encryption (WEP) for the data placed on wireless networks, it is a very weak security measure that can be cracked within a few minutes by anyone with the AirSnort program running on a Linux-based computer. The answer to the weakness of the WEP feature is to use additional virtual private network (VPN) software to restrict access to the network and encrypt the data you place onto and take from the wireless network—so that even if someone gets your data, he needs to have the same VPN software and access codes to be able to use it. VPN software is a must among roving corporate users accessing the company network from the variety of dial-up, DSL, cable, and wireless Internet access methods available.

Certainly in very secure environments, from military posts to private research facilities, security experts do not trust any data leaving the immediate area, however well encrypted it may appear to be.

Who Will Design, Install, and Maintain Your Wireless System?

With the plethora of wireless products available in computer stores, it may appear as easy to install and implement a wireless network as it is to replace a computer mouse. Indeed, some products, especially all-in-one client network cards and access point kits, make the process very easy. But as you get further into the subject matter and start to expand the network with products from different companies and use different software, you will find nuances in firmware used in the network equipment, differences in terminology for the same items, different software, and occasionally different channel changing capabilities for different products.

Your best bet is to select a reputable, qualified vendor who can give you references to other customers, who will use high-quality equipment from major manufacturers for dependability and consistency, and who will intentionally design and implement your network for a bit of overcoverage to ensure reliability. The vendor you select should be able to accommodate different types of PCs and operating systems, work with different types of wired-network equipment and your servers, and most importantly, be attentive to your business and users' needs.

Your vendor should be willing and able to do a site survey before, during, and occasionally after your installation to ensure reliability and spot potential problems before and as they occur. The survey process should characterize the building structure to assess obstructions and reflections, and assess the environment for potential sources of interference, as well as interference your network may cause.

Implementation should consider security, vulnerability, and installing measures in addition to WEP. Ongoing maintenance should include changing security codes as employees come and go, just as you would change passwords to e-mail and network servers. You can enhance network security somewhat by using access point equipment that allows you to limit wireless access to only the specific wireless client cards you specify in the access point configuration. To do this, use their media access control (MAC) address—a unique number that identifies each and every network connection. Combining 128-bit WEP encryption between wireless equipment, MAC address control of which equipment can connect to an access point, and a secure VPN application between clients and networks is about as much as you can do to secure your network.

As part of your vendor selection process, you will also consider the cost of implementing your wireless system—pitting one vendor against the other and the cost of wireless versus wired.

The Cost of Wireless

Adding wireless to or using it as your home network might be more expensive than a few cables and conventional network adapters and a hub—a novelty or luxury. But going wireless at a workplace or places where construction or other issues make installing wires prohibitive may be the only way to go.

Let's compare the costs of installing wired and wireless networks in a typical small- to medium-sized office with 50 people/computers, even without considering whether or not cabling can be installed because of physical constraints.

TABLE 2.1

Cost Comparison Between Wired and Wireless Networks for 50 Systems

Equipment and Labor	Wired Network Cost	Wired Totals	Wireless Network Cost	Wireless Totals
Network Card (50)	$100	$5,000	$100	$5,000
Jacks and Cable Installation (50)	$50	$2,500	0	0
Patch Panels (3)	$400	$1,200	0	0
Patch Cables (100)	$5	$500	0	0
Hub/Switch (2–3)	$400	$1,200	0	0
Access Points (2)	0	0	$400	$800
Workstation Setup (1 hour)	$50	$2,500	$50	$2,500
Total		$12,900		$8,300
Difference				$4,600 less

The simple comparison in Table 2.1 shows you come out way ahead in cost savings when you go with a wireless network solution upon initial installation. With the money you save, you can expand your network by 50 percent for free versus a wired infrastructure. Long-term savings are also cumulative in that you do not have to do as much maintenance when users or systems move from one location to another—no patch cable changes at each end and far fewer bumps on the head from crawling under desks.

The initial and long-term savings could easily pay for VPN software to secure the network if needed. There is also long-term convenience to users, who can move about freely with laptops and take their data with them into conference rooms, meetings, and presentations without worrying about network cables or transferring files to another system or a server and retrieving them on another system later.

Multiply the savings by 2, 10, 20, or 100 times for larger scale implementations and the savings begin to add up to some significant money—enough that your CEO and CFO could be so impressed you could move up closer to CTO, if that is where you are headed.

LAN implementations are not the only place significant savings are apparent by going wireless. Consider simply connecting two nearby office buildings together when your company expands, typically done by running the equivalent of a T-1 carrier circuit or fiber optic thread through an underground trench. The permits and cost of trenching alone are almost prohibitive—well into thousands of dollars of heavy machinery work. Add a couple thousand dollars for burial cable or fiber and about a thousand for interconnect equipment at each end. Compare trenching with about a thousand dollars worth of wireless equipment for both ends and there is no comparison—you are going wireless. In some cases, you may even be able to interconnect directly with a branch office a few miles away via wireless—. something that would cost a couple thousand dollars for a Frame Relay or T-1 circuit installation and a recurring monthly cost of $1200 per month. Wired is obviously very expensive.

There are unseen costs of wireless—depending on what your vendor may charge for site surveys, interference checks and remedies, determining reflection and absorption that may affect signals, additional access points to improve coverage, and recurring security maintenance—but they may not be an issue at all in a clean environment and could be absorbed in the overall cost savings versus wired networking.

Summary

If the cost advantages of wireless networking excite you, then things are looking up. Certainly for a small, modest wireless LAN, the cost savings are obvious. Larger networks with more client systems may require different and more costly access point equipment. If your network spans a larger area than one access point or antenna scheme can cover, you will have to work out the design and costs of creating a contiguous, multi-access-point network. We still have a lot of work to do in considering network design, equipment selection, installation and setup time, and eventually performance tweaking. Before you can design, install, and set up a wireless network, you need to know a bit more about the various equipment and configuration options—from access points to antennas, cabling to client software—and that is covered in Chapters 3, 4, and 5.

CHAPTER 3

Wireless Network Basics

With your head full of jargon and technical details, you will want to put into perspective some of the components that make up a wireless network and how they work together.

For the most part, the components of a wireless network directly replace most of the common components of a wired network one-for-one, as shown in a simple configuration. Figures 3.1 and 3.2 show that a wireless network card replaces the wired network card; radio waves replace the Ethernet cabling, plugs, and jacks; and a wireless network access point unit replaces the Ethernet hub.

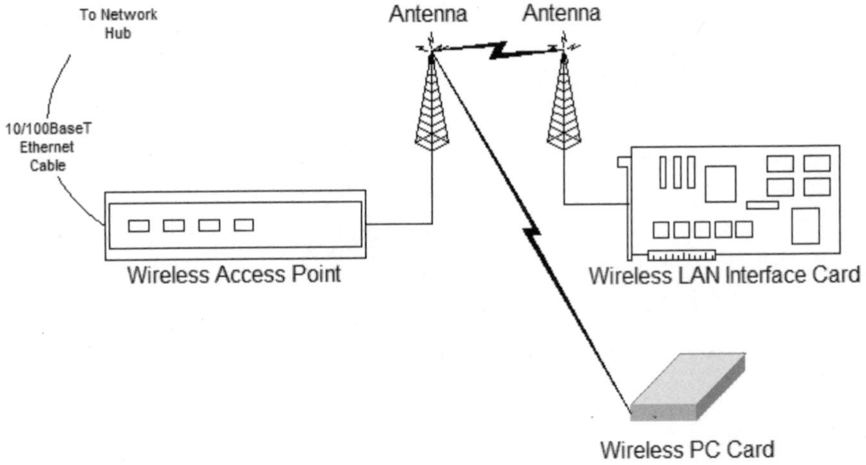

Figure 3.1 Basic wireless network components and their setup. An access point ties multiple wireless devices to the wired network and each other, as a hub does in a wired network.

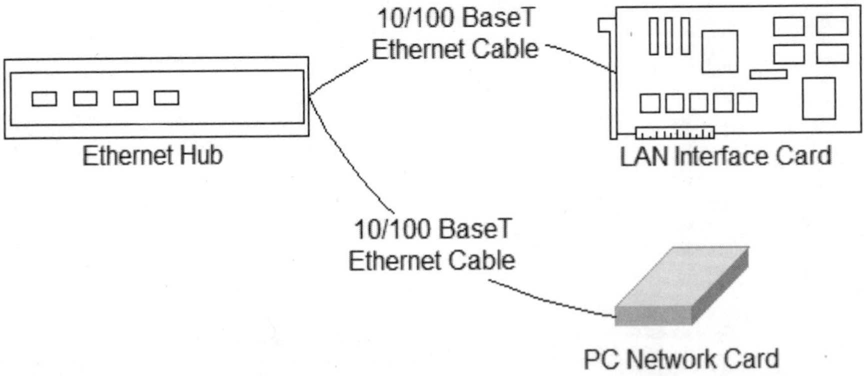

Figure 3.2 Basic wired network components and their setup. The Ethernet hub connects several different types of network clients to the network and each other.

The wireless interface card in the personal computer (PC) (running Linux, Windows, etc.) or Macintosh system that acts as a client on the network, and a wireless network device or base station known as an *access point*, connects multiple radios to the wired local area network (LAN) (or Internet) and each other. Although access points are more like hubs and are not considered repeaters, in a common LAN environment, they do extend the potential distance between client devices.

These figures illustrate the simplest possible plug-and-play (not to be confused with the plug-and-play interface standard) network configuration that can be achieved when these components are taken out of the box and set up.

Not shown are the network addressing and configuration details—the Internet protocol (IP) addresses, gateways, and domain name system (DNS) addresses needed to make the network devices be able to "talk" with each other, the LAN, and the Internet. These are parameters that must exist and be set up in any transmission control protocol (TCP/IP) network. In a wired network, these details are handled by either a digital subscriber line (DSL), cable modem, or router or another form of domain host configuration protocol (DHCP) server providing these services. In a wireless network, these details may be configured in the access point acting as a router, or left up to the modem or DHCP server.

The trade-off for not having wires for networking is the possibility of having to configure at least one and possibly two new parameters to connect to a specific wireless network system—the name of the wireless network and a security code. These allow you to connect and "talk" through a specific access point to other network devices and destinations.

If you take your laptop computer from your home wireless network to the local coffee shop with a wireless connection and hope to surf the Web, these last two details are essential because you will have to add the name and encryption code for the coffee shop wireless system to the configuration of your laptop. Once you are connected to the coffee shop network, their DHCP server will issue your computer an IP address on their LAN configured with the appropriate gateway and DNS addresses so that you can access the Internet and beyond. Fortunately, setting up the wireless network name and security code does not affect any wired network settings you may already have and does not require you to reboot your computer. This makes wireless more like adding a dial-up network connection than you would encounter by making major changes to an existing wired connection when you switch between networks.

The ability to change between different wireless networks at will, without complex configurations, allows you incredible freedom to roam. You retain your normal LAN workgroup or domain information and remote virtual private network (VPN) capabilities so you're never far from the office network—something your boss may really appreciate even if you do not.

Ready for Wireless?

Are you headed out to your local computer store to buy a wireless access point and PC card? Are you sure you are ready? Quite possibly you are not. My first attempt at recreating this simple wireless network scheme works fine when the access point is 5 feet away from a laptop in the same room, but fails miserably when the laptop is moved less than 50 feet from the access point. Failure at less than 50 feet away? Really? Why?

Figure 3.3 shows a not uncommon wireless system physical setup. The room containing the bulk of my computer equipment, wired network hubs, servers, and main Internet connection (also wireless) is in an office/recreation room separate from our house. The access point base station was sitting atop a shelf above one of the desks, about 5 feet off the ground. The place where I moved the laptop, a table outdoors on the other side of the house, is an otherwise "easy shot" under true line-of-sight conditions—though just a walk around the corner under normal circumstances, it is obviously (deliberately?) blocked by 30 feet of house—not line-of-sight.

The distance and the fact that I did not have optical line-of-sight between laptop and base station are compounded by the fact that both structures are traditional wood-frame construction with exterior walls of stucco (a form of concrete) bonded with wire mesh, commonly known as *chicken wire*. Although the office has large, single-pane window areas all around, they are covered by mini-blinds that have metal strips. The presence of these seemingly innocuous and, in the case of the stucco, invisible metal objects, is enough to reduce a wireless signal. You may have experienced a similar situation walking about your home with a 900 MHz or 2.4 GHz cordless telephone—static and fading signals. Same issues, different application. In the East, Midwest, or other parts of the U.S. and probably the rest of the world, this 50-foot span might not have been a problem because most exterior walls are either brick, wood, or vinyl. However, aluminum siding panels may present as much or more of a problem than the chicken wire in our stucco walls.

Figure 3.3
A not uncommon home wireless layout showing multiple signal obstructions. These obstructions can reduce a strong line-of-sight signal to barely usable across just a short distance.

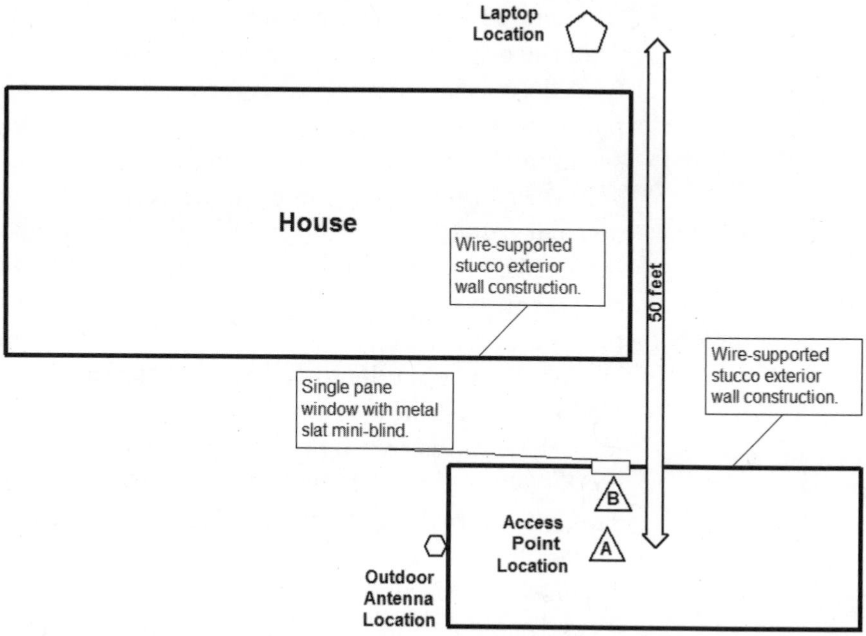

In technical terms, the received signal strength indication at the laptop, using the program included with the wireless PC card and the NetStumbler program (for Windows), showed a very weak –90 to –95 dB signal from the base station (access point). For reference, with the laptop sitting 1–2 feet away from an access point, the received signal strength is measured at –40 to –45 dB. So, our original access point signal weakened some 50 dB, or approximately 1 dB per foot, although signal attenuation over line-of-sight distance diminishes somewhat predictably, but not linearly, by the calculation:

$$96.6 + 20 \log(f) + 20 \log(d) \text{ dB} = \text{PathLoss dB}$$

where f is frequency in GHz and d is path distance in miles.

Field tests by researchers at the University of California-Berkeley (http://wireless.per.nl/multimed/cdrom97/contents.htm) indicate the signal loss should have only been –40 dB, if I assume that the concrete walls and concrete patio and driveway are equivalent to the indoor test case in the tests cited. I expect that the chicken wire and metal mini-blind slats added the additional 10 dB of loss. 10 dB of power loss reduces the signal to 1/10th the original signal strength, whereas 10 dB of power gain results in a signal 10 times stronger than the original signal. This is quite significant either way, especially at 2.4 GHz, where wireless networking signals are weak and low power to begin with, and attenuate rapidly at distance and with seemingly innocuous obstructions.

A very impressive, comprehensive on-line path loss calculation and path plotting tool is available at http://members.gbonline.com/~multiplx/wireless/wireless.main.cgi, with links to similar tools and documentation at http://www.qsl.net/n9zia/index.html. The path loss tool is a must-have reference for those digging into the technicalities of particularly challenging longer distance wireless projects. It will also show just how fragile the path of ultra-high frequency and microwave radio frequency (RF) signals can be. Once you get a grasp on the nominal signal levels, types of antennas, and surrounding terrain, such a tool will be invaluable to plan and troubleshoot wireless LANs.

I used the on-line tool and submitted very modest values for my access point and a PC card at the same elevation and a distance of 0.01 miles between them. The 40 dB loss I experienced was actually better than the results of the tool's calculations, which showed I should have seen 64 dB of loss. That a theoretical calculation appears to give a worst-case result than in my practice, shows that we need to consider that perhaps my link should not have worked. And to have a reliable link, I should take steps to improve the signal.

An important aspect to consider is that of *fade margin*—an extra amount of signal over and above the level you may obtain in an average experience. This additional signal level protects you if conditions change—like someone walking or standing between your computer and an access point—so that you can still maintain a solid communications link. Fade margin is extremely important over longer distances, especially those spanning varied terrain, over water, experiencing sun one day and rain the next, or through significant altitude changes where atmospheric conditions can affect a signal dramatically—such as a mountaintop access point at 3000 feet communicating with devices at 1000 feet or below.

My first attempt to overcome this short-path signal problem was to mount an omnidirectional antenna outside the office and connect it to the base station, to overcome the effects of the stucco and mini-blinds at that end. Because I used inadequate coaxial cable between the access point and the antenna—Times Wire and Cable model LMR-240 instead of the recommended larger model LMR-400 or Belden 9913F7 cable with less signal loss characteristics—using the outdoor antenna was no better than the local antenna on the access point.

Fortunately, the wireless PC card I am using in the laptop, an Orinoco Gold model, has a jack for an external antenna, and I have its complimentary decorative tabletop antenna. The antenna provides +2.5 dB of signal gain and more flexibility for where I can place it. Immediately upon plugging in the antenna, the received signal increased from –90 dB to a reasonably usable –80 dB level—what the card's diagnostic program calls "low." The reason the signal increased +10 dB by using only a +3 dB antenna has to do with physical placement of the antenna. I was able to position the external antenna about 8 inches above the edge of the laptop, roughly 15–18 inches higher than the position of the PC card inserted in the slot at the side of the laptop keyboard.

Looking for further improvement, I shifted the position of the access point without external antenna to a position just inside and directly up against a window, rather than sitting 8–9 feet away from the window on an interior shelf. This improved the signal another +15 dB, to a very solid –65 dB, a level the card diagnostics rates as "very good."

If I position the laptop card's external antenna on the same desk level the laptop is sitting, but with the laptop blocking the path, the signal drops –10 dB to –75 dB, rated again as "low" by the diagnostic. If I move the antenna to the base station side of the laptop, so the screen is not blocking the signal path, the signal goes up +8 dB to about –67 dB or to a "good" level. Disconnecting the external antenna results in only a –3 dB signal drop, adequate for truly portable use without the external antenna as another tether.

This experiment with a common scenario proves a very crucial point in your wireless network buying decisions. You will want to get equipment that accommodates external antennas, or at least their convenient placement. If I had simply gone to the local computer store, I probably would have ended up with a wireless card or an access point that did not have a connection for an external antenna. I was able to reposition the access point to give it a better view toward the direction I was going to be using the laptop—this time. When I move to the pool on the other side of the office, I would expect to experience signals problems there. Since it is obviously inconvenient to have to reposition the access point unit in different windows each time I change locations, and it is a nuisance to have to drag an external antenna along with the laptop, I have to consider better options. You will encounter these problems in office buildings and warehouses with metal framing, shelving, partition walls, and common office furniture.

The seemingly obvious option of choice for me, since I am not opposed to placing an antenna outdoors and running cable from it to the access point, is to purchase a suitable length of Times LMR-400 cable to improve the signal to the outdoor antenna and thus its effectiveness. Other options would be to get an access point with two antenna connections, install two pieces of coaxial cable, and position antennas near the windows on either side of the office; or purchase a second access point to feed signal from the opposite end of the house, since I do have a wired network running to most parts of the house, and all are likely places for positioning another access point.

Within the house proper, an apartment, or relatively open office space, you probably will not encounter this type of signal path obstruction, but it is worthy of note if you are trying to share your wireless network with your neighbors or working in an office complex with walls supported by metal framing.

While you might think that an antenna and some cable would be much cheaper than adding a second access point, think again. Antennas for wireless networking cost at least $40 each, more for more rugged outdoor antennas. LMR-400 coaxial cable costs about $1.50 per foot, connectors for the coax cost about $5 each, and you will probably need some form of pigtail or adapter cable ($20–40) to connect your PC card or access point to the LMR-400. To do a neat installation of the antenna, I would need at least 50 feet of cable, $75 worth, plus $10 for connectors, plus $30 for the adapter cable, plus the antenna—a total of $145 or more, versus a second access point cost of $150–160, and I do have the luxury of a nearby wired network from which to connect the second access point.

Certainly your buying decision needs to be based on what you know, or can know, about the type of signal coverage area you will have in your particular terrain and what you can do within the terrain to establish and maintain good signal strength. Cheaper or easily available is no match for starting out with the right equipment. A good friend of ours lends us the saying, "The price of quality only hurts once." I'm going to take that little jab of pain quietly and head to the local ham radio equipment store to get some LMR-400 and connectors.

Oh, yes, I do have the luxury of access to nearby well-stocked electronics stores to obtain cable and connectors—items that most computer stores do not sell, even as accessories for wireless networking. Unless you are in a major metro area like San Francisco, Los Angeles, Dallas, Houston, Chicago, Milwaukee, Atlanta, New York, or Washington, D.C., you will probably have to order your cables and antenna pieces online. A list of Web sites of popular wireless equipment vendors is provided in Chapter 4 and in the appendices.

How Did Wireless Suddenly Come to Involve Wires?

Quite simply, wireless-anything involves some of that "magic" described in Chapter 1. OK, some folks call it physics, with a lot of atmospheric and random physical variations thrown in—er, "magic."

> Any sufficiently advanced technology is indistinguishable from magic.
>
> *Arthur C. Clark*

Through no fault of equipment or operation, a signal that left the access point at a whopping (relative to low-power radio signals) –40 dB signal (as measured in side-by-side access point-to-PC card comparison) dropped an amazing 40 dB or more through 50 feet of space and common construction and trim materials. This is what wireless is all about—a technology designed and intended to reach no more than 300 feet (100 meters) clear line-of-sight between devices. If you expect no more distance between devices than that, and understand the conditions that have an effect over this performance, you will not be dissatisfied. With a variety of signal enhancement techniques, you may be pleasantly surprised to obtain a working distance of a mile or more.

But understand the limitations, be prepared to experiment, and accept that there will be some failures through no fault of your own.

In a situation where you cannot achieve line-of-sight from a single access point, barring moving the location of one or both devices, or removing the obstructions, sometimes you can augment the signal conditions with wires—at least wires between devices and antennas—to optimize the signals.

Summary

We saw two things that have an effect on our wireless system, obstructions that diminish range and optimizing antenna placement to overcome diminished range. Obstructions are somewhat variable. Improving your signal with antennas is a somewhat limited solution in that you cannot build and install antennas with enough gain at both ends to overcome all loss and still have a workable, mobile wireless device in which one end may have some wires involved and the other end typically does not.

We will see cases where using antennas separate from the wireless devices, or even moving the entire wireless access point to a more optimum location, attached by wires to the wired network and a power supply, is desirable and optimal.

Wires are an important part of many wireless networks, as are antennas. Chapter 4 introduces and familiarizes you with these two critical elements of wireless networking.

CHAPTER 4

Antennas and Cables

Antennas and cabling go with wireless networking like milk goes with cookies, beer with sausages, or Scotch whiskey with cigars—neither are mutually inclusive, but on some occasions, the combination is inevitable. For the modest home or office wireless network, your network may be fine with the antenna included with your access point and the one built into the wireless card in your computer. If you have a large home, a large or complex office layout, or you want to provide neighborhood Internet access throughout your neighborhood, a college campus, or large metropolitan area, you will encounter the need to select antennas and special antenna cabling to establish and maintain signal presence throughout an area larger than the 100-meter distance most wireless equipment is designed to support.

As we have seen in Chapter 3, it may be difficult to get and maintain adequate wireless signals over a distance as short as 50 feet (16–17 meters). Thus there are times when your wireless equipment needs some help doing what it is supposed to do—either that or install more wireless equipment, which also involves more wires. In this chapter, we discuss some basic principles and types of antennas, as well as the cabling and connectors used to interconnect antennas with wireless equipment.

Antennas

Every wireless device has one—an antenna that is—a mysterious construction of wires, metal, and insulators that somehow converts radio frequency (RF) energy from a wire into a wave of energy thrusting into or plucking out a signal from the atmosphere. Antennas are essentially resonators—like piano wires or flute reeds—tuned to the frequency of the signal we want to transmit or receive.

My experience with antennas, as an amateur radio operator of over 30 years, is fair at best. My understanding of the theories and practices of antennas is limited—physics and electromagnetic waves are not my strong suit. I experiment like everyone else, and when something works, I leave it alone and just use it. Fortunately, using the analogy of a piano wire or musical reed seems to strike a chord (pun intended) of reasonable understanding.

In the case of getting a middle C from a piano to a listener's ear, the string has raw energy applied to it and just happens to resonate at 880 Hertz. It fluctuates the air molecules at that rate and pushes other air molecules further along until the note is heard or the signal weakens and is lost. Understanding how music gets from a piano to our ears is easy by comparison to a radio signal. A radio signal is conveyed along a wire until it exits a resonant device (such as an antenna) and becomes an electromagnetic wave that does not affect air molecules, but pushes onward to be imposed onto another antenna, converted from electromagnetic waves to electricity, and then detected by another device.

What most people involved with radio know about antennas are:

- The elements of the antenna must be of the proper dimension to be resonant at the RF at which they are to be used.
- The elements of an antenna are constructed of conductive material.
- Antennas are typically constructed of wire—smaller or larger—but other metal objects such as rods or bars may be used, depending on design.
- The antenna connections must form or be made of properly matched components for the most efficient transfer of electrical energy from its source to the atmosphere when transmitting, and vice versa for receiving.
- They must be mounted far enough away from obstructions and other metal objects so that frequency resonance and optimum radiation of signal is not affected.
- The amount and direction of radiation of the radio energy can be manipulated by various antenna design and construction methods.
- Size matters—the size of an antenna depends on the RF used and what the antenna is used for.

Most of those tidbits probably seem obvious—judging by the appearance of different antennas on top of houses, on fenders of cars, and those strung between trees, poles, or on radio towers. You will notice a variety of antenna styles, each representing a different type of antenna used for different purposes. In the context of wireless networking, you will find five different antennas in use:

- Omnidirectional quarter-wave and collinear antennas
- Directional Yagi or beam antenna—both ring and wire element styles

- Directional waveguide or slot antenna
- Directional helical antenna
- Directional parabolic dish

I will defer details about antenna theory, design, and construction to the American Radio Relay League's (ARRL) *Antenna Handbook*, and concentrate on antenna basics related specifically to wireless networking systems. Of the five antenna types listed above, each has distinct benefits.

Omnidirectional Antennas

The radio antenna on your automobile is a common example of an omnidirectional vertical antenna—a simple wire or rod oriented vertically to match the RF radiation polarity of most radio broadcast stations.

An ideal but theoretical omnidirectional vertical antenna would radiate 360 degrees from a point in a spherical pattern, as shown in Figure 4.1.

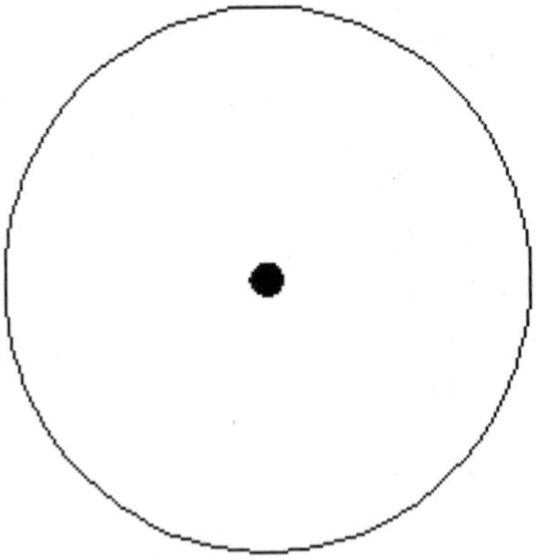

Figure 4.1 The theoretical isotropic radiator (center dot) takes up no space and has an Ideal spherical radiation pattern (outer circle).

This theoretical antenna employs an isotropic radiator—having no depth, width, or height—and is used only as a reference to calculate antenna performance—gain or loss—expressed in decibels as *dBi* or *decibels relative to an isotropic radiator*. Your neighbors and most environmentalists would really appreciate isotropic radiators—if they existed.

Because there is little need to do so, and it is nearly impossible to achieve this spherical radiation pattern, most antennas are based on a simple dipole—an antenna made of two equal elements, one pointed up and the other pointed down. The signal radiation pattern from a dipole is nearly spherical, with slight dimples of little or no radiation directly above and below the tips of the elements—envisioned as two circles of signal radiation emanating from the center of the dipole radiator. Dipoles may also be set horizontally, as in the case of long-wire antennas. A dipole is the typical real-world reference antenna from which measurements and antenna gain, expressed as *dBd* or *decibels referenced to a dipole* (see Figure 4.2).

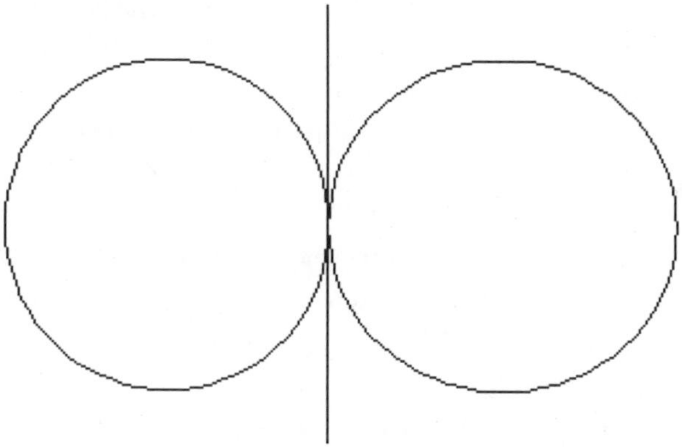

Figure 4.2
A practical, realistic vertical dipole has a donut-shaped radiation pattern.

Dipoles are fine in many applications, but more often, vertical antennas have an underlying counterpoise or ground-plane surface, causing the signal to radiate in a semicircular pattern above the ground or horizon plane beneath it. This design is more practical for use on cars and other objects where mounting and supplying signal from the bottom, rather than the center, is more practical (see Figure 4.3).

Antenna purists will argue adamantly about the differences in antenna performance and the theoretical isotropic versus the realistic dipole reference points. My view is to make sure that whatever antenna performance numbers you use employ the same reference point—dBi or dBd—or correct the difference to a reference value of your preference and work from that. In reality, we can easily compare a real-world dipole to other antennas, so using dBd as a real-world reference makes more sense than shifting numbers around to dBi—because an antenna that performs two, three, five, or ten times

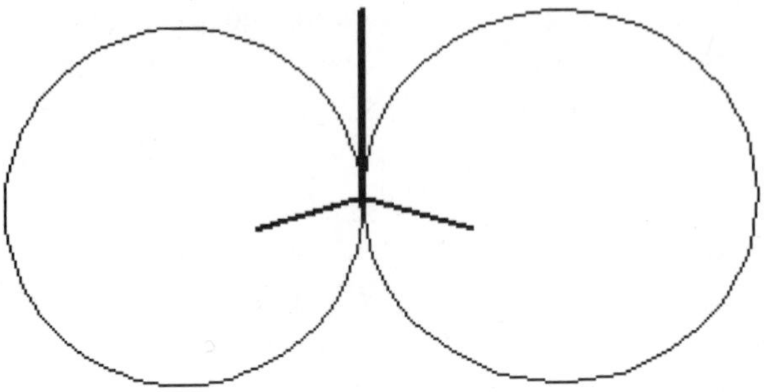

Figure 4.3 A vertical ground-plane antenna also has a donut-shaped radiation pattern, though somewhat flattened or having less signal immediately below the plane than above and across it.

better than the lowly dipole is something we can realize. No one will ever know how much better or worse something real performs against something that does not and cannot exist in real life (like the isotropic antenna) and the difference is generally only a single decibel or so—not worth arguing about in practical cases where signal levels typically vary by 10 to 30 dB.

With a vertical antenna, there is a single main radiating element offset by a comparably sized or larger surface area or ground-plane beneath it. This acts as a signal counterpoise or return point for the energy flowing from the radiating element—in effect, a modified dipole.

Omnidirectional antennas are typically oriented vertically, perpendicular to the Earth, so the signal they radiate spans out and around across the horizon. If the antenna were oriented horizontally, parallel to the Earth, much of the available signal would be lost, radiating into the Earth and up into the atmosphere. We want our wireless signals more earthbound, but not wasted into the ground either.

A basic vertically polarized omnidirectional antenna, as shown in Figure 4.3, has a radiating element equal to 1/4 of the wavelength of the signal frequency of interest. As antenna theory and measured reality goes, the maximum amplitude (voltage), current, and, as a result, power of an alternating wave signal is at 90 (1/4 wavelength) and 270 (3/4 wavelength) degrees from the starting zero-amplitude point of the waveform. A 3/4 wavelength would be just as good, if not a little better, because there would be two maximum signal points in the wave. However, 1/4 wavelength is an accepted practical reference point.

A 1/4 wavelength is equal to 234 / operating frequency in megahertz (MHz) = 1/4 wavelength in feet, or 2808 / operating frequency in megahertz (MHz) = 1/4 wavelength in inches. At the frequencies used by 802.11b wireless devices—effectively 2400 MHz—1/4 wavelength = 1.17 inches, making for a *very* short antenna!

A 1/4 wavelength antenna offers no signal gain and is also known as a *unity-gain antenna*. Ideally, all of the power applied to the antenna is radiated as is in all directions above the ground-plane uniformly.

Making antennas with multiple 1/4 wavelength elements—odd multiples (3, 5, 7, 9, etc.) of 1/4 wavelength elements coupled together—causes their radiation patterns to accumulate and compress, so that the antenna signal gains by forcing the radiation pattern into various shapes of focused patterns—patterns of concentrated radio energy.

Since a 1/4 wavelength antenna for 2.4 GHz is so small, and offers little or no advantage to signal transmission or reception, you will find that most omnidirectional antennas for use at frequencies above 150 MHz are built and configured with multiple elements to provide signal radiation gain. Other variations exist as well—combining 1/4, 1/2, 5/8, or 3/4 wavelength elements in deliberate configurations to achieve typical signal gains of 5–6, 8–9, or 10–12 dB or more.

Most base station antennas for wireless services—2.4 GHz 802.11b or 5.3–5.8 GHz 802.11a—are high signal gain devices. Their physical size is several times longer than a 1/4 wavelength for their respective frequencies. Beware that higher gain in antennas means the radiation pattern from these antennas is not as full (wide or tall) as omnidirectional antennas, as shown in Figure 4.4. The radiation pattern shrinks from a full, wide "flood" effect, extending from horizon to the sky, down to a narrower beam only 10 to 30 degrees wide. The bottom of this beam is elevated up from the horizon and pushed down considerably from "reaching the sky," thus forcing the power to only a few degrees above the horizon.

You can visualize this signal pattern compression much like deflating a basketball and pushing down on the top of it to reshape the ball pattern into a donut or bagel shape. In extremely high-gain antennas, the pattern resembles a flattened bagel, or in extreme cases, a nearly flat pancake-like signal pattern. This pattern shaping creates an umbrella under which there is low- or no-signal below the horizontal plane of the antenna and a void of signal above the antenna. The pattern compression as you use higher gain antennas *must* be taken into account when selecting and installing different antennas specific to your desired coverage area. More discussion of this consideration is provided in the "Deploying Antennas and Feedline Cables" section.

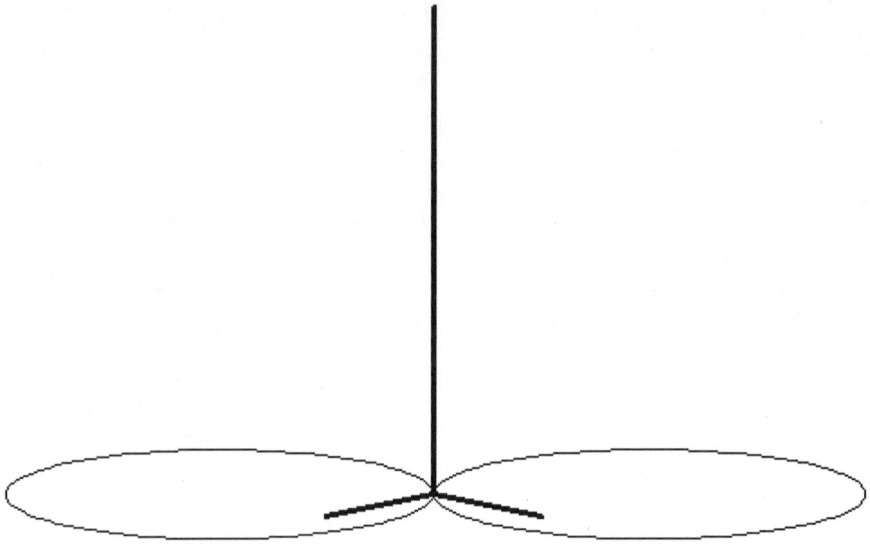

Figure 4.4
The vertical radiation pattern of a high-gain vertical ground-plane antenna. The radiating element is comparatively longer, providing more radiating surface area. Note the compressed or elongated shape of the pattern—less signal is available above and below the plane of the antenna versus a 1/4 wavelength unity gain antenna—as more signal is forced out to the sides.

Directional Yagi Antennas

The popular and unsightly rooftop TV antenna is a common example of a Yagi antenna (named after Hidetsugu Yagi, a Japanese electrical engineer who came up with this type of antenna) or beam antenna (so nick-named because it concentrates the RF signal into a beam of radiated energy).

A Yagi antenna enhances the normal 1/4 wavelength dipole antenna by adding a reflecting element behind a dipole antenna and several directing elements. This creates a concentrated beam pattern of radiated signal in a single direction, with minimal signal radiation to the rear and sides of the antenna's designed directionality (see Figure 4.5).

The common home rooftop television antenna is intended as a receiving antenna covering an extremely broad range of frequencies—from 50 MHz on up to nearly 1 GHz in one physical framework—which accounts for the various sized and positioned elements. It is effective, but not the efficient design you would find for a specific application, such as two-way or amateur radio or wireless networking. When you can design and create an antenna for a very specific frequency range, and especially for a range as high as 2.4 or 5.8 GHz, you can take advantage of the shorter wavelength, which affects not only the dipole antenna size but also the length, spacing, and number of directing elements, to create a lot of signal gain in a very small mechanical package.

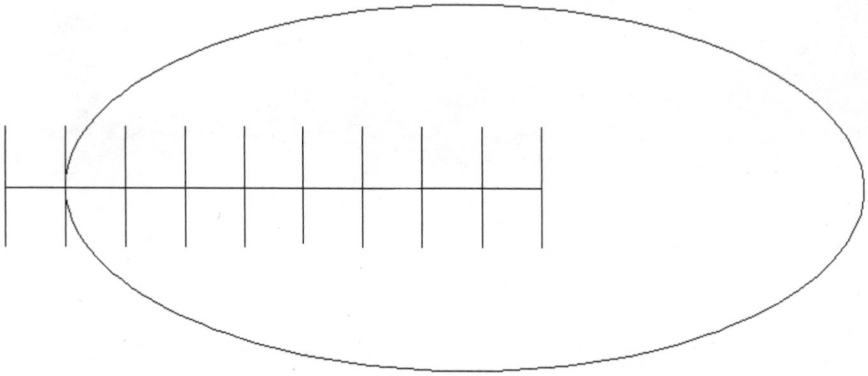

Figure 4.5
The basic form and radiation pattern of a Yagi-type antenna.

As with omnidirectional antennas designed to provide signal gain by forcing the radiation pattern into a narrower shape, Yagi or beam antennas do the same thing plus add the advantage of concentrating the signal radiation into a specific direction. The simple rule is more gain—less pattern area but stronger signal in the direction of the pattern (see Figures 4.6 through 4.8).

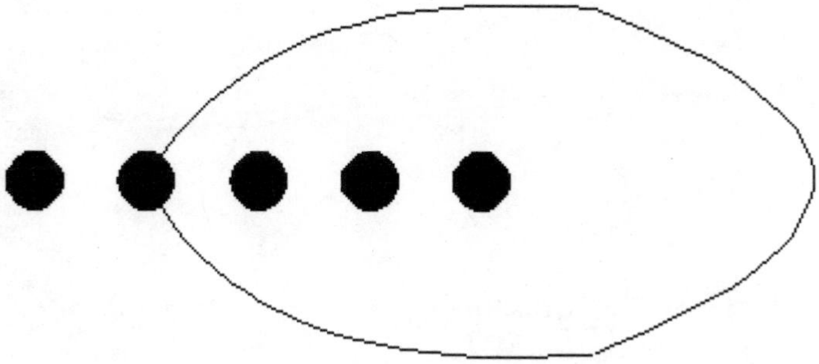

Figure 4.6
A top view of the typical radiation pattern from a relatively short vertically polarized Yagi antenna. The dots represent the individual vertical elements of the antenna.

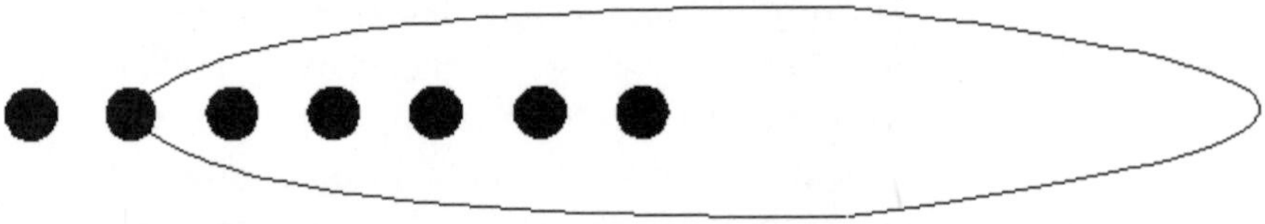

Figure 4.7 The narrower radiation pattern of a larger Yagi antenna.

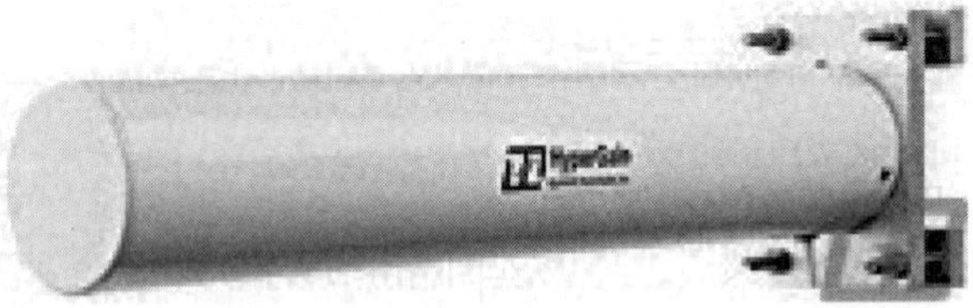

Figure 4.8 A picture of a commercially manufactured ring Yagi antenna. (Photo courtesy of Hypertechlink.com.)

One of the simplest and most impressive antennas you can use with wireless networking can be built with about $10 worth of parts from your local hardware store and two easy to find electronic pieces. The project is known affectionately and specifically as the Pringles Can Yagi—yes, the Pringles potato chips in the tall red can available at grocers nearly everywhere! See Figure 4.9.

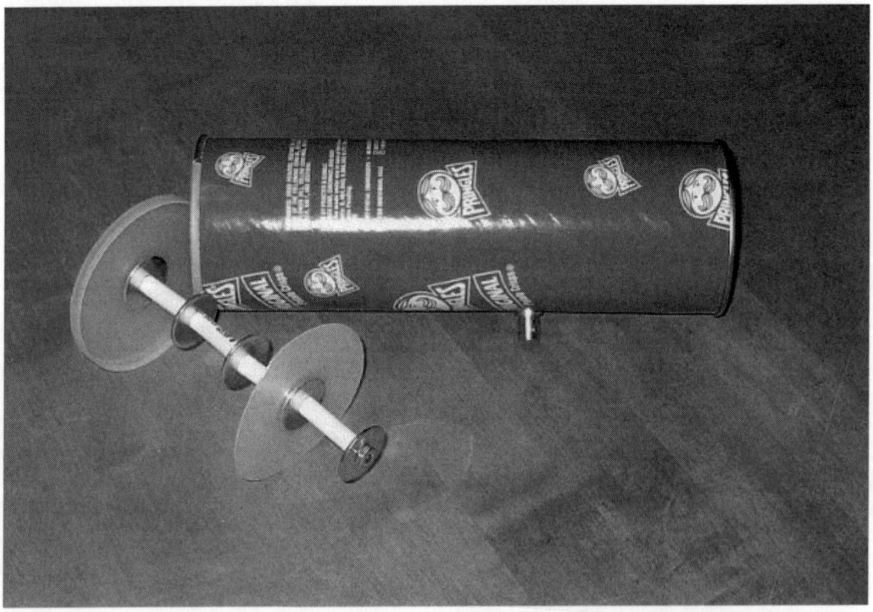

Figure 4.9 A photo of a homemade Pringles can ring Yagi antenna.

First created by Rob Flickenger and documented at http://www.oreillynet.com/cs/weblog/view/wlg/448/, the Pringles can antenna is a wireless tinkerer's pet project. Because the main structural material is merely cardboard, and the electrical characteristics of this implementation are not ideal for 2.4 GHz, the "can-tenna" offers an opportunity to play with antenna construction and toy with the magic of RF, but it is not suitable for long-term or reliable commercial use.

In antenna terms, the "can-tenna" is really a ring Yagi, imitating the antennas many precable TV services installed on home rooftops to receive broadcasts of select movie content, and those initially deployed by WavePath, a wireless Internet service provider in the San Francisco area, before being acquired and replaced by Sprint Broadband wireless services. The feed element is a 1/4 wavelength wire. The RF signal applied to the wire is induced into the elements of the antenna. The wire also picks up RF signal induced upon it from the elements.

This antenna, as constructed, does not stand up to the laws of physics and accepted antenna design because the elements—the hardware store washers—are too small (less than 1/4 wavelength) for use in the 2.4 GHz 802.11b band, but it does provide positive results. What does this criticism mean?

At the very least, a properly designed ring Yagi for 2.4 GHz would have circular elements approximately 1-1/4 inches in diameter, rather than smaller 1-inch washers. Using smaller than appropriate washers means the signal pickup and radiation are inefficient at best, and may have a negative influence on the transmitted or received signal at worst.

Specifically, if the antenna elements are not the proper size for the frequency used, it will not be resonant. When an RF signal is applied a nonresonant antenna, some of the power that is applied reflects back to the source (in the case of transmitting) or is lost to the atmosphere (in the case of receiving). Reflecting power back to a transmitter can damage the radio's internal circuits—they are designed to put out power, not absorb it, and as a result, excessive heating and high electrical currents can break the transmitting circuits. Broken transmitting circuits can often cause excessive power to get into the receiving circuits and desensitize or damage them as well.

Radio people—from CBers to ham radio operators to commercial radio engineers—know that power reflected back to a transmitter, reflected power, measured in relative dB, watts of power, or standing wave ratio (VSWR) is a bad thing. Those of us who work with radio signals strive to make our connections and tune our antennas for minimal power reflected back into transmitting devices. In most cases, we have test equipment to measure either the forward transmitted and reverse/reflected power, the standing wave ratio, or both—to help us assess our antenna systems. Unfortunately, such equipment is too expensive for one-time use in setting up a wireless network system, and the transmit power levels for wireless networking are so low (30 to 250 mW) that accurate measurement is difficult at best anyway.

Simply, a mismatched antenna, or any mismatched connection at an RF transmitting device, can harm your radio—in this case, your access point, add-on power amplifier, or PC card. Thus, long-term use of a less than optimal antenna like the Pringles can project is not recommended. Receivers generally do not care too much if they are connected to a nonresonant antenna, but they will not work as well and may be subject to significant amounts of interference from other signals.

Any one of several widely available, commercially made antennas specifically designed for wireless networking service is preferred and highly recommended for serious long-term wireless networking use. Although premade antennas are seemingly expensive for the small amount of metal involved, a few dollars spent on the correct antenna, and the correct cable to connect to it, can save you a lot more than the cost of having to replace damaged access points and PC cards because you used the wrong antenna or went the budget route on cheap cable and connectors.

The ring Yagi antenna, as demonstrated with the Pringles can project, is but one of several types of directional antennas that make our desired wireless signal stronger by focusing it in a specific direction. The ring Yagi is a variation on the dipole and more conventional wire element Yagi as you see in TV antennas, on the roofs of ham radio operators, or in commercial service on some buildings and radio towers. Experimentation and engineering have produced flat panel-style antennas that provide signal gain and directivity, as well as a strange variation on the Yagi called a *helical antenna*.

A *helical antenna* is no more or less than a precisely wound coil of wire tuned to the RF of interest. The radiation pattern is very focused and, as may be anticipated, radiates out in the direction of the helical turns, rather than essentially flat in either the horizontal or vertical plane. Signal gain is determined by the number of turns, again in proportion to a number of 1/4 wavelengths. Physically, this antenna resembles the commercially made Yagi shown in Figure 4.8.

Other types of directional antennas include the *parabolic* or *dish antenna*, as you might see used for satellite television reception or with a grid reflector, as shown in Figure 4.10.

Figure 4.10 A commercially made parabolic dish antenna with wire grid reflector. (Photo courtesy of Hypertechlink.com.)

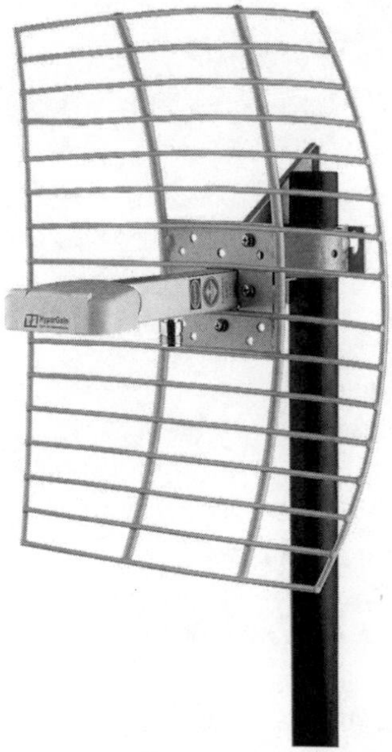

Parabolic or dish antennas provide much higher signal gain, typically +24 dB, within a reasonably compact size, compared to using extremely long Yagi antennas. They have little or no signal radiated to or picked up from the back side of the antenna.

Flat panel antennas offer high directional gain and little or no signal presence to the rear of the antenna and have a visually less obtrusive appearance—suitable for melding into building architectures, made into decorative ornamentation, and attaching easily to walls, rather than being mounted on gnarly bits of piping and brackets (see Figure 4.11).

Figure 4.11 The author's flat panel antenna mounted on a tall mast. This antenna is used for Sprint Broadband wireless Internet access, affectionately known as a "pizza box."

Antenna Radiation Polarity and Diversity

An important aspect of antennas to consider is the *polarity* of the signal they radiate. A radio wave signal may be imagined to be two-dimensional—having a length or extension in the main direction of signal focus and either a height or width, depending on whether the signal is vertically or horizontally polarized or aligned. Horizontal polarization is typically better to reduce received noise, as most man-made and environmental noise is vertically polarized. But it is easier to use vertically polarized antennas for omnidirectional coverage so that half of the signal is not wasted by radiating into the Earth, floor, or ceiling.

The radiated energy from all vertical omnidirectional antennas is, as you might imagine, vertically polarized. For maximum signal coupling or transfer to and from wireless devices, antenna polarization should also be vertical. Matching polarization is easy if both devices use vertically oriented antennas. Signal polarity matching is more challenging using PC card wireless adapters that may sit horizontally or vertically in their respective computers, or if one device uses a ring Yagi or a helical antenna and the other uses a vertical antenna. The labeling on panel antennas may indicate how they are to be mounted to achieve horizontal or vertical polarization.

In practice, wireless networking systems are built with little or no awareness or regard for polarity, mostly because at 2.4 and 5.8 GHz, the wavelength is so short that the radiated signals may twist along their path as they encounter reflections and atmospheric changes. These signals may change polarity several times, making polarity alignment difficult, if not impossible. You may find that tilting or rotating the antenna on some devices will improve reception as a means of improving your system coverage. This is especially easy on devices like the LinkSys WAP11 access point unit that employs two antennas on gimbal mounts.

The dual antennas on some devices typically act as *diversity reception antennas*, where the signal may arrive stronger at one antenna versus the other, depending on signal reflections, polarity twist, and the position of devices that communicate with a particular access point.

When selecting and deploying higher gain antennas, remember you *must* adhere to RF power limitations versus increased antenna gain, as covered in Chapter 1. You must also adhere to the Federal Communications Commission (FCC) rules governing 47 CFR Part 802.11 systems and not modify cables or connectors provided with manufactured equipment.

Wires

Every antenna is connected to its corresponding radio transmitter or receiver by some type of wired connection. In very small wireless devices, the antennas may be directly connected to the transmitter/receiver unit with virtually no wire separating them. Radio energy is not unlike the AC energy that powers lighting, appliances, and computers, or the DC energy that powers flashlights, cellular telephones, or laptop computers—that energy must flow in a complete circuit from source to load and back.

In most cases, the wire used to connect radio antennas to transmitters or receivers is actually two wires, constructed in a specific coaxial form to create a cable. Coaxially constructed cable is most practical and common in most applications. A coaxial cable is a concentric unit comprised of a center conductor wire, surrounded by a dielectric insulation, surrounded by an outer or shield wire—both of these wires run continuously from end to end of the cable. The insulating material used between the center conductor and the shield wires provides a specific spacing between the two wires and prevents electrical shorts, as well as fills the void to keep contaminants from destroying the effects of the cable. The type of insulating material also affects the efficiency and power handling capability of the cable. Polyethylene foam is the most common insulator used, but Teflon® and even air may be used in high-power applications. The size of the center conductor wire and the distance between it and the outer shield wire form are proportional, and these dimensions define the *impedance* of the coaxial cable.

Impedance is an important factor in properly matching the radio source or detecting unit and the antenna. Impedance is also a characteristic of antenna elements. Impedance is expressed using the term for electrical resistance—*ohms*. In most all radio applications, 50 ohm cable is used for a variety of traditional and practical reasons. You will not be able to measure this 50-ohm impedance with traditional electrical instruments, as the impedance is related to RF signals rather than common AC or DC voltages. At this point, I will again defer discussion and details about cables and their electrical and design issues to the ARRL's *Antenna Handbook* and similar expert reference material on the topic.

The basic things most people involved with radio know about wires—especially those related to antennas—are:

- The basic rules of electricity—voltage, current, and resistance—apply to antenna wires.
- Wires transfer energy differently along their core and surface, depending on the RF applied.
- The higher the frequency, the more loss experienced from the cable.
- Size matters—long, thin cables have more signal loss than short, fat cables.

Smaller wires cannot handle as much current or transfer as much power as larger ones. Less distance between wires reduces the amount of voltage that can be applied to avoid arcing and shorting-out of the applied energy. Also, the closer the inner wire is to the surrounding shield, the more capacitance the cable has, which diminishes performance at higher frequencies. Larger diameter cables typically impose less loss between radio and antenna and can handle more power. Less loss means more of your radio signal gets to where it is going.

Higher frequency signals travel on the outside surface or skin of the wire, rather than through the inside body of the wire. So whether the center conductor wire is hollow, solid, or stranded matters little, though higher frequencies tend to move better over a smooth, solid wire surface than a twisted stranded wire. And if the outer circumference of the wire is larger, there is more area for energy to flow across. Unfortunately, we cannot reduce the frequency of our wireless devices in order to use cheaper, more practical cables. Also, if we reduced the frequency, then antennas would get impractically larger and more cumbersome. We would also encounter more competing signals, leading to interference issues at lower frequencies.

Basically, you want a short, fat, smooth cable—and preferably none at all—connecting your wireless device to the largest antenna possible, and have that antenna pointed right at the other end of the connection. Because cables are rarely connected directly into the equipment we are using, they need connectors that allow us to remove the cable from the device as needed. Connectors, in nearly a dozen unique sizes and styles, introduce additional signal loss and are another potential point of failure, but they are necessary.

The cables used in wireless applications are common coaxial varieties available from many electronic suppliers, all radio/communications equipment suppliers, and vendors focusing on wireless networking equipment. Smaller, more flexible cables are used for the last foot or two of connection between access points and antennas and the larger, more efficient feeder cables.

The cables of choice are usually the following:

- CommScope WBC-100 and WBC-195—a very thin (1/8 inch) cable found on "pigtail" cables for wireless cards. The signal losses or attenuation of these smaller cables in the wireless range of 2.4–2.5 GHz are 20–40 dB per 100 feet—which means that any signal gain you would get from using a 10–18 dB antenna is lost (okay, buried deeply) in that length of cable, but short runs of 2–6 feet do not impose enough loss to be of concern.

- Times Microwave LMR-400 and Belden 9913F7—relatively thick (1/2-inch) cables used for longer cable runs between access points and fixed position antennas. Plain LMR-400 is not as flexible as the more expensive UltraFlex version of it or Belden's 9913F7. The signal losses (or attenuation) of these cables are about 6–8 dB per 100 feet—which means that you would only lose roughly half or slightly more of your signal over that length of cable, rather than 90 percent of it with smaller cable. Obviously, keeping the length of any cable you have to use to a minimum is best (see Table 4.1).

TABLE 4.1

Coaxial cable attenuation values listed in decibels for common wireless cable types. Smaller cable sizes and longer cable lengths impose higher signal losses.

Cable Type	Loss per 100 ft @ 150 MHz	Loss per 100 ft @ 450 MHz	Loss per 100 ft @ 2.5 GHz
WBC-100	8.90	15.8	39.8
WBC-195	4.40	7.80	19.0
WBC-200	4.00	7.00	16.9
WBC-300	2.40	4.20	10.4
Belden 9913F7	2.00	2.80	8.50
WBC-400	1.50	2.70	6.80
WBC-600	1.00	1.70	4.40

To illustrate the effect of cable length and loss on typical access point setup, let us start with a typical –40 dB local signal received from an access point by a wireless card in a laptop, install an outside antenna for the access point, and subtract out the losses (see Table 4.2).

The resulting values in Table 4.2 clearly indicate that only one-quarter (6 dB of gain results in four times the output level, while 6 dB of loss results in one-fourth of the level) of the original signal will reach the antenna using standard wireless network products and cabling. Using the most economical antenna, one with only 2.5 dB of gain, still leaves us with only 30 percent (4 dB less) of the signal we started out with, radiating into thin air to reach our client systems.

TABLE 4.2

Cumulative gain and loss effects of installing an antenna 50 feet from an access point (or client adapter).

Component	Reference Value	Cumulative Loss/Gain	Effective Radiated Signal
Starting received signal level	–40.0 dB	0.0 dB	–40.0 dB
3-foot WBC-100 pigtail cable to adapt access point connection to antenna cable	–1.2 dB	–1.2 dB	–41.2 dB
Antenna cable connectors (1 each end, –1 dB each)	–2.0 dB	–3.2 dB	–43.2 dB
50 feet of WBC-400 (or LMR-400) cable	–3.4 dB	–6.6 dB (cable loss)	–46.6 dB (end of coax)
Omnidirectional antenna with +2.5 dB gain	+2.5 dB	–4.1 dB	–44.1 dB (radiated)
Replace +2.5 dB antenna with +8.5 dB antenna	+8.5 dB	+1.9 dB	–38.1 dB (radiated)

To get more signal out of the access point, we must use an antenna with higher gain. In this example, an antenna with 8.5 dB of gain results in a radiated signal that is 1.9 dB higher or almost two times (twice the power would actually be +3 dB) stronger than the signal level from which we started. Higher gain antennas are often an inexpensive way to get signal where you want it, unless you want it in many directions without constraints. One problem with obtaining a signal increase by using antenna gain is that the pattern or radiated signal becomes narrowly focused, reduced from an almost spherical pattern around a simple antenna element that provides no gain, to a signal that is "squished" into a narrower or flatter shape, as discussed in the previous section.

Obviously, if we use the larger 600 grade cable, our cable loss would be less, and more signal would be available to and radiate from the antenna. Using smaller 200 grade cable, we would experience severe loss and have almost no useful signal available to radiate from the antenna or could not practically have enough antenna gain to overcome the loss imposed by the smaller cable. These considerations and tradeoffs will be a significant part of designing wireless networks—essentially, use the cable with the lowest loss possible in all cases, but consider the mechanical aspects of flexibility and weight or forces the cable may apply at the connection points.

In very special cases, you may find a use for a cable called RADIAX™—a unique type of cable that, unlike normal coaxial cable, keeps the signal contained within the outer shield until it gets to the end. RADIAX is designed to radiate signal by letting some of it leak out intentionally along its length. RADIAX is intended to increase the distribution of RF signals throughout a large area, without using multiple antennas or access points.

Manufacturer links:

- Times Microwave—www.timemicrowave.com
- Belden—www.belden.com
- Andrew RADIAX™—www.andrew.com
- CommScope WBC Series Cables—www.commscope.com

Vendor links:

- Ham Radio Outlet—www.hamradio.com
- Amateur Electronic Supply—www.aesham.com
- HyperLinkTech—www.hyperlinktech.com

Connectors

Every cable must have a connector on the end, unless it is directly attached to an electronic circuit, which is rare. Connectors come in different sizes, styles, polarities or genders (male or female center conductor), materials (brass, silver, or nickel), insulators (nylon, Teflon, or polystyrene), and fastening types (push and turn bayonet style, threaded, or push/snap-in).

Contrary to the cable sizes they are made for, some of the smallest connectors have the best performance (e.g., less signal loss, good power handling capability). But larger connectors are required for mechanical stability with larger cables and to preclude the use of small connectors for attachment to PC cards, to avoid breaking the cards.

In wireless networking applications, you will find that most of the RF connectors used are identical to industry-standard RF connectors, with one annoying, significant difference—the male and female center connector pins are female in the otherwise male connector body and male in the usually female connector body. This prevents using commonly available off-the-shelf RF connectors for at least one very important reason. By FCC rules, 802.11 devices are intended to be sold and installed as a system—the antenna or cable provided with a specific piece of wireless equipment is to be used only with that equipment. If standard connectors are employed, there is less assurance that you will have a proper single-source system, and thus you may violate the FCC rules.

Since wireless networking frequency allocations are shared with amateur radio operators, and amateur radio operators are allowed much more latitude in terms of customizing and experimenting with radio devices, they may be considered exempt from the FCC's regulations for unregulated 47 CFR Part 15 equipment. This means they can break up a system into different components and may exceed the transmitting power and antenna gain limitations of Part 15. This is a terrific reason to get your "ham license," but it also means that you cannot resell or profit from any wireless networking operations when implemented or modified as a ham radio station.

Type N

The type N connector was developed in the 1940s and named after Paul Neill of Bell Labs. It was designed as the first true microwave capable coaxial connector (see Figure 4.12).

Figure 4.12 Male and female N connectors.

TNC

TNC stands for Threaded Neill Concelman, the connector's namesake, developed in the late 1950s (see Figure 4.13).

Figure 4.13 Reverse male and female TNC connectors. The standard TNC connector has the male pin in the male connector body with rotating female-threaded shroud.

SMA

SMA stands for subminiature version A. The connector was developed in the 1960s (see Figure 4.14).

Figure 4.14 Male and female R-SMA or reverse-polarity SMA connector uses the same external body style as the standard SMA connector, but the male and female pin are on opposite bodies.

The material from which connectors are made also plays an important role in connector performance. Gold will, of course, oxidize the least and is one of the best conductors of electricity, making for very low-loss interconnections. Silver is a more economical alternative, and though it oxidizes, more silver-oxide actually increases conductivity. Copper, while a good conductor of electricity, corrodes rapidly like silver, but copper oxide is less conductive. Brass, nickel, and chrome may be very appealing to the eye, but like copper, are not suitable for exposed, outdoor use without significant waterproofing, as should be done for all exposed connections.

MC and MMCX

MC and MMCX are micro-miniature connectors developed in the 1990s to save space and weight in microwave equipment. They simply snap into their mating connectors and allow swiveling for flexibility in attaching to PC cards.

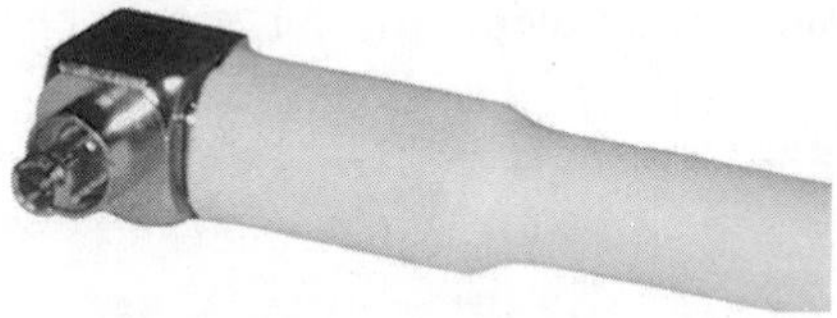

Figure 4.15 The MC-Card connector is a special implementation of the MC connector for quick connect and disconnect applications for PC card use.

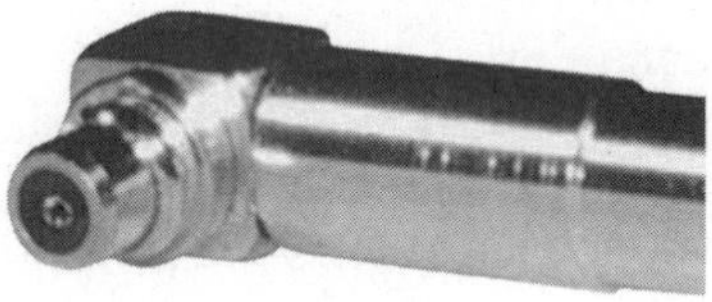

Figure 4.16 The MMCX connector is an alternative to the MC connector for quick connect and disconnect applications for PC card use.

Remote Mounted Access Points and Amplifiers

Based on the antenna and cable data presented earlier, you may find it prudent to try to place the access point or use an inline amplifier directly at or very near your antenna to minimize cable runs and signal losses.

Mounting your access point with an integrated antenna, or an additional antenna in a location that provides a clearer signal path to other access points, is an excellent option, provided you can weatherproof the device and provide power and wired Ethernet connections back to the wired network. There are many types of adapters intended to allow you to carry power to remote access points over the same Ethernet cable that carries your network data—a method called power over Ethernet (POE).

If you wish to place an amplifier near your antenna, but keep your access point separate, you may also use special adapters that can mix and then isolate DC power and RF to supply power to the amplifier through your coaxial cable.

No matter where the equipment is located in proximity to the antenna, you are still under regulations to know and maintain proper power levels radiated from the antenna.

The addition of signal amplifiers is permitted within the regulations, as long as you do not exceed the appropriate radiated power levels for the antenna you are using. Keep in mind also that wireless equipment is sold as stand-alone or as a deliberate system that must not be altered. The addition of 250 mW, 500 mW, 1, 5, or 10 W amplifiers is probably *illegal* in your circumstances.

A most important consideration for total radiated power levels is limiting human exposure to higher power RF signals. Applying 1, 5, or 10 W to a high gain parabolic antenna will create a very strong RF signal near the antenna—strong enough to act like a microwave oven and cause heating of tissue and fluids. This could cause an excessive RF exposure hazard to nearby users or workers on antenna towers. You must perform the radiated power calculations, as well as know and maintain the legal power limits and safe exposure distances for your implementation. Failure to do so could result in serious health and legal issues that you do not want to encounter, for the sake of moving a little data back and forth.

Summary

With a basic understanding of antennas and cabling, you are almost ready to begin enhancing and expanding the coverage of your wireless network. To do that, you will need to be familiar with antenna and cable selection for your particular application, installing the antennas and cables you will be using, performing a safe and professional installation, and testing the performance of the overall system you created. When you are comfortable with these steps, or if you know you will not need complicated antennas and cabling, you can move on to Chapter 6 to learn about the various types of wireless networking equipment and their application in your system.

CHAPTER 5

Common Wireless Network Components

Now that you have some idea of what wireless networking is about, including a briefing on rules and regulations, safety and interference, and how antennas and cables play in the wireless world, it is time to describe the bits and pieces and start putting them together.

The two most common pieces for a typical home or small office environment are the adapter that attaches to a laptop or desktop system—the client device or *wireless network adapter*—and the wireless device that interconnects your wireless clients to your wired network or some form of Internet connection—the network, or in some cases a server-side device called an *access point*.

These two essential pieces are most commonly found at retail computer stores or sold through mail order outlets and will work together to create a simple wireless network in the shortest time. Add more client adapters, another access point or two, maybe an external antenna, and you find yourself able to support more people over a greater area, creating a true wireless local area network (WLAN) architecture.

If you are building your network from the ground up, you may find that a combined access point and gateway/router—called a *wireless gateway/router*—is simpler than using separate components to connect to your digital subscriber line (DSL) or cable Internet service.

As your future needs expand, or if your present need is to span a network beyond a building or two, and adding wires or fiber optic links is not feasible, you will find that there are other types of wireless equipment available to tie two separate network segments together, such as a wireless access point-like device called a *wireless bridge*.

If you have a serious access control and network security requirement, you may want a separate virtual private network (VPN) firewall, an integrated secure wireless access point, access control software, or to configure your network with IPSec encrypted TCP/IP packets, which we will cover in Chapter 9: Wireless Network Security.

Client-Side Wireless Adapters

There are four basic styles of interfaces intended for use in your client systems—a personal computer (PC) card or PCMCIA card typically for laptops, a peripheral component interconnect (PCI) bus interface card, a few industry standard architecture (ISA) bus cards for older/legacy PC systems, and universal serial bus (USB) port adapters. Each of these acts like any other standard network interface card that you would use in an IBM-compatible/DOS- or Windows-based PC, Apple Macintosh, or UNIX (Linux, FreeBSD, Sun Solaris) system.

The basic difference between a typical Ethernet card and a wireless networking card is that the wireless card connects to a network over radio waves, instead of a cable with twisted pairs of wires. Detailed differences include the ability of a wireless card to detect and offer connections to different wireless networks, and for the software and drivers for the card to display them to you so that you can choose the appropriate one for your needs.

While the following devices are considered client-side, rather than server or network access-side devices, with the right software or operating system configuration, you can use these interfaces and software as the primary network access interface between wireless and wired networks—creating gateways, routers, and firewalls—as you might do with any other network interface card.

If all you need or want to do is establish a simple peer-to-peer network between two different computers, these adapters will do that for you also. All it takes is a simple software configuration change, and you are able to select network or peer-to-peer connectivity within a few clicks of your mouse.

PC Card

The acronym PCMCIA, short for Personal Computer Memory Card International Association, an industry trade association which initially established the PC card standards, has given way to the more generic term "PC card," since devices for this type of bus interconnection system are no longer restricted to memory cards. Almost every laptop built since 1995 has a PC card slot available to accommodate memory cards, modems, network adapters, external disk drive interfaces, and other input/output (I/O) devices.

The PC card wireless network adapter is probably the most common adapter for client computers—makes sense: laptops are the portable devices for which we most often want portability and net connectivity.

PC card wireless adapters usually have built-in antennas and no connectors for external antennas. Manufacturers of these devices range from Apple, Belkin, Cisco, D-Link, LinkSys, and Lucent/WaveLAN/Orinoco to SMC, Symbol, and US Robotics, among others. Wireless aficionados typically choose adapters that use Intersil's Prims 2 wireless chipset so that they can take advantage of various commercial, shareware, and freeware software tools to exploit the features of the chipset. These software tools include the venerable AirSnort program for Linux that allows you to find and decode wireless security encryption keys, to NetStumbler and similar programs that can detect and show all of the possible unhidden wireless networks near you.

CF Card Adapter

CF stands for Compact Flash, an interface port found on many handheld personal digital assistant (PDA) appliances. The CF port is typically used to expand the memory or allow portable storage between devices, but it can also be used like the PC card slot on laptop computers to support modems and wireless network cards. CF cards offer the same features as other client-side adapters, but may not have external antenna connects, and the robust software tools you might use on a PC, Macintosh, or Linux system are generally not available for PDAs.

PC Card PCI Bus Adapter

Of course you want your desktop or tower PC (Windows or Linux) or Apple Macintosh systems to be able to participate on your wireless LAN, especially if you intend to use one of those systems as an access control point, a router, or bridge to interconnect wired and wireless LANs together. For these purposes, you can find PCI adapters that accept PC card interfaces. This lets you select one PC card interface make and model for all of your systems, and apply them to nonlaptops as needed. You also get all of the same hardware (antenna) and software (tools) features of whichever PC card you select.

PCI Bus Card

If you are not picky about having the same PC card interface in all systems, you can get dedicated PCI card wireless adapters. These come in two forms—those that have a PC card permanently attached to them and those that are specifically built from discrete components and chips onto a PCI bus card. These cards typically have an external antenna jack on the back panel and come with a specific antenna attached to them.

As with the PC card, CF card, and PC card/PCI adapters, the hardware and software features are similar to the PC card adapters.

ISA Bus Interface

If you absolutely must connect older PC systems—those pre-PCI/pre-USB era systems that have only a 16-bit I/O bus—to your wireless network, a few vendors still offer ISA bus wireless network adapters.

USB Interface

Someday the 8- and 16-bit ISA bus, along with serial and parallel ports, really will be obsolete, I promise. PC vendors have been telling me so for years. By then, the limited number of PCI slots your system has will be taken up with a fast video card and some other special widget, and you will be left with only your USB or FireWire/ IEEE-1394 ports to connect new devices. Many users probably do not want to remove the covers from their computers to install a PCI card, or cannot open their systems in the case of laptops and notebook. In these cases, several vendors are offering wireless adapters that connect via external USB ports. The USB port is one of the most efficient, temporary, as-needed, automatically configured Plug-and-Play I/O ports, and using this interface method is highly recommended, especially if you are screwdriver-phobic or all of your system's PCI or PC card slots are occupied with other devices.

Network-Side Wireless Equipment

To gain all of the flexibility you have in roaming about with your laptop and a PC card wireless network adapter, you will need an access point or similar network function every step of the way. The basic access point appliance makes it simple to connect a wireless computer to the conventional wired networks of the world—a function you could do with almost any wireless network interface and Windows 2000 Server or Linux software features. But why bother when someone has built a device to do it for you?

There are reasons for everything, and for almost any way of doing something the harder or more expensive way, there is a piece of ready-made hardware available to do the job for you.

Access Points

Access point devices are the wireless equivalent to a combined hub and bridge/gateway/router in wired network. They accept, and to some extent manage or sort out the wireless connection from a few or many wireless client devices. They also may dole out domain host configuration protocol (DHCP) settings for your network, or pass that chore on to another device or server, and otherwise convert wireless signals into wired network signals and vice versa.

You can create an access point using a client-side adapter and Windows or Linux server software configurations, or you can buy a specific access point device from any number of vendors.

The access point is the core of a wireless network. Many vendors make access points. LinkSys makes a popular device called the model WAP11, SMC has the SMC2755W, D-Link has a model DWL-900AP+, Cisco has several models in the 340 and 1200 series, and Apple makes the AirPort base station. Without these devices and the firmware they provide (embedded, highly specific software), wireless networking would truly be left to the backroom late night antics of a select few thousand Linux or Windows gurus trying to make wireless network adapters and server software do some otherwise simple networking tasks.

The primary purpose of an access point is to allow selected client devices to connect to a wired network, and conversely to disallow unwanted clients access to the wired net. This is accomplished by using system ID (SSID) names and wired equivalency protocol (WEP) security keys to control network access. Of course you can turn off the security features and allow anyone and everyone access to the wired network, but at great risk of providing access to hackers or bandwidth thieves.

An optional secondary function of an access point can be to provide, or pass on to another server or device, dynamic host configuration protocol (DHCP) requests to give wireless clients a modest handful of necessary parameters (IP, gateway, and DNS addressing) to make and use a connection to a wired network.

You may still find the need to use server-based software to provide network access control, network security, specific firewall features, or network traffic routing between wireless clients talking to a wireless access point, before getting to the wired network or Internet. How you design and manage your network is up to you, but more often than not, using ready-made off-the-shelf devices will save you a lot of time and money.

Wireless Bridges

A wireless bridge acts as a repeater of signals between one wireless network segment and another—extending the range of the two wireless devices at either end of the bridged gap between networks.

Some bridges, such as the SMC 2682W, may also perform access point functions, making them ideal to be positioned in the middle building between two other buildings to provide wired network access through that location.

The LinkSys WET11 is a simple wireless bridge designed to convert any existing 10BaseT Ethernet device to a wireless client—making it a useful addition to Ethernet-ready printers, scanners, or laptops. It is also possible to use this device as an access point to a wired network by connecting it to a wired network interface card on a server or any other workstation with software capable of passing traffic from one wired network connection to another—a function called *routing*.

These simpler bridges have small antennas included with them that must be used with the unit. The units will accommodate, but cannot be run legally, with separate external antennas that could extend their range and effectiveness further. Another limitation to these devices is that all of the wireless network traffic between two different locations goes through the single channel the bridge uses, rather than repeating signals from one channel to the other to obtain full 802.11b bandwidth.

Products such as the Orinoco (now Proxim) Point-to-Point Backbone Radio Kit are more like a true duplex repeater in that they take signals from a wireless device on one channel in one direction and rebroadcast them on another channel to another wireless device in another direction. This is effectively high-end, powerful bridging to get wireless signals between two points a significant distance apart (up to 6 miles), because it is sold as a system with high-gain directional external antennas. This type of bridge is very effective, not only for its long range capability, but also because it uses separate radio channels for each direction of communication so that you get full 802.11b bandwidth. This also gives you more security and control over access into and through the bridge.

Wireless Gateways and Routers

Most of us started out with a wired network and migrated to including wireless components. We probably have a router or firewall of some kind guarding our workstations and servers from open traffic on our DSL or cable Internet connection, and then we add a wireless access point.

To save money and complications with separate equipment, wireless gateways and routers are access points with firewall and router capabilities, providing these two functions in one more affordable unit. These are intended to be used by people who will mostly or only have wireless devices on their networks and do not need wired network connectivity; they will send wireless traffic directly to and from the DSL or cable modem connection.

Orinoco (now Proxim) makes three products in this category—the RG-1000, RG-1100, and BG-2000. The RG-1000 is especially attractive for home and small office use because it has a built-in 56k modem that can be used for Internet access, so even homes without high-speed Internet access can enjoy wireless networking. The BG-1000 accepts high-speed Internet access and distributes it to your wired and wireless clients at the same time. These units are similar to the LinkSys BEFW11S4 and the SMC Barricade series of gateway/routers for home and small office use. Larger offices and enterprises will want to consider more robust, manageable products such as Orinoco's AP-1000 or models from Cisco.

Wireless Signal Power Amplifiers

If anything is going to extend the range of your wireless system beyond adding proper external antennas, it will be providing more radio frequency (RF) signal by increasing the power output of your access point to the antenna. To get more power, you need power amplifiers.

Hyperlinktech is one of a few vendors that offers affordable power amplifiers in 100 mW (milliwatt), 250 mW, 500 mW, and 1 W models—http://www.hyperlinktech.com/web/amplifiers_2400.html. These products are Federal Communications Commission (FCC) certified, and the vendor provides a cross-reference to specific wireless interface products. Thus, you get the right amplifier to suit your needs and your antenna, so that you can operate your system legally.

Most wireless network adapters provide between 30 and 100 mW of RF power output—barely enough juice to get across the street in some cases, especially if you use an external antenna fed with a long run of high-loss coaxial cable. Your resulting RF signal could fizzle out of the antenna at a mere 1 mW or less.

It may seem simple to just hook up a power amp into your antenna line and call the system good—and for the physical part of the task it is—but in doing so, you might exceed FCC regulated power levels or expose others to dangerously high levels of RF.

When considering an amplifier to boost signals, you have to know where to put the amplifier in-line with your antenna so that your wireless device does not overload the input to the amplifier, rendering it useless. Thus, you have to know the gain of your antenna; you need to know the loss factor for the cable you are using to properly factor losses and gain, keeping in mind dBi versus dBd reference values; and you have to know how to set the power level of the amplifier so that the resulting radiated signal is not too strong. If the output level of the amplifier cannot be set or measured properly, you should assume you are putting out maximum power and introduce some loss by adding an appropriate length of coax to reduce the maximum radiated power to within legal limits.

In Chapter 1, we summarized the FCC rules and made some rough comparisons to power levels and antenna gain. In Tables 5.1 and 5.2, we present a more practical cross-reference to follow when applying different power amplifiers to different antennas in different wireless configurations.

Power Limitations for 802.11b Systems

It is highly unlikely that anyone will ever know or report that you are running your wireless system in excess of the legal limits, but remember that if your wireless system causes interference to other devices, you are putting your operation at risk. You also have to be careful where you place your antenna if you are going to run full power. Tucking it into a flower box behind a frequently used bench on your patio will likely overexpose anyone who sits there to dangerous levels of RF.

It is also unlikely that you have handy the proper test equipment to measure the power output of your wireless devices to know whether or not you are in compliance with the power limits.

The most practical and accepted way to ensure that your system is within legal limits is to keep track of the power levels, gains, and losses of your system components in decibels, then add or subtract accordingly, to arrive at a calculated RF radiation level.

Point-to-multipoint configurations. For point-to-multipoint setups (access point to clients), you are allowed up to 30 dBm or 1 W of transmitter power output (TPO) feeding into a 6 dBi antenna, 36 dBm total, equal to 4 W of effective isotropic radiated power (EIRP) (above the reference level of an isotropic antenna). The TPO needs to be reduced 1 dB for every dB of antenna gain over the baseline 6 dBi antenna gain. In this system, it is fairly easy to work back from that 36 dBm maximum output level, substituting power output levels, antenna gain, and feedline loss values, to determine if your system output is above or below the legal limits. Table 5.1 shows typical RF output levels versus antenna gain to achieve, but not exceed, the maximum output level.

TABLE 5.1

Amplifier power output and antenna gain values to maintain legal limits in point-to-multipoint wireless configurations. These values assume that there is no feedline loss between the wireless device and the amplifier, nor loss between the amplifier and the antenna.

Amplifier Power Output	Max Antenna Gain	Radiated Power Level
1 W = 30 dBm (18 dB gain)	6 dBi	4 W = 36 dBm
500 mW = 27 dBm (15 dB gain)	9 dBi	4 W = 36 dBm
250 mW = 24 dBm (12 dB gain)	12 dBi	4 W = 36 dBm
100 mW = 20 dBm (8 dB gain)	16 dBi	4 W = 36 dBm
30 mW = 12 dBm	24 dBi	4 W = 36 dBm

You will probably discover that the antennas you can buy do not have exactly 6, 9, 12, 16, or other integer-level gains—5.7 dBi, 8.2 dBi, 13.7 dBi, etc., may be more common—but it is easier to round up, to be on the safe side.

To your advantage, as a margin of error, you can add 1 dB of loss for each connector in the path between wireless device and antenna. So in the case of a wireless device plus amplifier plus antenna, there are at least four connectors in the line (one at the wireless device, one at the input to the amplifier, one at the output of the amplifier, and one at the antenna), giving you 2 dB of loss to the amplifier, which may reduce its output slightly, and 2 dB of loss from amplifier to antenna.

Connecting an antenna or amplifier to most wireless devices typically requires a pigtail or jumper wire to convert from an MC card or MMCX to a Type N connector for connection to a length of feedline. Thus, you experience at least another 2 dB of loss for each of these connectors and the length of the jumper cable. But this only affects the power level driving the amplifier, not the amplifier output, which may still be at the advertised level.

With all of this factored in, you could still use a 24 dBi gain antenna with a 250 mW amplifier. With direct connections or no loss between device, amplifier, and antenna, this would give you 48 dBm radiated output, which you have to reduce 12 dB to stay at or below the 36 dBm limit. To accomplish this, you might insert about 100 feet of coax between the wireless device and the amplifier to reduce the signal to the amplifier to 6 dBm. With its +12 dB gain, you would have 18 dBm output into the feedline connected to the antenna—giving you a 42 dBm antenna output that you would have to reduce another 6 dB to limit the radiated level to 36 dBm. So, you have to add another length of coax with 6 dB loss. In this scenario, you may be better off using a 100 mW amplifier, so that you can use less coax and have less overall loss in the system. Remember that the coax loss also affects the received signal. 12 dB of coax loss decreases your –77 dBm receiver sensitivity to –59 dBm, which makes for a pretty deaf receiver for 802.11b use.

You may need or want to use a lower gain antenna to obtain a wider antenna radiation pattern that covers more area, in which case you can increase the power output to a maximum of 1 W. For instance, you can apply a 9 dBi gain antenna to a 250 mW amplifier, which results in 33 dBm radiated output, or roughly 2 watts—still not too shabby to cover an office or small park area.

Remember, too, that antenna gain works in both directions. And since most client systems will be using lower power 30–100 mW adapters, and their signal needs to get to the access point as much as the access point's signal needs to get to them, a higher gain antenna—but not so high as to overshoot the area you want to cover—may be necessary.

This is where that "magic" stuff comes into play in real life—balancing the advantages and disadvantages of various antenna types, coax losses, and power levels for the area and range to be covered.

Point-to-Point Configurations

Point-to-point wireless configurations are typically used to bridge two different network segments over a long distance—or make a short-distance link very robust and reliable. Since the FCC encourages the use of directional antennas to minimize interference to other users, you have much more control over the radiation of signal. To reward this type of configuration, the FCC is more lenient about power levels and antenna gain—you do not have to reduce your power as much if you use higher gain, but directional antennas.

Instead of having to reduce the power to the antenna by 1 dB for every 1 dB of antenna gain in excess of 6 dBi, you only have to reduce the power 1 dB for every 3 dB of antenna gain. In this case, if you have 1 W (30 dBm) RF output and a 24 dBi antenna, instead of a 6 dBi antenna, you need only reduce the power to the antenna by 1/3 of the difference (18 dB) or only 6 dB. This is easily accomplished with a length of coax to achieve the desired amount of loss needed to stay within maximum power limits (see Table 5.2).

TABLE 5.2

A listing of how much loss you need to place between a 1 W (30 dBm) RF output (wireless device or amplifier) for various antenna gain values to stay within the maximum radiated power level for point-to-point links (e.g., 30 + 6 – 0 = 36, or 30 + 9 – (1/3 of 3 dB) = 38).

Amplifier Power Output	Antenna Gain	Differential Loss Required	Resulting Allowable Radiated Output
1 W = 30 dBm	6 dBi	0 dB	36 dBm
"	9 dBi	1 dB	38 dBm
"	12 dBi	2 dB	40 dBm
"	15 dBi	3 dB	42 dBm
"	18 dBi	4 dB	44 dBm
"	21 dBi	5 dB	46 dBm
"	24 dBi	6 dB	48 dBm

In RF terms, 6–12 dB is a significant amount of power increase (4 to approximately 12 times). So thanks to the FCC rules, in this case, it is truly to your advantage to use an amplifier and high-gain antenna to get the strongest signal going at each end between points A and B of your point-to-point setup.

Once again, be aware that you have to watch out for excess RF exposure levels. 42 dBm is 16 watts of radiated power at the front or "business end" of the antenna, and 48 dBm (6 dB or 4 times more than 16 watts) is 64 watts of radiated power. *This is more than enough power to start warming soft fleshy parts of the body and likely cause some tissue damage.*

802.11a Point-to-Multipoint

The FCC is more stringent concerning power limitations in the 802.11a or 5 GHz band. For strictly in-building use, you are limited to 50 mW total power output in the 5.15–5.25 GHz or lower portion of the 802.11a band, and you are prohibited from using any antenna that is not built into the device—period—no exceptions.

For exterior point-to-multipoint operations, you may use the 5.25–5.35 GHz middle band, with the same gain restrictions imposed for 802.11b point-to-multipoint service, but a maximum power of 250 mW, or the high band of 5.725–5.825 GHz with a 1 W power limit for a 6 dBi antenna—or 36 dBm/4 W radiated power level.

For point-to-point links, you may use the 5.725–5.825 GHz or high band with more generous power and gain limits—1 W maximum RF power using up to 23 dBi antennas with no power reduction for the increased gain, as with the middle band and 802.11b point-to-point links. The absence of power/gain limitation is probably because there is an additional 7 dB of path loss to be factored in when using 5 GHz versus 2.4 GHz.

Summary

We have introduced you to the essential components of wireless network systems—from the client computer to the fixed-station equipment that converts wireless to wired networks, and the RF signal components that your clients rely on for a solid RF signal connection. We have stressed again, and likely still not enough, the legal regulations and safety concerns surrounding the RF equipment involved in getting a strong RF signal connection. Our next step is to take these components and start building wireless networks. Then, once we have them built, we will go through the steps to configure the client-side and network-side equipment, so that they can communicate properly and securely.

CHAPTER 6

Typical Wireless Installations

Where are you going to install your wireless network—home, office, warehouse, coffee shop, or campus? What are you going to use your wireless network for—personal/family, profession, or public Internet access?

The application of your wireless network should dictate the equipment, configuration, security and access control software, equipment location, and installation procedures you will use.

Most home, personal, or recreational and small office wireless networks will probably fare well with almost any off-the-shelf product available. Wireless products from LinkSys, SMC, D-Link, and Belkin are commonly found in local retail outlets such as Best Buy, CompUSA, Fry's, and Circuit City. The access points available from these companies will accommodate up to 10 and sometimes more individual clients/users on the network, and have user-friendly installation and setup software.

Apple makes a series of products it would prefer you to use with its Macintosh computers, available from Apple stores and other retailers. Off-the-shelf equipment is usually quite economical, priced as a suitable alternative to the complexity of Ethernet cabling and hubs. With a combined wireless access point, hub, and router/firewall product, you can easily accommodate wired and wireless users and share your broadband Internet connection through the house or office.

For your medium to large office, enterprise, and college/university campus applications, you will probably be accommodating more than 10 users. A higher quantity of users dictates that you will need commercial grade products from Proxim, Orinoco Wireless, 3Com, or Cisco because off-the-shelf products often limit the number of connected computers to 10–25 systems maximum. The setup and installation software and processes are more complex, including very robust security, often integrated with existing server and firewall systems and requiring intermediate to advanced network engineers or administrators to install and set them up.

Large corporate wireless local area networks (WLANs) may also require wireless bridges so that the entire wireless network can span multiple office buildings. You may even implement *mesh routing* to provide multiple access points to the best wired network connections and enhance the overall reliability of the network. Mesh routing means that most or all of the access points your clients connect to have one or more wired or wireless backbone connections between different access points. This technique is used to ensure the access points are always able to provide the best connection to the wired network, in case one or more paths to the main network from other access points fails.

If you intend to become a wireless Internet service provider (WISP), your equipment, configuration, security, and access control requirements will be the same as for corporate environments, with the possible addition of some form of user-friendly web- or client-based sign-on software to limit and control subscriber access to the service.

Equipment costs are somewhat higher for commercial applications and related products than for home/small office products because they are more capable in terms of numbers of clients that can connect through them; availability of different antenna systems; and network management, security features, and software to better control these networks. If you are building a corporate network from the ground up, rather than adding wireless to an existing network, the costs are comparable—perhaps higher for equipment and initial installation, but far lower in terms of hub or switch-to-cable-to-desktop maintenance.

ISPs and those developing wireless networks to span wider and more public areas will incur incrementally higher costs for additional access point, bridging, and routing equipment—but certainly nothing like the expense of building out, deploying, and maintaining DSL or cable systems.

With all these systems in mind, let us begin to draw upon them and lay out some typical wireless networks. After the basics are laid out we will cover each of these in their own chapters, and include typical software and security configurations.

Wireless at Home

Wireless networking is perfect for home networking. In fact, you could say that this application is what wireless networking was designed for—to avoid the hassle of finding just the right place to locate a hub central to a bunch of wires, avoid drilling holes, keep you from having to crawl around in attics and under homes to run cables, and prevent the hassle and mistakes of connecting tiny wires to nearly impossible phone-style jacks.

A typical 600–1,000 square foot apartment or 1,200–2,500 square foot home is an ideal place to install a wireless access point and outfit each desktop and laptop system with a wireless adapter card and then go about enjoying the Internet, local file, and printer sharing quite easily.

Figure 6.1 shows a typical home wireless network setup with one desktop personal computer (PC) or Macintosh system using a wireless network adapter—connected internally to the PC bus or externally through the universal serial bus (USB) port. The desktop system is used as the host for sharing a common printer. The access point could be a LinkSys WAP11, an Orinoco RG-1000 or similar, or a LinkSys BEFW11S4 that provides router/firewall functionality, a dynamic host configuration protocol (DHCP) server, and a 4-port Ethernet hub for wired connections. The printer could also be networked wirelessly using a wireless bridge device LinkSys PPS1UW for a USB printer, or LinkSys BEFW11P1, which includes a wireless access point, router for cable digital subscriber line (DSL), and a print server interface.

Figure 6.1
A typical apartment requires only one access point.

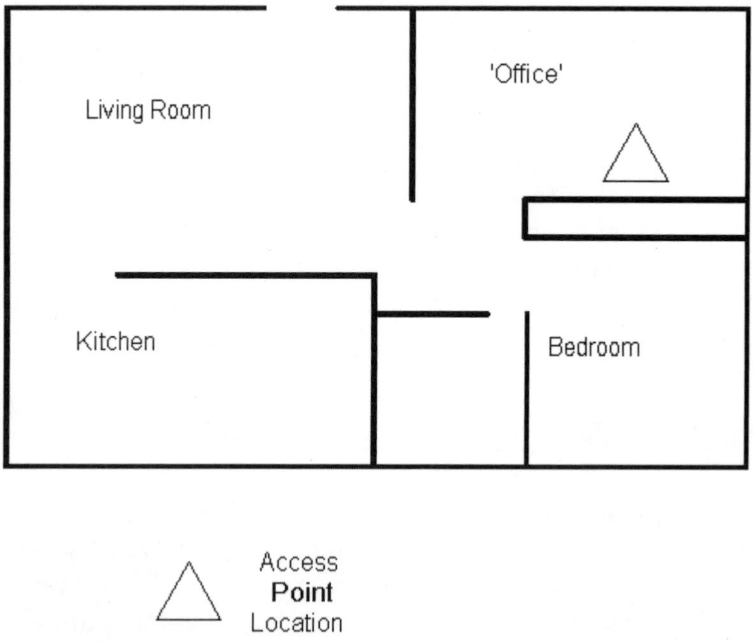

The LinkSys BEFW11P1 may be the ideal solution for apartment or home use—just one device is needed to interconnect printers, laptops, and desktops. Location of the access point within a small apartment, as shown in Figure 6.2, is not critical to provide adequate wireless signal coverage.

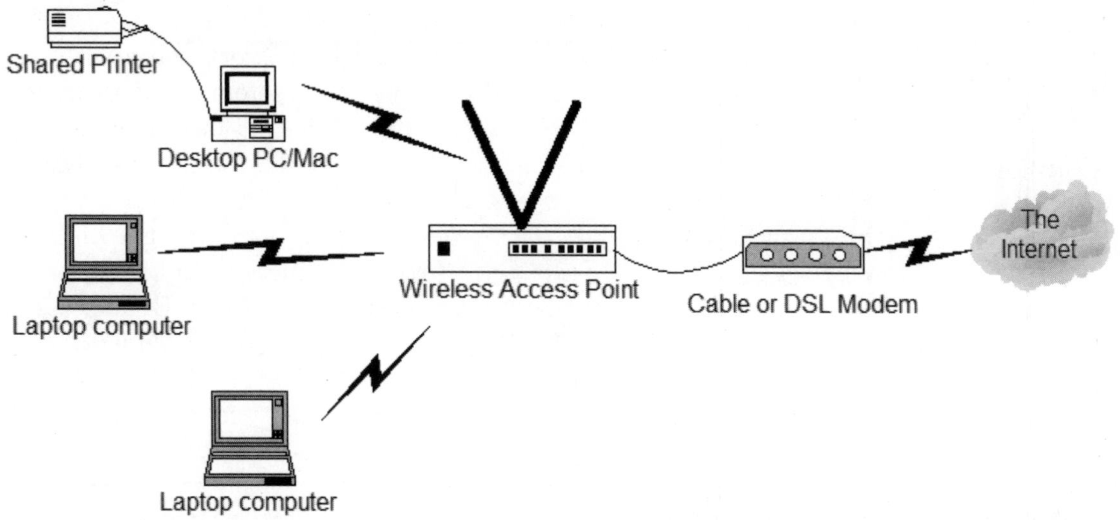

Figure 6.2 Wireless networking in a typical apartment requires only one access point to cover the entire area, and will probably provide coverage for your neighbors 2–3 apartments away as well.

Covering the average 1,200–1,800 square foot one- or two-story single family home, duplex, or even a four-plex is possible with one access point. A larger sprawling ranch-style home with detached garage/rec room/granny quarters may present a new challenge—especially if the exterior walls are of stucco or other metal-containing construction (don't forget that foil backed insulation in the walls and ceilings can also inhibit wireless signals). When you need to cover a larger area or two buildings, as shown in Figure 6.3, two access points and one Ethernet cable between buildings may be required.

In closer, higher density quarters, such as apartment buildings or multi-family structures, you should consider the security of your network an absolute must. You should use 128-bit wired equivalent privacy (WEP) encryption and consider changing the WEP key often to avoid having your network compromised and your cable or digital subscriber line (DSL) bandwidth stolen. Several precautions against file sharing or using password protected file sharing are worthy of note as well. If you live in a house with stucco or aluminum siding or a metal-sided mobile home, security may be less of an issue because your wireless signal will not go far beyond the exterior walls, but be careful.

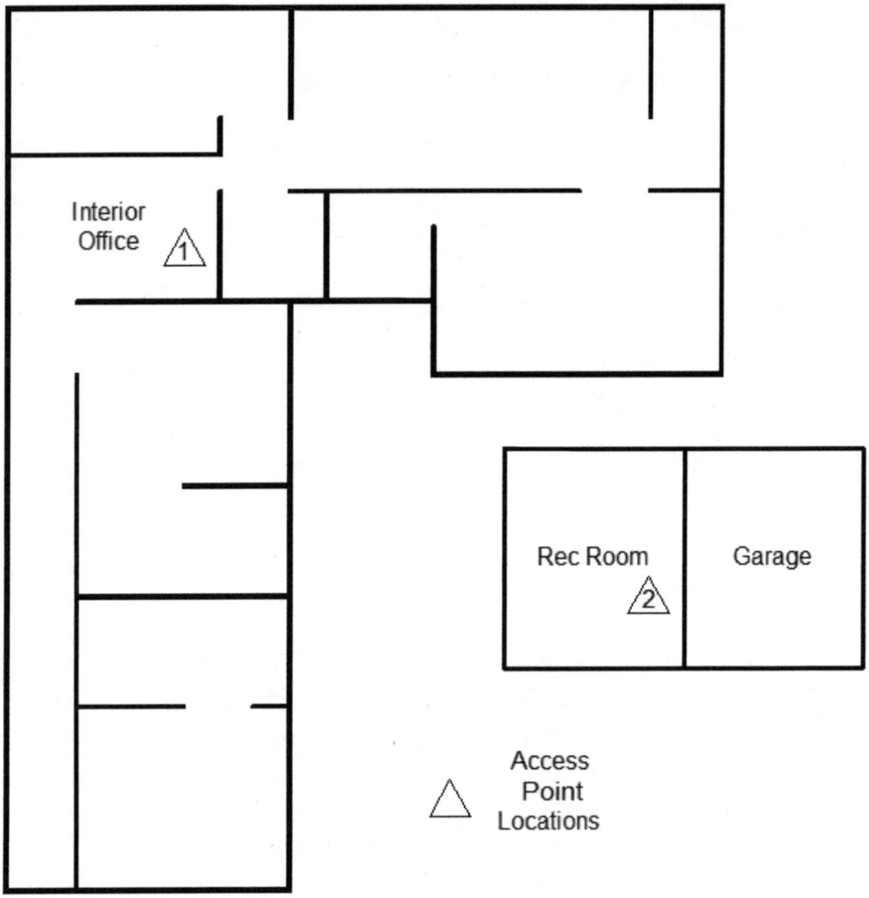

Figure 6.3
An expanded ranch house with a large area to cover may require two access points.

Wireless at Work

Your workplace may be a typical 2,000–10,000 square foot office or office and warehouse facility for 10 to 20 people, with 2 to 10 office areas, engaged in sales, light research or manufacturing, warehousing, or retail sales. Such facilities are not unlike a large home for coverage area, and even though you may be able to easily wire such a place with easy access over walls through a suspended acoustic ceiling, you may not want to bother with the trouble and expense of installing or reinstalling wires and jacks. In many high-tech areas, occupant turnover is quite high, and many tenants leave behind their own mess of uncertain wires and routing to their ideal common hub and server placement.

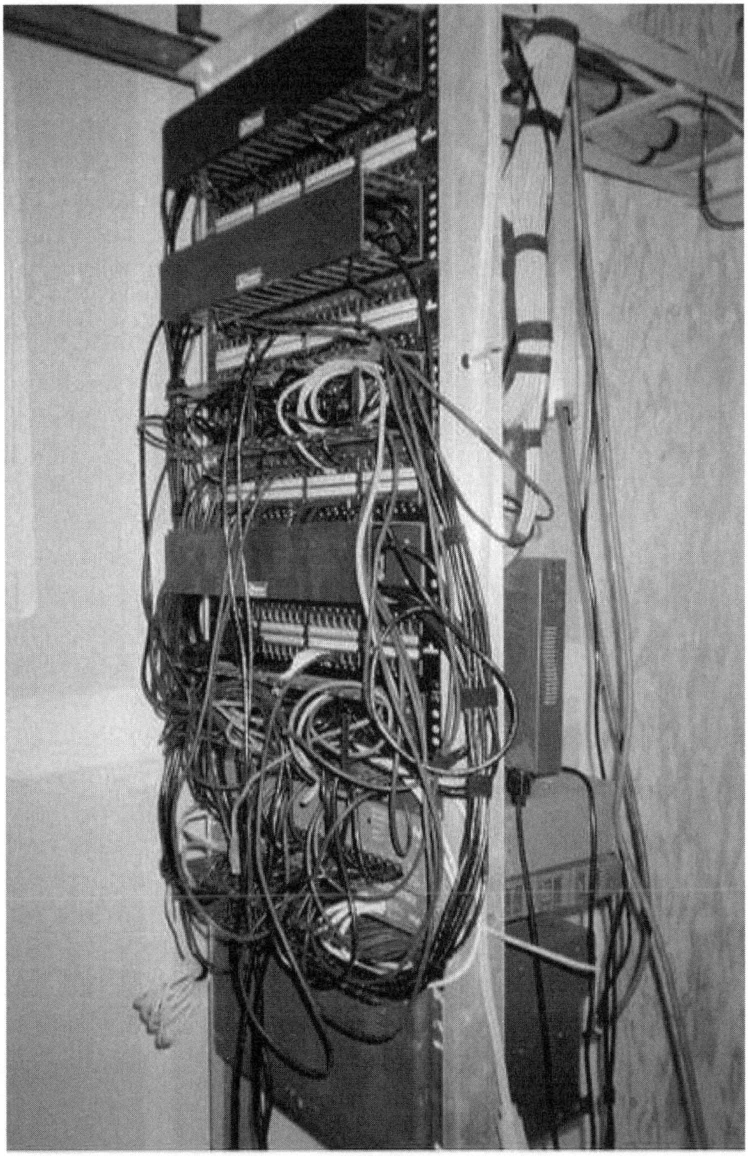

Figure 6.4
If you inherited a network wiring rack like this, or worse, and an equally messy telephone wiring rack, you can imagine the mess that is in the ceiling of your office.

The network diagram for a small office facility may be as simple as that shown in Figure 6.1. If you have a server for a Web site, e-mail, printer, or file sharing, you would probably wire it directly to a router/hub appliance, rather than connecting it through wireless—if only because that is one of the most critical components of your business, and losing connectivity due to a weak signal or interference could be costly.

When your office space begins to grow to or exceed 10,000 square feet, as shown in Figure 6.5, you will probably encounter more walls, pipes, electrical wiring, and plumbing that can interfere with the wireless signal. In such cases, you probably need to use two or more access points hardwired with an Ethernet cable to your main network/Internet connection point. Figures 6.5 and 6.6 illustrate the possible physical placement of two access points to cover a larger area, and the simplified diagram shows how they are connected to the network.

Installing two access points is not a significant mechanical or wiring consideration unless you cannot get a wire and power to the desired access point locations. Even if you need to install an access point in bridging mode to span some distance, you still need to get power to the access point so that it will work. This is when power over Ethernet (POE) comes into play. POE uses special adapters at each end of a standard Ethernet cable—one to supply power to unused wires in the Ethernet cable, and one to take that power and apply it to the access point unit.

When you introduce a second access point, you also introduce more complexity into the configuration of each access point and the client systems, because as the client systems must be able to roam freely and maintain a reliable connection as they move between access points.

One of these complexities is ensuring that each access point is configured to use different channels—preferably nonoverlapping channels of which there are three that do not share any of the 802.11b frequency spectrum—1, 6, and 11. That there are only three such channels limits your configuration options when you have more than three access points that may have overlapping coverage areas. However, three access points should be more than adequate to cover any one working area. If you need four or more access points, you can share channels with the access points that are farthest from each other.

Figure 6.5
A mid-sized office space with several walls and signal obstructions.

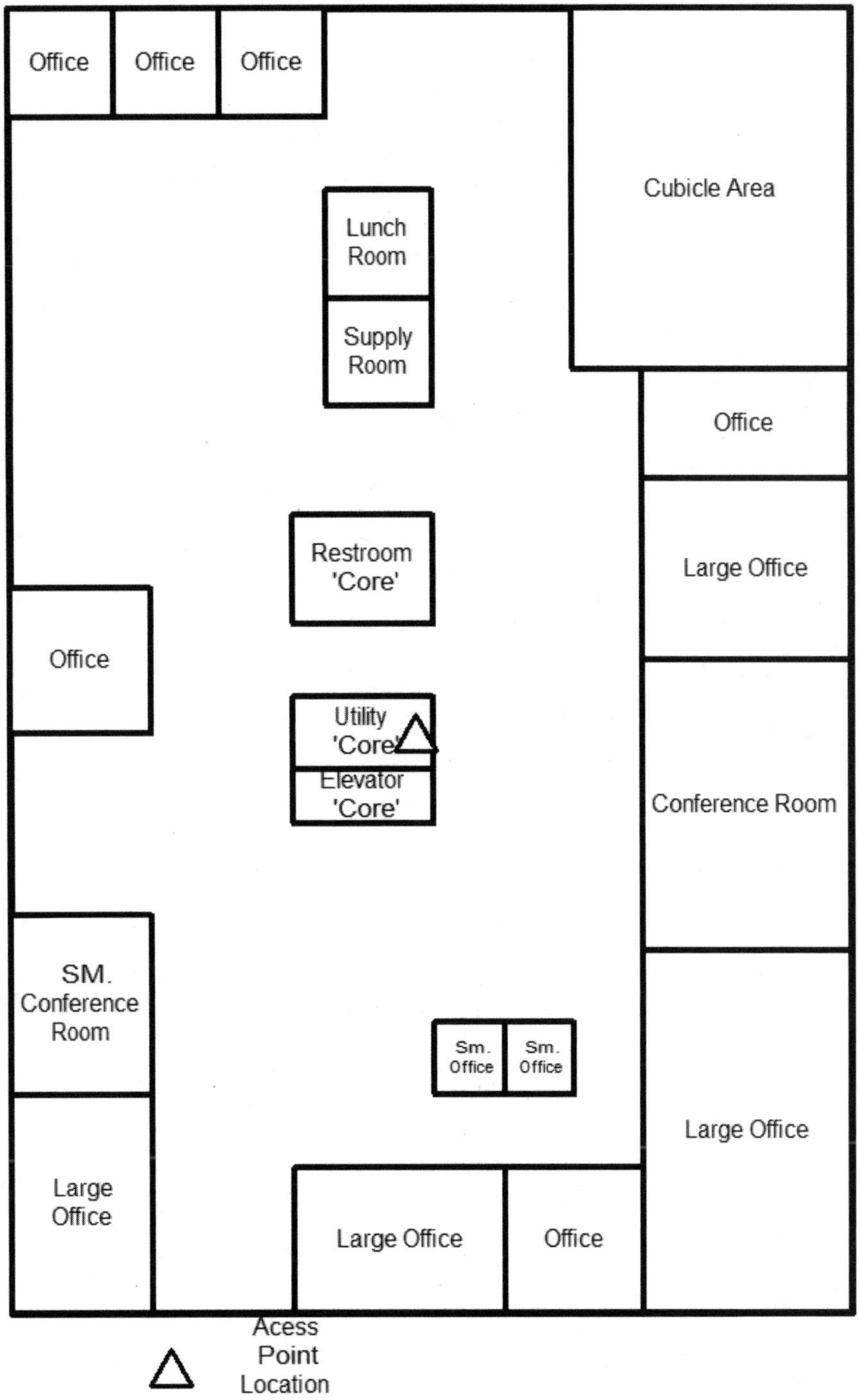

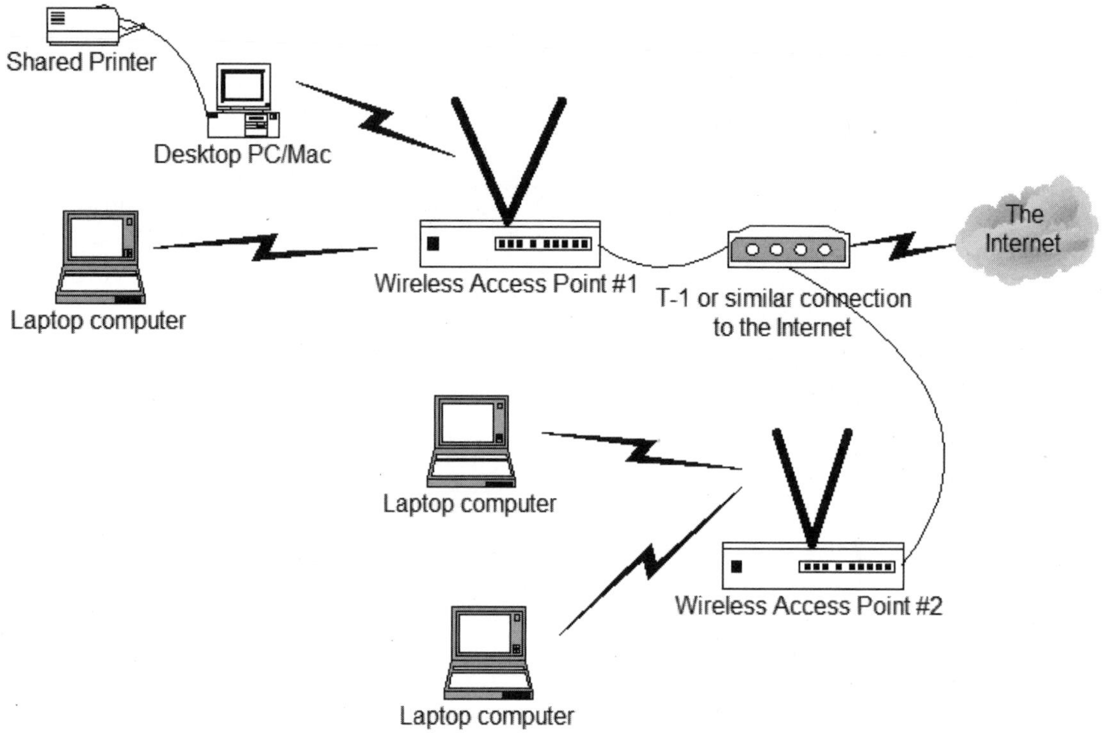

Figure 6.6 *Dual access points.*

Another of these complexities is selecting the proper service set identifier (SSID) for each access point. Theoretically, more than one access point can share the same SSID if they are using different channels; the radio portion, rather than the software portion of the wireless setup, negotiates a change of access points for the clients as they move between the different coverage areas for each access point.

Unless the facility is an older structure with wood framing, the walls will be set with steel frame pieces, as shown in Figure 6.7, which will reflect or block your wireless signals. Most office ceilings are suspended across steel frameworks hung on steel wires, as shown in this above ceiling photo in Figure 6.8, providing even more random points for signal reflections and multipath distortion of the desired signals.

Figure 6.7
Metal framing is quite common and prolific in new commercial construction. Metal framing contributes a new point of signal reflection every 16–24 inches.

Figure 6.8
What lurks above the ceiling could be your worst enemy for wireless networking. This picture of a new office facility, though not cluttered by an accumulation of old unused wiring, presents many opportunities for unwanted signal reflections.

If framing, ceilings, and old wiring are not your enemy when going wireless in a facility, there are still unseen hazards to be realized. The utility core of most buildings, as shown in Figure 6.9, holds much of the wiring, plumbing, and ventilation pieces needed to keep the place usable by human occupants—steel, copper, tin, cast iron, rust—all forms of electrical and electronic variables at play behind the wall-papered, painted, and tiled walls of rest rooms, closets, offices, and elevators. Speaking of elevators—you may notice your wireless signals changing sporadically throughout the day—timed oddly enough with the familiar whoosh and ding of an elevator going up and down and stopping on various floors. Could it be that large metal box moving into and out of the path of your wireless signals? Elevators are not made of wood or plastic—they are made of steel—which blocks radio signals!

Figure 6.9
Your office's utility core contains myriad hazards to wireless signals—water and waste plumbing, ventilation, and electrical components that act as unseen reflectors or blocks to your wireless signals. (Using your laptop in the "john" is probably not a good idea anyway!)

If, as in the case of our office example, you expand to another floor or part of one, at least one more access point may be needed to cover the other floor, especially if the flooring construction is poured concrete on metal suspensions as shown in Figure 6.10. Concrete contains a great deal of steel reinforcing bar, and is often poured over a corrugated metal pan to hold it in place. Radio signals barely penetrate dense rebar, and will not get through steel plating reliably at all.

Figure 6.10
Only Superman, a cutting torch, and jack hammer or nuclear radiation are getting through this steel floor pan with steel-reinforced concrete poured on top of it to reach the wireless clients above.

Wireless on Campus

A campus can be anything from a two- to three-building office complex and research park to the local community college. Pick any cluster of separate buildings and try to interconnect them—a challenge unless they have been designed with interconnection in mind. You would think that property owners and building contractors would have thought ahead and provided conduits for cabling between buildings. But more often than not, this is not the case.

The buildings in most multistructure complexes are built as stand-alone units, provided with separate addresses, power meters, and main telephone cabling and no common facilities except the driveway and parking lot. Interconnecting buildings by cable or fiber optics typically involves expensive trenching, piping, and repaving—

costing $10,000 or more. Then, even if you pay for this improvement to the facilities, the phone or power utility may claim ownership or some control over the pipe you paid to have installed, and suddenly your pipe is no longer your pipe to use as you please.

Faced with these costs, you can readily see that installing a few hundred dollars worth of wireless equipment can save you thousands of dollars immediately. You may even be able to extend your telephone system to the other building by using voice over IP (VoIP) technology routed through wireless network bridging equipment.

As your future wireless network is planned and grows to cover multiple buildings, and your network complexity increases proportionately, you will probably shift your equipment acquisitions from off-the-shelf consumer/home/small office products to larger scale network products from Proxim, Orinoco, 3Com, or Cisco. I don't know of any business with over 50 users that has not transitioned its network to use higher-end professional grade equipment. Boards of directors, investors, CEOs, and CIOs seem to like putting their money and careers on the line with a real networking infrastructure versus stuff the neighborhood computer geek uses at his or her house. If your business is on the line, invest heavily in the right infrastructure components to make sure your business and its employees stay on-line.

When your business expands beyond 50 to 100 users, and you have 3, 4, or 18 buildings to hook up to corporate networks, servers, and handle mega-packets of Internet traffic, you will not be surfing the Web over a garden-variety DSL circuit or even a single mere T-1. You may be obtaining Internet access over a larger T-3, DS-1, or high-capacity asynchronous transfer mode (ATM) circuit or multiple circuits. You may be providing separate Internet access circuits for each building so that all of your Internet traffic is not committed to just one circuit, and you can have backup routes, should one circuit fail. In this case, you may be employing primary and alternate traffic routing, and might be dabbling in a technique called *mesh routing*, where multiple Internet access ports are shared between network distribution points. Here, each building is provided with two or more ways to get to the Internet or just the main corporate network servers.

Wireless networking is an ideal technology to mesh with mesh routing. In fact, it is almost unavoidable, given that access points are typically located within sight of each other and can conceivably, if not practically, share bandwidth between them.

When you look at building up a wireless network for two- to three-buildings or more, obviously the complexity of the network increases, and the cost and quality of equipment you will use gets kicked up a notch or two. Even with better equipment and more deliberate planning, you cannot ignore the base-level problems inherent in any wireless network. Larger buildings with plumbing, electrical services, and other metal components present the same signal blockage and reflections as smaller facilities, no matter how professional the equipment. In fact, larger facilities have more rest rooms, thus more plumbing, and certainly more electrical circuits within them than smaller ones. Site surveys to determine any radio frequency (RF) signal issues—interference and optimum access point locations—will become more involved, as it should be when you are going to service many more users.

You will probably find yourself planning for many more access points, more bridges or Ethernet cable to connect to the access points, and more wireless bridges—altogether creating more setup and configuration complexity. You could find yourself building multiple wireless networks if you want to try to carry a VoIP phone system from building to building without using expensive dedicated T-1 carrier circuits between each building—one network for data and another for voice services.

In its simplest form, a multibuilding network expands from our simple two-access point 10,000 square foot single-story wireless network, to replicating that setup on multiple floors of some buildings, or to additional square footage on the same floor of a building, and then interconnecting the work areas of the buildings with wireless bridges as illustrated in Figure 6.11. The network topology for connecting multiple buildings may be similar to the illustration in Figure 6.12.

Bridges typically use directional antennas to create a stronger and more limited signal, to keep it within the bounds of the path between the buildings. As with access points, the bridge units themselves may be mounted at their respective antennas or use low-loss coax to interconnect them.

You can create a failsafe or fallback path between all the buildings, in case one of the bridge paths should fail, by adding an additional set of bridges between buildings 2 and 3, and configuring them or their respective routers.

Figure 6.11 The physical topology of interconnecting separate buildings with wireless bridges. Directional antennas are used at each end of the wireless link to provide a stronger, more reliable signal and prevent broadcasting interbuilding traffic around the entire area.

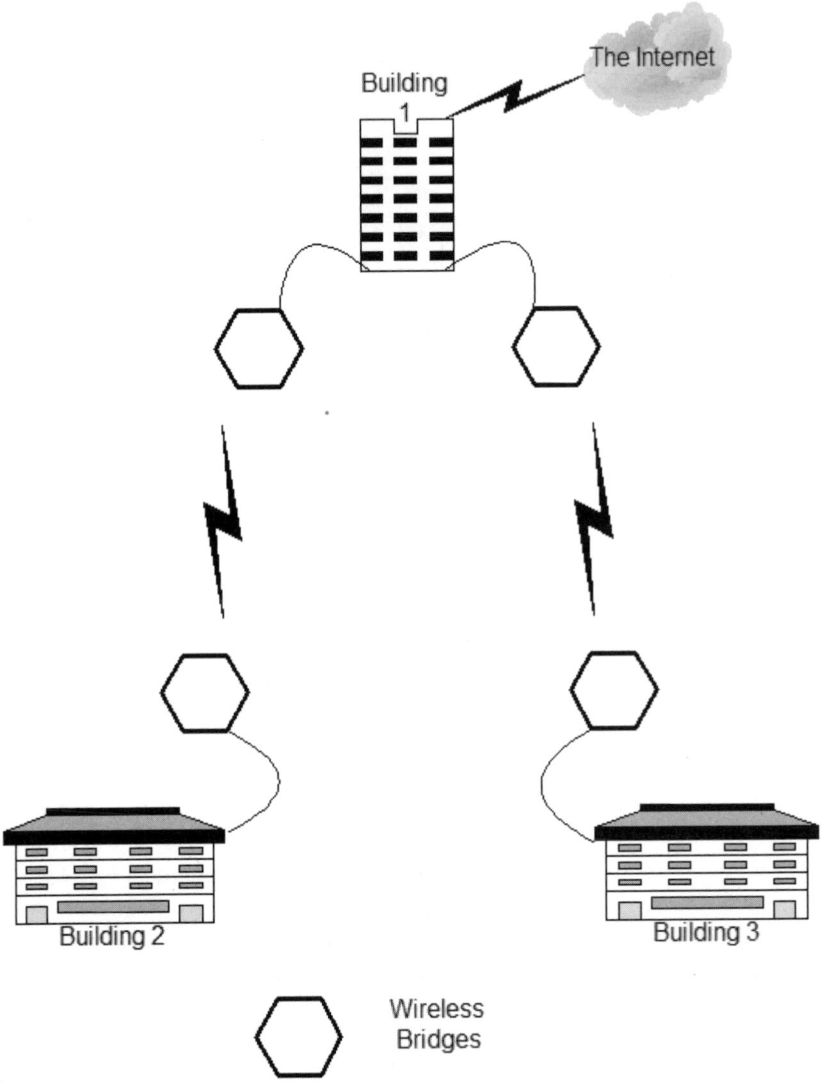

Wireless in the Community

Community wireless systems have literally sprouted from grassroots efforts. Like-minded individuals begin with the desire to connect to the Internet from almost anywhere they happen to be within a given geographical area, a bit of bandwidth to the Internet to spare, a few pieces of wireless equipment, and the technical know-how to begin

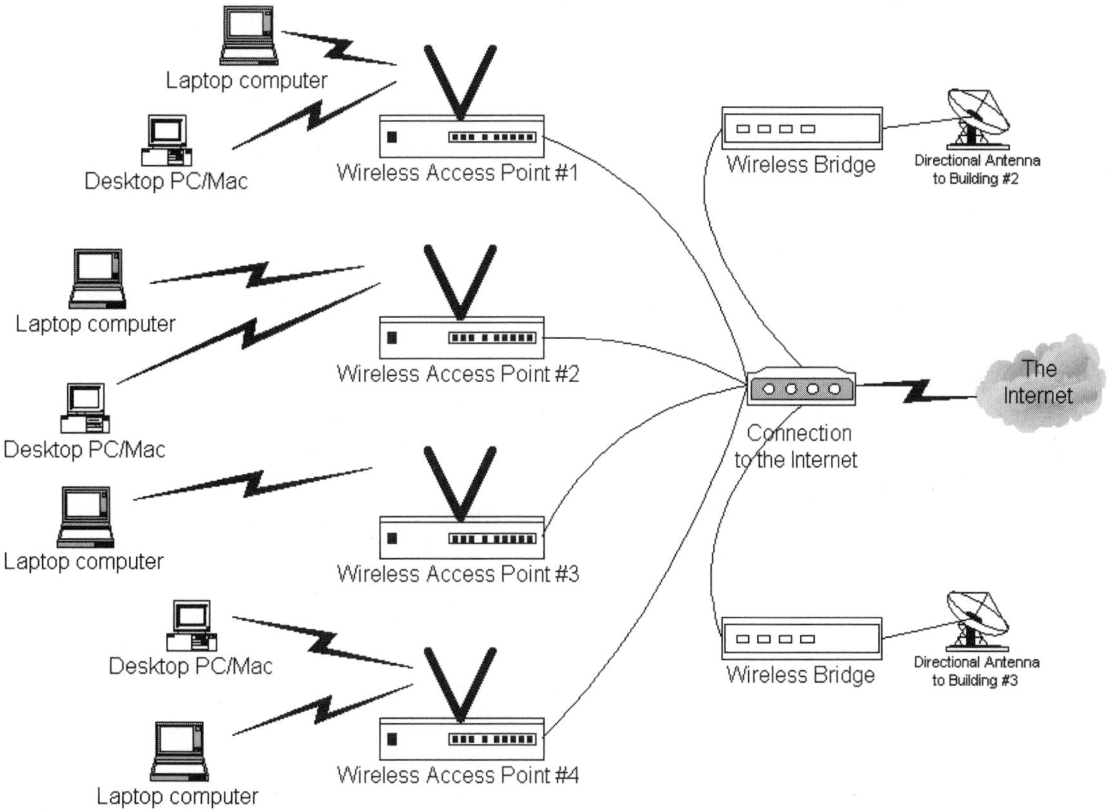

Figure 6.12 A simplified network diagram for Building 1 showing the interconnection of two bridges within the main network. The interface to the Internet would be protected with a firewall or router.

interconnecting, configuring, and securing different networks, all while allowing them to appear as one common source of Internet access. What may have started out as sharing bandwidth with a few neighbors grows to serving anyone in the community with a wireless network card and either plain curiosity or a reference list of "hot spots" of freely available Internet access.

You will find such community networks thriving in Seattle, New York, San Francisco, and other "hip" locations—only coincident perhaps with a high percentage of coffee shops—with a lot of people who like to lounge and surf at the same time. If one coffee shop with a DSL connection and an access point is a good thing, why not many? Why limit wireless access to coffee shops? Perhaps libraries, parks, and other public gathering places would be good locations for wireless Internet access!

In a corporate or personal network, you have greater awareness of the obstacles your network and wireless signals will face, and tighter control over access point location and client system configurations, so that you can implement and manage your network. An urban or suburban area in general brings into play many more and different concerns about signal strength, integrity, access control, and security—not the least of which is that you have no idea who will be using your public network, much less have any control over their system configuration, to help them resolve any problems they may encounter.

There is certainly no assumption or expectation that you will be building a community wireless network, much less troubleshooting and repairing one or directly supporting the users of the network, at least not without discussing the matter with others who have done or are currently doing so.

What you will find in such networks may be a deliberate, systematic deployment and configuration of equipment and access controls, so that your community appears to be served by a homogeneous entity, or simply a cooperative group of free service providers associated by a bond of common interests and the dissemination of "hot spot" locations and access information. Common gateways, shared or backup routings to avoid service outages, and deliberate engineering and installation of high-grade equipment and antenna systems to optimize coverage and performance may or may not be part of your community network. Again, these are typically grassroots efforts—or they began that way—albeit provided by very capable, professional, and responsible people for the love of doing so. Without significant funding, and the deliberate management, design, and deployment of such projects typically associated with huge corporations (like national ISPs or telecommunications companies), community networks will likely not follow any specific design or deployment patterns.

Wireless Internet Service Providers

Chances are, if someone has started a community wireless network where you live, someone else stood back, watched the opportunities develop, and took the big step into building a wireless network for hire. This means someone is charging users a per-day or per-month fee to share access to wired bandwidth—either a cable or DSL circuit at home or a larger T-1, T-3, DS1, or ATM circuit fed from a large Internet access provider data center.

Probably hundreds or thousands of folks get by on a neighborly handshake agreement. Party A charges Party B $10 to $20 per month to share the use of Party A's existing Internet access. Party A gives Party B the SSID and WEP key of their access point, and Party B inexpensively surfs away in their backyard, front yard, den, etc. Party A could decide to extend this offer to other neighbors as well, covering the full cost of their cable or DSL service, but spreading the bandwidth pretty thin among the paying users. There is no contractual service agreement, bandwidth performance guarantee, 24/7 technical support, etc. Party A's off-the-shelf access point or cable/DSL access could fail at any time, leaving Party B stranded. While indeed a wireless Internet service provider, or WISP, you can hardly call these backyard shade tree operations anything to build a business on, much less define a technical architecture or infrastructure around it.

If you do not have a kind neighbor willing to share costs and bandwidth, you may find new locally owned ISPs who charge higher fees for the flexibility of wireless networking capabilities beyond your own backyard. With a more structured, professional business environment, you get access through commercial-grade equipment that is installed at advantageous antenna sites and backed up with performance assurances and tighter security. Although this support is not as easy as peeking over the back fence, these ISPs may be better equipped to handle your needs. These businesses provide one or more levels of access control or subscriber authentication, much as you would find user name and password requirements to log on to a dial-up account. They may even have their system configured for the more robust IPSec level of transmission control protocol/Internet protocol (TCP/IP) security, rather than leaving your data to the whim of whoever can crack your WEP codes.

So far, there is no single nationwide wireless provider, like Earthlink, AT&T, AOL, or Sprint. But like the early days of the Internet circa 1994–1995, these companies are probably waiting out the evolution of wireless services from mom-and-pop-do-Internet until just the right moment to swoop in and run the market just as they do dial-up and, to some extent, other broadband services.

Instead of Earthlink and AOL, Boingo (www.boingo.com), HereUAre (www.hereuare.com), and Sputnik (www.sputnik.com) have created revenue sharing opportunities for local wireless providers. Those with bandwidth to spare, in a location with several potential users who are hoping to offset the cost of that bandwidth and their wireless equipment, can sign up with one of these companies as an affiliate or partner, use their software for access control to their network, and let their users enjoy their subscription to the service in other areas of town or other towns.

All of this has to do more with service or business models than with how these systems are deployed and what equipment they use—placing them more on the application side of wireless implementations. But this information is still food for thought if you want to build a larger wireless network outside the bounds of building frameworks and obstructions.

To build these networks for local use is as simple as positioning your access point with antenna, or just the antenna, in a location that exposes the signal from your access point to your intended audience of users. For a coffee shop, simply locating the access point or antenna in a central, unobstructed location—similar to a home or small office deployment—should work well enough, though signal coverage may drop off in the back store rooms or rest rooms. For an outdoor café or city park environment, the network architecture may be best served with two or more access points bridged together into one network backbone, with external antennas strategically placed to cover the entire area of interest.

Wireless ISPs have the more significant challenges of finding available building, radio tower, and similar space that provides them the ability to bridge their subscriber access points together (to avoid wireline charges to create the network). Each WISP site costs about $5,000 to build, and from $150 to $600 per month in space rental fees—that is, after spending $2,000 or more in time to experiment to see if a particular site will work well or not.

Building a reliable, Internet-everywhere wireless ISP service to serve indoor and outdoor users will take as many nearby sites as are used for cellular telephone services. Deutsch Telecom, doing business as T-Mobile in the U.S., is building up such a wireless infrastructure, but it will not be able to guarantee interference-free, totally reliable services because of the nature of 802.11a and 802.11b services, equipment, and regulatory issues.

Indeed, this may be a lot cheaper than renting or building and securing a reliable data center facility, buying into dozens of T-1 phone lines, expensive telephone system equipment, banks of modems, and dozens of servers, to run a conventional dial-up ISP. However, there are a lot of things working against wireless ISPs using 802.11a and 802.11b—mostly in the areas of being limited in how much area can be covered reliably with the fragile signals, and being able to withstand potential interference from other forms of radio use that share the same frequency bands.

Summary

We have covered several typical installation types in this chapter. From this, you should be able to determine which wireless network style or topology best suits your needs. As you venture into this new technology, it is suggested that you begin small—create a simple home or office network; get familiar with the equipment, setup software, different parameters, and security levels; and most important—the signal coverage and potential interference issues in your surroundings.

Both the wireless software configuration and the signal characteristics can be a bit daunting at first, though most of us are impressed by our first "on the bench" installations and tests ("it really works!")—until you encounter a new brand or version of wireless network card, or begin to move your laptop around the house or office and find out how wireless really behaves.

Once you have your network up and running and discover signal fades and dead spots, you will probably start experimenting with antennas—different locations and orientations—to improve the signal coverage. If you cannot cover the entire area of interest with one access point and antenna setup, you will advance to installing two or more access points and discover how to configure the access points and client-side software so that you can move between access point coverage areas seamlessly. Your confidence and expertise will increase to installing a wireless bridge or two and expanding your network to other buildings or separate coverage areas. In all cases, please remember you are dealing with the magic of radio waves and a technology originally intended only to replace the cumbersome Ethernet wiring that hampers or prevents some network installations and use a portable computer in a truly portable way.

CHAPTER 7

Hardware Installation and Setup

Have access point and wireless network card—will roam and surf at the same time. That is the promise and, for the most part, reality of wireless networking. Simple installations are simple—experiencing how they work is both amazing and enlightening. I cannot emphasize enough that you are dealing with the magic of radio waves—some of it black magic and some of it white magic. The black magic aspects cover both how amazingly well wireless works in cases where you think it could not, and how poorly it works in places where you would expect it to be flawless. The white magic is that, in most cases, it simply does work well—as designed and expected.

You will encounter some new widgets and software parameters when you are installing, connecting, and configuring your new wireless equipment—none of it too complex, but sometimes frustrating, as you encounter parameter differences between makes and models of equipment.

You will encounter all of those mysterious hidden signal reflections and shields as you wander around with a laptop, testing out the wireless signal coverage. From this, you will begin to experiment with antennas and different locations to optimize the signals.

As you work with different antennas, you will begin to visualize how each of them works, projecting in your mind the signal radiation patterns as you aim in one direction or the other. You may even pick up some orienteering skills as you figure compass angles and directions to unseen target locations. As you install antennas and feedline, you will begin to appreciate loss and how "fatter pipes" work best. Perhaps the trickiest part about doing your own antenna work is installing the special connectors—requiring dexterity, good eyesight, a few tools, a meter to check your work, and some patience.

Single Access Point Installations

Nothing could be simpler than installing your first access point—generally a plug-and-play process, with off-the-shelf products and step-by-step instructions. Go to the computer store, buy an access point and a wireless card or two, remove the products from their boxes, review the instruction manual, uncurl the cabling, plug the unit into power and your network, install the software, configure, and you are "good to go." Right?

The first step—going to the computer store—should not be done without a short shopping list of criteria for what you want to accomplish, the types of systems you will be setting up for wireless networking and their capabilities, and some research on available products. The first thing to look for is which units have detachable antennas. Even though you are not supposed to remove them and add external antennas, many people are going to want to do that. In reality, for an indoor system, that is probably OK. It is difficult to find a place that has both an AC outlet for power and an Ethernet connection. If you are going to run a wire or two, it may as well be antenna feedline, rather than stringing an unsafe AC extension cord into the attic or atop the ceiling tiles.

The second thing to look for is how you configure the unit—via Web browser through the Ethernet port, or separately via universal serial bus (USB) or serial port. The latter has a security advantage in that no one can configure your access point unit from the network side of things—although many commercial access points support configuration via simple network management protocol (SNMP) over the Ethernet port.

Another consideration is whether the unit also supports providing or passing through domain host configuration protocol (DHCP) services—which determines where your wireless client will get its Internet protocol (IP) addresses. You also need to know which level of wired equivalent privacy (WEP) encryption the unit supports—40-bits (5 ASCII characters) or 104-bits (13 ASCII characters) + 24 bits of encryption key data = 64-bits or 128-bits protection level. 104/128-bit encryption is obviously somewhat better than 64 bits, but you must have wireless cards for your client systems that support that level of WEP to be able to use it. It is still possible to purchase, new or used, cards that support only 40/64-bit WEP.

If you have to buy a cable digital subscriber line (DSL) or generic broadband router with some form of built-in firewall protection, DHCP server, and network address translation (NAT) in addition to the wireless access point, you might consider buying an access point with all three functions integrated into one unit. There are cost and convenience advantages to this, versus the technical flexibility of being able to change out either functional unit later.

What the Instruction Manual Will Tell You

Once you have made your buying decision, let's take a step back to that part about reading the instruction manual. This part cannot be emphasized enough. Get a cup of coffee, a soda, or a fresh bottle of water, and take a short break to ensure you know what you are going to be doing. Let's make this first installation a successful one, so that we can concentrate on the fun aspects of wireless and become accomplished at it.

You should already know whether your access point is configurable by a web browser once it is hooked up to your Ethernet connection, through a serial port, or USB port and special software. Without this information and following the appropriate steps, installation of your access point's configuration software could fail and have to be re-done, or you could end up losing control of your access point and have to reset it to factory default values and start over. We still live in an age where "plug-and-play" is not a 100 percent reality, and many USB devices and Windows operating systems still require that you install software before connecting devices.

If your access point is configurable over Ethernet via web browser or SNMP, you will need to know its default IP address or if it gets an address from a DHCP server. The former is more common, so you can determine a starting point to configure the server.

Once you have determined software installation and connectivity, and gained control of the access point with configuration software, you will have to know how you want to set it up. Factors to consider include:

- Whether to change the default IP address to an address compatible with your network
- How DHCP is going to be handled for clients
- What channel the access point uses and if you need or want to change it to a nonoverlapping channel (1, 6, or 11)
- What security method, if any, you are going to use
- If security is used, establish a security key to be set into the access point and all of your clients

Table 7.1 provides a handy worksheet to make note of the default values that come preset in your access point and wireless cards, and your customized values.

TABLE 7.1

A handy worksheet to make note of your access point and wireless card configurations.

Parameter	Default	Customized
Product Make		
Product Model		
Firmware Version		
Configuration Interface	USB ___ Ethernet ___	
Configuration Method	Software ___ Web ___ SNMP ___	Software ___ Web ___ SNMP ___
Configuration Password	Read Access Write Access	Read Access Write Access
SSID		
Access Point IP Address		
Access Point Gateway IP		
Access Point DHCP Source		
WEP Level	Off ___ 40/64-bit ___ 104/128-bit ___	Off ___ 40/64-bit ___ 104/128-bit ___
WEP Key 1 Value	ASCII: Hex:	ASCII: Hex:
WEP Key 2 Value	ASCII: Hex:	ASCII: Hex:
WEP Key 3 Value	ASCII: Hex:	ASCII: Hex:
WEP Key 4 Value	ASCII: Hex:	ASCII: Hex:
DNS Servers		
Access Point Mode	Access Point Bridge	Access Point Bridge

I suggest starting out with no WEP security key, just to get your clients onto the wireless network for a brief testing period. Once you determine the wireless portion works, then turn on security. This requires that you know if you are going to use 40/64 or 104/128 bit security level, if your key will start out as a string of ASCII characters or hexadecimal (Hex) characters, and which key of the four available you are going to use. These last two items are a source of great confusion when using equipment or operating systems from different vendors.

Windows XP numbers the four available WEP keys 0, 1, 2, and 3, while most wireless devices number them 1, 2, 3, and 4 (equivalent to Windows' 0, 1, 2, and 3, respectively). LinkSys wireless devices and Windows support entering the key in ASCII or Hex, but some operating systems and devices require the key be entered as Hex. There are simple conversion programs that let you enter either ASCII or Hex, and then present the converted values for you—one is available for use on-line at www.powerdog.com. Once you have determined a key value to use, make note of both the ASCII and Hex versions of the key so that you can easily apply the appropriate one in your configuration.

Once you have all of this information mapped out for your access point, check the parameters and configuration capabilities of your wireless adapters. You may have to mix and match WEP key levels and wireless channels to establish a common set of parameters you can use throughout your entire network and for anyone visiting your network.

Hardware Configuration Concerns

Repeat the process of becoming familiar with your access point for each of the wireless adapters you will apply to your client systems. Be especially aware of USB driver installation requirements before connecting external client adapters to the systems that will use them.

Also, be aware of any input/output (I/O) port address or interrupt request (IRQ) configuration problems with either industry standard architecture (ISA) or peripheral component interconnect (PCI)-based plug-in adapters for desktop systems. The concern over I/O conflicts is especially critical with some system boards that have built-in audio, certain audio cards, and the Linux operating system. You may have to disable or reconfigure your audio device manually to work around conflicts presented by the wireless adapter.

Windows 98, 98SE, Me, and XP support true plug-and-play for most compatible PCI devices, and recent versions of Linux do as well, though embedded audio and video chipsets may be stubborn about their plug-and-play capabilities. Check for updates to your system board basic input/output system (BIOS), and be sure the BIOS settings have plug-and-play enabled. Reset the PCI/non-volatile random access memory (NVRAM) configuration if necessary to get all the devices properly recognized and reconfigured with any new hardware you install.

If in doubt about I/O port issues—what they are and how to resolve them—check out my PC configuration book, *IRQ, DMA and I/O* (3rd edition, IDG, 1999). It may be a little hard to find a copy, but you will swear by it once you get your hands on it!

Connecting and Configuring Your Access Point

Now that you know what to expect of your access point and client adapters, it is time to start hooking things up. Start with the access point since it is core to your network and central to all the clients that will connect to it over wireless.

If the access point uses a USB interface for configuration, install the configuration software and hook up this—and only this—interface first, leaving the Ethernet connection unplugged until you have configured the access point for your specific network parameters. If Ethernet is the only way to configure the access point, use your browser or configuration software to begin the configuration. Start with providing or changing the configuration password to prevent anyone from tampering with your access point.

Security Note: *Anytime there is a security option to control access to a device or its configuration, enable security immediately and change the default password to a unique passphrase. This applies to access point configurations, wireless network service set identifiers (SSIDs), WEP keys, and your local network's workgroup or network identification (typically 'Workgroup' in Windows). In Windows, disable or password-secure any and all file and printer sharing services to prevent hacking and information theft.*

Establish a strong password policy—typically at least 6 to 8 characters long containing a mix of letters, numbers, and punctuation characters if allowed—avoiding common names of family members, pets, etc. Get creative and take a simple word like "plateglass" and substitute a few characters to make it hard to decipher, like pl8gl@ss for example.

Connecting and Configuring Your Client Adapters

As ubiquitous and promising as (true) plug-and-play is supposed to be, your installation of a wireless adapter may not be as smooth as you would like. I have encountered the simple failure of drivers to install, whether or not a new PC card or USB device has been installed and detected—making for some careful doctoring of Windows and a couple of laptops to make everything well. Still, over 80 to 90 percent of today's PCs will accept any wireless networking adapter you can toss into them, and they will work fine!

Barring problems getting your computer and its operating system to recognize the availability of a wireless network adapter, you need to become familiar with the normal parameters a wireless network setup requires. If your system does not initially recognize your chosen wireless adapter and accept the driver installation for it, we probably have a fix for that before you get to the network parameters.

PC card wireless adapters. The PC card adapter is the most common wireless network device—catering to the laptop market, but also commonly used in special PCI and ISA adapter card slots for desktop systems. Connecting a wireless card, having the operating system recognize it, and installing the driver software is a simple and successful task over 90 percent of the time.

Occasionally, the card manufacturer's driver software will not install in the operating system, and your only solution is likely to be hooking up to the Internet to search for, download, and try a different version of the software.

Apple Macintosh users will find their choice of wireless adapters limited to the Apple Airport wireless product line. Only a few of the typical wireless network equipment vendors supply wireless card products for the iBook series. While limiting your options, this makes setting up an Apple product for wireless quite a bit easier.

If your operating system preference runs to Linux or FreeBSD, you may find drivers for your card provided by the manufacturer, generic drivers with the operating system, or drivers available from open-source contributors.

USB wireless adapters. Installing a USB wireless adapter is as simple as installing a PC card—no I/O issues, but you must check the instruction manual to determine if you must install the driver software before connecting the device to your computer. This is a typical scenario for Windows 98SE, Windows Me, and 2000, while Windows XP should be able to accommodate post-connection driver installation.

PCI and ISA wireless adapters. There are two types of internal plug-in wireless adapter cards: those that have the wireless interface components built onto the board, and those that contain an adapter or slot for a PC card adapter, thus creating a PC card slot in a standard desktop PC I/O socket.

The PCI-based cards will be plug-and-play compatible, identified by the system BIOS and then the operating system. These cards cause the operating system to look for a driver, and prompt for a driver to be installed if none exists.

ISA-based cards are typically not plug-and-play compatible, so you must know how to install the driver software and configure the card manually.

If you are going to experience any hardware configuration issues installing a wireless adapter, they will be with PCI and ISA cards—the classic, legacy headaches plug-and-play is designed to avoid, if plug-and-play cooperates.

ISA device conflicts are common and well known. You must know what your current PC configuration is and be able to select unique I/O address and IRQ settings that do not conflict with other devices already in the system, or go through an entire system reconfiguration to make all the pieces work together.

PCI devices very rarely conflict with other devices because they are configured automatically by the plug-and-play process and proper driver software. However, you may find some manufacturers' products that do not reconfigure themselves based on normal plug-and-play rules, and you will have to manually reconfigure or disable them. Users of Linux have encountered problems with some system boards that have built-in audio chipsets that will not reconfigure properly when plug-and-play detects a change in hardware. In these cases, you must check the system BIOS settings and change the configuration for the audio components or disable them to get your wireless adapter to work.

The key to success with legacy ISA and some PCI devices is to find, record, and check your system's current hardware configuration against the requirements of any new devices you are installing. If you know the settings you have to work with, what the proper settings should be, and what you can change the settings to, you can make your new device work just fine. Given a choice for legacy/ISA cards, I would suggest using I/O address and IRQ settings typical for those of a normal wired network card—either address 280h or 340h and IRQ 5 or 10—as plug-and-play can work around these fixed settings in most system configurations.

Configuring Your Wireless Adapter

Once your wireless adapter is in place and your operating system recognizes it, you will have access to the wireless network setup parameters to specify network identifications, WEP keys, wireless channels, and standard transmission control protocol (TCP/IP) network parameters. For Microsoft Windows and Apple OS X users, these parameters are available in the operating system's network setup screens. Linux users will have to tinker with specific configuration files.

Windows XP

Microsoft Windows XP was made to be wireless-aware and about as wireless-friendly as possible. When installed, most wireless adapters are at least recognized by plug-and-play in Windows XP followed by a prompt to provide the adapter's installation CD for driver installation.

Although XP will usually automatically install the drivers to make the card functional, often you will have to supply the driver CD for XP to complete the tasks. XP's driver installation process does not install any of the adapter maker's special software for configuring or monitoring the adapter's status. Once XP is done installing drivers, it is recommended that you run the setup program from the driver CD to gain the full benefit of the card.

Once the card, its drivers, and its software are installed and ready, XP becomes wireless aware, requiring only that you provide the specific parameters needed to connect to a wireless network. From there, XP's built-in wireless-aware network support can present you with a new set of network status and configuration screens.

When you go to Start, select My Computer, then My Network Places, and select View Network Connections, you will see a listing for "Wireless Network Connection" with an odd-looking antenna icon, similar to that shown in Figure 7.1. Older versions of Windows have no such distinction in their network properties dialogs.

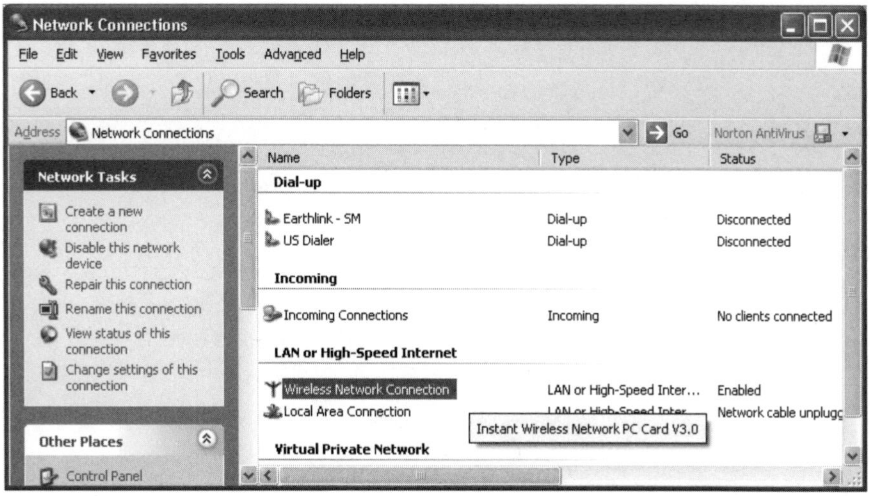

Figure 7.1 XP Network Connections window reveals its support for wireless with a distinctive antenna icon representing an installed wireless adapter.

Right-click on the new icon, select Properties, and then the new Wireless Networks tab appears—see Figure 7.2. From here, you can preconfigure a known wireless connection or reconfigure an existing connection.

To configure your system for a new wireless network, select the Add button to bring up a fresh Wireless Network Properties dialog—see Figure 7.3.

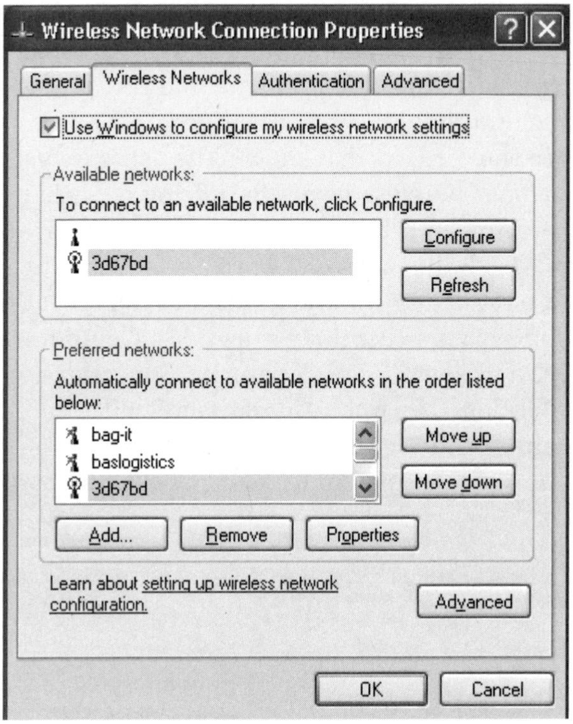

Figure 7.2
Known active and previously configured wireless networks appear in the Wireless Network Connection Properties dialog.

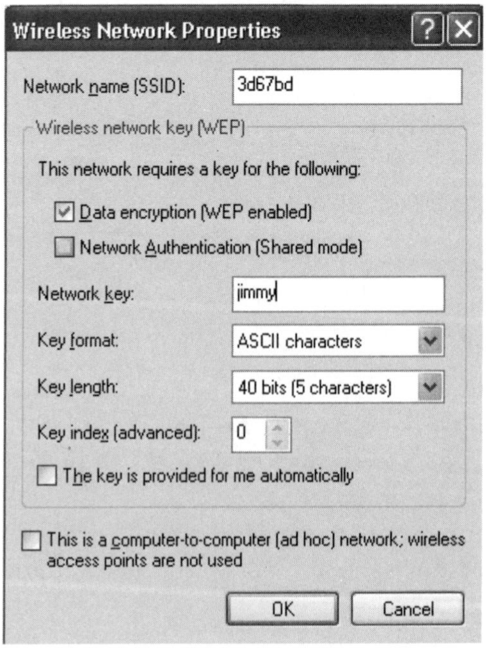

Figure 7.3
The Wireless Network Properties dialog is the place to preconfigure the SSID and encryption parameters required to connect to an access point, or set up an ad-hoc, peer-to-peer wireless connection.

In most cases, you will enable data encryption (WEP enabled), not select Network Authentication (Shared mode), and provide a Network key in normal text (which is then converted into Hex) of the appropriate length for a 40/64 or 104/128-bit WEP security level. Unless you or your network administrator elect to change which key is used on a daily or weekly basis, you will most often leave the Key index (advanced) setting at 0 (zero).

Note: *Windows XP numbers WEP keys from 0–3, while the configuration programs of most access points and wireless adapters number the keys 1–4. If your network is not using the default first key, your network administrator should specify if you are to set the Windows XP key index or the wireless adapter configuration, and which specific key number to use in each case. You may assume if the network administrator says to use Key 0 he is making reference to Windows XP (the adapter configuration has no key 0). If the administrator says to use Key 4, he means the key setting for the wireless adapter (Windows XP has no key 4). For Windows 98–2000, you will use the configuration program provided with your wireless adapter to make these changes.*

When you have completed the settings, click OK a couple of times in succession to close the dialogs, and then observe the wireless icon in the tool tray and the pop-up status flags that appear as your network adapter finds and connects to your access point. Double-clicking the icon will present a dialog similar to Figure 7.4, showing you the relative signal strength and data packet activity of your connection.

Right-clicking the wireless adapter icon in the tool tray will present a small menu with choices to Disable, obtain Status, Repair, View Available Wireless Networks, and Open Network Connections. Selecting Disable will disable the network interface at the software level, blocking all network traffic through the adapter. Status is the same as double-clicking the icon, and will present Windows' Signal Strength and Packet Information dialog. Repair will invoke the Windows `IPCONFIG/ RENEW` process to make this connection and try to find a DHCP server to obtain new IP address settings in case you have lost your connection. View Available Wireless Networks will present the Wireless Network Selection dialog, shown in Figure 7.5. Open Network Connections presents the complete Network Connection window, as shown in Figure 7.4.

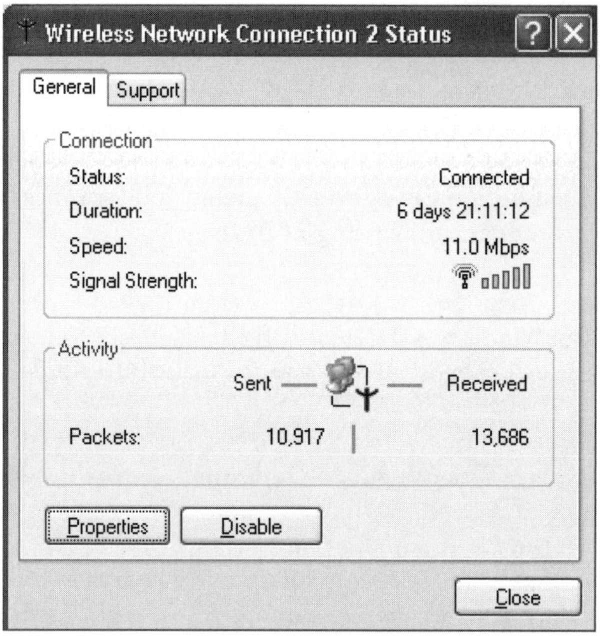

Figure 7.4
Click on the wireless network icon in the tool tray to see the status of your wireless connection—at least whether or not packets are flowing in both directions. Select the Support tab to see your TCP/IP address information for this connection.

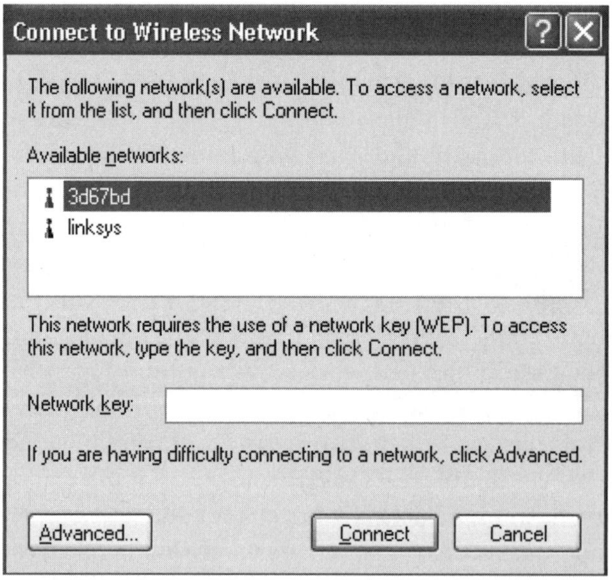

Figure 7.5
The Connect to Wireless Network dialog shows you the SSID of which wireless networks are available. In this view, two network SSIDs are shown.

If an access point is not broadcasting an SSID, because keeping the SSID hidden adds an additional level of security, you have to know and preconfigure the SSID for this specific connection to use it.

If you have not previously connected to either of these networks or have not configured a connection, and know the SSID and WEP key for the network, simply type in the key information, then select Connect to make your connection.

Clicking the Advanced button shown in the Connect to Wireless Network dialog causes this dialog to close, then opens the Wireless Networks tab dialog from the wireless adapter's properties so that you can reconfigure parameters as needed. Here, Microsoft has provided easy access to various complicated settings that were otherwise buried or accessible only from leaving one context and navigating through another.

Your wireless adapter probably comes with a program to give you more information about your wireless connection. Figure 7.6 shows the media access control (MAC) address of the access point for which this connection is associated, its transmit data rate, channel, packet throughput, link quality, and the signal strength of the active connection.

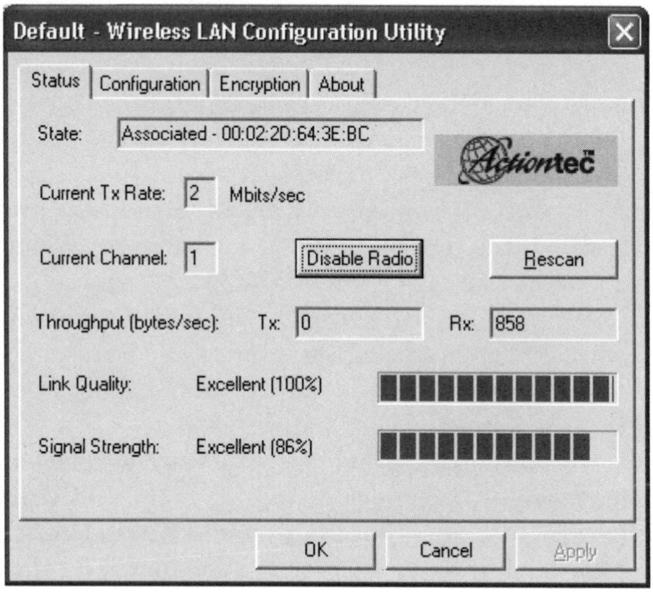

Figure 7.6 The status screen from an Actiontec wireless adapter connected to an access point. Through this dialog, you also have access to the adapter's configuration parameters.

The Link Quality indicator gives you a relative indication of the connection's ability to provide full data bandwidth. The Signal Strength indicator tells you how strong the wireless signal is. You could use this program to reconfigure your card under Windows XP, but since XP has its own wireless support, this program's reconfiguration features are mainly for use with Windows 98–2000.

Windows 98, 98SE, and Me

Earlier versions of Windows also support plug-and-play, but since they are not wireless-aware, you must install the software that comes with your adapter so that you can configure and check the status of your wireless card. For this section, I set up a Windows Me desktop system with a LinkSys WMP11 PCI bus wireless adapter. No matter what type of wireless adapter you have—PC card, PCI, ISA, or USB—the setup and configuration involves the same wireless network settings and parameters.

Because much of the hardware you buy today was not known to these earlier operating systems, you must install the driver and configuration program from the CD-ROM that comes with the card. This can be done before or after you physically install the card in your PC system chassis.

Once the card and software are installed and you restart the system, a resident configuration program is left running in your Windows desktop tool tray. Double-click the icon for the program to access the wireless network configuration. These screens, as shown in Figure 7.7, can be very helpful in determining connection status and quality, especially if you are running Windows 98–2000, which do not have built-in wireless support.

Figure 7.7 is the status screen for a LinkSys WMP11 PCI desktop wireless card showing the connection state to a nearby access point and its MAC address, the associated access points SSID, channel, transfer rate, and signal strength.

The programs provided with your wireless adapter may be the only way you have to configure them and wireless networking in general, which is the case with the WMP11 card under Windows Me. Figure 7.8 shows the Configuration dialog for the adapter.

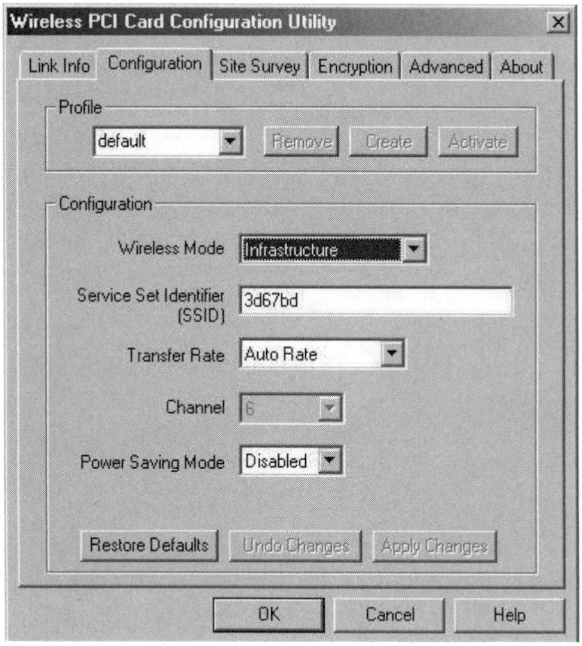

Figure 7.8
The Configuration dialog for the LinkSys WMP11 gives you access to a few basic parameters—the most important of which is the SSID of the network to which you wish to connect. You will have to interact with this dialog, or the Site Survey dialog in Figure 7.9 and the Encryption dialog shown in Figure 7.10, each time you change networks.

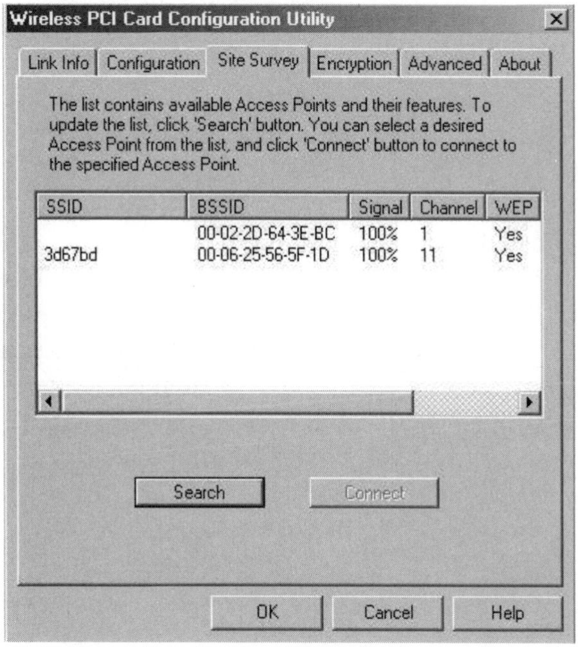

Figure 7.9
The Site Survey dialog for the WMP11 shows you all available wireless networks, similar to Windows XP's View Available Wireless Networks dialog. From the list, you can select a different network to connect to.

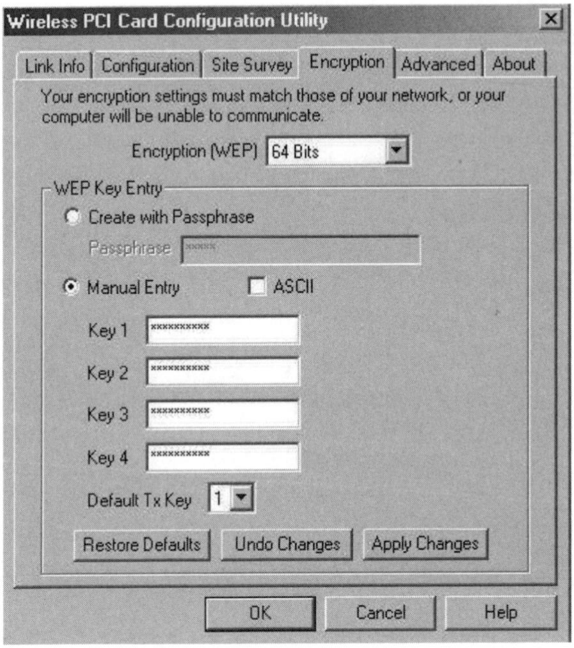

Figure 7.10 The WMP11's Encryption dialog is where you set up the WEP encryption level and keys. Expect interaction with this dialog to change your WEP key values when you change networks.

Setting up a wireless network can involve a few other parameters, as shown in Figure 7.11, such as Fragmentation Threshold and RTS/CTS Threshold, which are usually left at their default and enabled values—including automatic determination of network access authentication methods. These values are provided to customize client and access point interactions for performance, but changing them can reduce your performance when connecting to other networks.

As you work with your wireless adapter and those in other systems, it is suggested you keep track of the default and customized settings for each client system so that you have a record of what works best and can easily get the client operating properly if something changes. Like the table provided for access point configurations, Table 7.2 is a handy tool to use to record your client adapter settings.

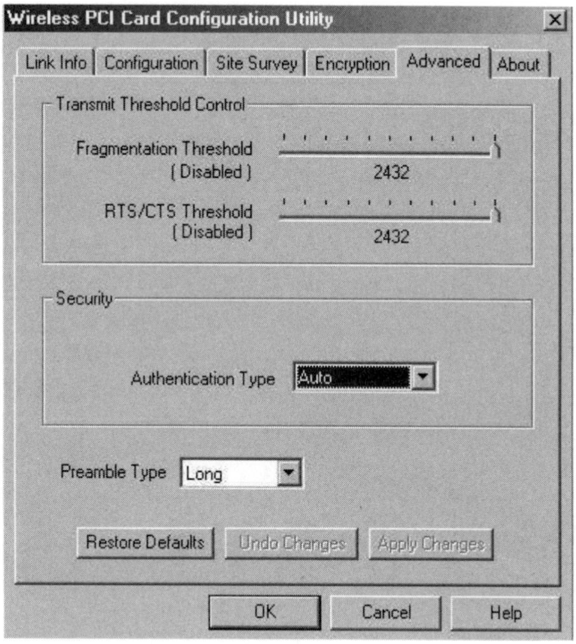

Figure 7.11 The Advanced settings dialog for the WMP11 adapter provides access to some of the more arcane and seldom changed parameters for wireless networking.

With your access point and client network card installed, configured, and ready to try out, you are of course anxious to see if you can indeed connect your computer to your network without wires, and hopefully keep it connected. If you are just setting up a client adapter to connect to someone else's network, the next section applies to you as well.

First Connect Problems

The first successful wireless connection you make may be very exciting—expectations are high, you are tentative, filled with anticipation, and then—*"it works!"* You are elated, taken aback slightly, pulse racing—sort of like getting a tingly shock from touching the tip of a glowing magic wand—then off you go merrily surfing the Web in every corner of the house. You will be toting your laptop out the front door, down the driveway to the mailbox, through the garage to the backyard, past the doghouse to the swingset, into the back door, around the kitchen, plop down on the sofa to surf for

TABLE 7.2

A handy worksheet to make note of your wireless interface configurations. Keep one of these charts handy with data for each network you connect to so that you can remember the SSID and WEP key settings.

Parameter	Network 1 (Home)	Network 2 (Office)
Product Make		
Product Model		
Firmware Version		
Configuration		
Interface	USB ___ PC Bus ___	
SSID		
WEP Level	Off ___ 40/64-bit ___ 104/128-bit ___	Off ___ 40/64-bit ___ 104/128-bit ___
WEP Key 1 Value	ASCII: Hex:	ASCII: ASCII: Hex:
WEP Key 2 Value	ASCII: Hex:	ASCII: Hex:
WEP Key 3 Value	ASCII: Hex:	ASCII: Hex:
WEP Key 4 Value	ASCII: Hex:	ASCII: Hex:
Wireless Mode	Infrastructure ___ Ad Hoc ___	Infrastructure ___ Ad Hoc ___

sports stats during the ball game—then you see your neighbor out watering the lawn and you rush out to show him just how cool this wireless stuff is.

About the time you get halfway across the street or around the neighbor's hedge, the signal drops off and you wilt sheepishly, feeling like your lush green lawn just turned to crabgrass and dandelions before your eyes. You lost the signal—oh darn—you begin thinking about adding more power, external antennas, multiple access points.... This will be described in Chapter 8.

What if your first connection attempt is not successful? What went wrong? Suddenly wireless networking begins to feel like the first time you installed an Ethernet card in your desktop PC to begin enjoying that broadband cable or DSL connection. You might have spent all night fighting addressing and IRQ issues, futzing with straight versus cross-over cables, resetting your router, and changing IP addresses and workgroup names, only to find that you had the Caps Lock key on when you typed in your password or something equally simple to recognize and fix.

You have almost all of the same issues getting your hardware installed: you may have to adjust some networking parameters in your operating system, and then you have all those new wireless parameters to worry about. What is the problem? Which parameter do you tweak first to correct the problem?

If you had no trouble at all installing the drivers and the hardware, and you saw what appeared to be all the right plug-and-play (Windows: "New Hardware Found" and new hardware ready to use) indications. In Windows, chances are that a new icon or two have appeared in the Taskbar Tool Tray, telling you that the status and configuration of the card are running OK. So, let's not suspect hardware, software, or drivers just yet. You have a few new parameters to get used to for both your access point and client configuration before a successful connection all the way through to a network or the Internet is possible.

Common Connection Problems

Once the hardware is installed and configured, there are only a few types of problems you could have in establishing a wireless connection—most of these in the few variables that must align correctly between any access point and client adapter to establish a connection between themselves and with a specific network.

Some of these problems are easily solved, and some require specific knowledge and technical intervention—with more hardware or sense of the magic of wireless signals. Below is a mini troubleshooting guide, including the following problems:

- **SSID**—Is the SSID known and properly configured at the client system?

- **Improper WEP key**—ASCII alphanumeric or Hex key? 40/64- or 104/128-bit?
- **DHCP configuration**—Access point preventing client IP setup from DHCP server.
- **Wireless signal quality**—Nonexistent to intermittent to weak/poor signal.
- **Wrong WEP Key Index**—0–3 or 1–4 correlation?

SSID. If you do not know the SSID of the network access point you wish to connect with, you will have a very tough time making the connection—even if you know the WEP key information.

Many private networks are secured first by configuring the access points so that the SSID is not broadcast. This will not prevent someone from deciphering the SSID and WEP key out of the data that exists on the wireless signal, then trying to make a connection, but it makes the task more difficult.

The lesson is that you need to know the SSID of the access points you want to connect to—whether or not the SSID is broadcast—and place this information into the SSID entry point for the client-side wireless connection configuration. Otherwise, your client-side has no idea which network to try to connect with.

WEP key. Using the wrong WEP key to attempt a connection to an access point that requires one is another show-stopper. Unfortunately, this element of networking takes us back to the early days of DOS, and perhaps before, with hand-coded PCs—when users typically knew how to deal with a lot of cryptic alphanumeric representations of bits of text or even nonsense text characters. The WEP key is passed between the client and the access point in hexadecimal form, but there is a provision in most devices to provide the key information in plain ASCII text, which is then converted to Hex format.

If you are lucky, your client or access point setup program will reveal the Hex format of your ASCII test WEP key entry—and if so—record both the text and Hex versions of the key, in case you run into a configuration program that will accept only one or the other.

If you feel the need to experiment with different ASCII and Hex sequences, or figure out one from the other, a quick pass by the www.powerdog.com Web site should satisfy you. If you would rather try the conversions yourself, check the ASCII–Hex conversion chart in the appendices.

Indications that you are using the wrong key are usually on-screen messages indicating some form of failure to authenticate—either by a recurring login dialog; an apparent authentication, but no data packets flow back and forth (monitored by the network status screen of Windows or your adapter's configuration or status program); or the lack of a valid TCP/IP configuration, covered next.

The remedy for your WEP key woes is to determine the correct WEP key in ASCII and Hex—for all four key index references, whether the key required is 40/64-bit or 104/128-bit—and which of the four key index references is being used at the access point. Provide that information at the appropriate point in your client configuration, and if you do not have any of the other common problems, your connection should come together just fine.

Dynamic client configuration. There is nothing worse than making sure you can establish a solid connection with an access point and then failing to connect with the network beyond the access point. If you think you have connected to an access point but cannot access any network resources or surf the Web, then your client configuration has probably not been provided useful TCP/IP address information for the network you are using, or the client configuration you are trying to use has the wrong information.

Most wireless networks are set up so that either the access point provides DHCP services with supplied TCP/IP information, or the access point passes through DHCP requests and configuration to the client-side adapter—so you do not have to be bothered with knowing the network information for every wireless network you use. Without the right TCP/IP information, your client system might as well not be connected to the network at all—by wires or wireless—as the network's router will ignore or block your data.

When DHCP is used to configure network clients automatically, your client device may receive an address within the host network's preassigned IP address range, or a private nonroutable IP address beginning with 10.x.x.x or 192.168.x.x addresses. Within this automatic configuration scheme, your client device will also receive a gateway address and probably a DNS server address or two.

A typical failed automatic TCP/IP configuration results in your client system being assigned a default and little used 169.x.x.x-range private IP address, and you will see no gateway/router or DNS addresses being assigned to your client.

You can view these settings using the WINIPCFG program in Windows 95 Me; IPCONFIG program in Windows NT, 2000, or XP; IFCONFIG program of most versions and variants of UNIX (each easily summoned from the command-line or the operating system GUI Run feature); or through the TCP/IP control panel under the Macintosh operating system.

Microsoft Windows operating systems provide the means to get your network drivers to look again for a DHCP server and request a new automatic configuration. Under Windows 95 Me, you can try the Renew button within the WINIPCFG program dialog. If you use Windows NT, 2000, or XP, you can issue the IPCONFIG/Renew command in a command prompt window to try to renew the connection. In Windows XP, the wireless networking status menus and dialogs provide a Repair button to perform a new DHCP query. In either of these cases, the program will either succeed and your system will be reconfigured, or you will receive an error message, indicating a failure to renew the network information.

A failure to get or renew the information means that you do not have a valid connection to the network you are trying to reach (see the SSID and WEP key items above), there is no working DHCP server on the network (ask the network administrator, or if you are that person, check your DHCP server!), or the access point is not configured properly to allow clients to access DHCP services and requires reconfiguration.

If the network does not use DHCP, then you will have to ask the system administrator to assign you specific TCP/IP information, or you must define the appropriate network parameters yourself and then assign unique IP addresses to each client system. You need a unique IP address for your client system, a gateway address so that it knows which device handles traffic outside the immediate network, and probably a DNS server address or two so applications can look up the IP addresses of the host names for which you want to connect.

Sometimes simply rebooting your client system will establish or renew the client configuration and get the right information from a DHCP server.

Signal quality. You probably knew I was going to have to mention the magic of wireless networking somewhere in this context. And in doing so, we are referring to black magic that consumes any and all form of radio energy, interferes only with the radio energy you want to use, or seems to shrivel up and fade away like the wicked witch in the *Wizard of Oz* when hit with a few drops of water.

With radio signals, the number and variations of potential problems are almost infinite, but fortunately not infinitely difficult to overcome, though you might wish for a wizard's hat, some pixie dust, a magic wand, or a pair of ruby slippers of your own to apply to these mysterious problems.

The first thing you can do to avoid signal problems is pick a clear channel that does not overlap other channels—this means 1, 6, or 11—leave 2 to 5 and 7 to 10 to someone else (of course you will all be doing this now!). Or, try channels farthest away from other 802.11b traffic you detect. Detection is easy using NetStumbler or a similar tool, as shown in Figure 7.12, to see what other devices are "on-the-air." Switch to another clear channel and recheck your results. The channel with the best connectivity is the one to use.

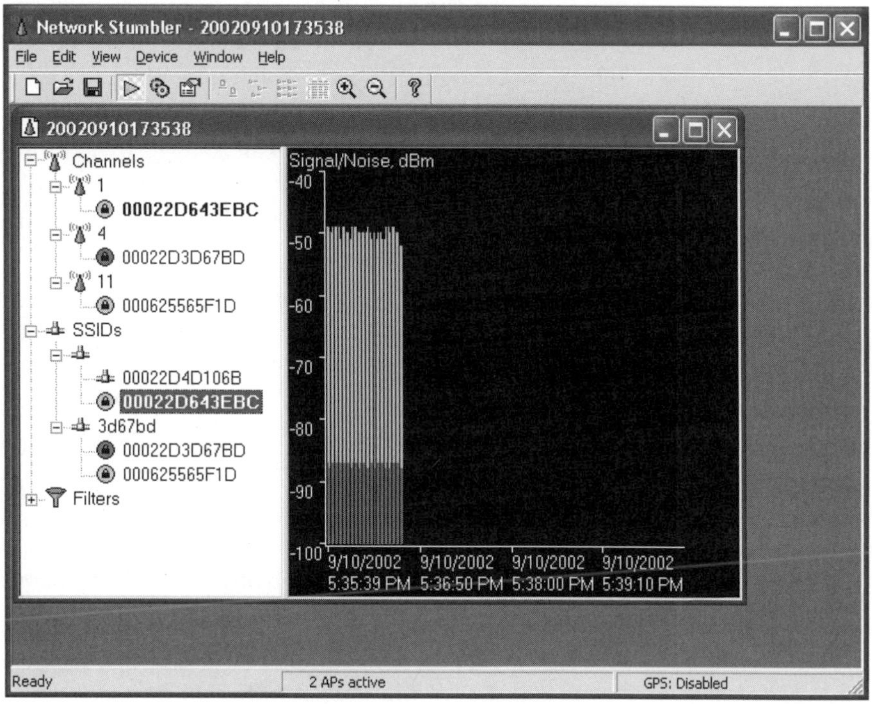

Figure 7.12 NetStumbler program display showing multiple active channels, some with and some without SSID.

While you are using NetStumbler to observe the signals that appear on different channels, watch for unusual spikes or drops in signal strength. This may be an indication that a non-802.11b signal source may be nearby, affecting the 802.11b signals you want to use. If you observe such events, you have just experienced a very crude, but effective tool you can use to get closer to the offending device—that is, if you are using a laptop. With a laptop, wireless card, and NetStumbler, you can tote your system around to different locations, watch the signal strength changes, and see if the changes get more or less pronounced. More dramatic changes mean you are getting closer to the offending source; less means you are getting farther away.

You will be looking for a microwave oven, 2.4 GHz cordless phone, or some other seemingly obscure unsuspected piece of equipment sharing the 802.11b spectrum. What you choose to or are able to do about the interfering device depends on what it is and who owns it (refer to Chapter 1 for information about our neighbors in the wireless world). If you cannot sniff out the cause of the interference, it may be time to call in a more knowledgeable resource, armed with a spectrum analyzer and a network sniffer to determine the nature of the interference source, and perhaps suggest alternatives. Of course, if you are trying to network with 802.11b equipment and cannot avoid the interference source, you have the option to set up 802.11a equipment—avoiding spectrum shared by so many neighbors.

If an interfering signal is not the culprit, and you likely will not be using 802.11a equipment, you might find the mysterious signal changes as you move about the access point coverage area to be caused by obstructions or reflections within construction materials—wires, pipes, framework, etc. The only way to avoid these, short of tearing out the walls and creating a truly open-office layout, is to reposition the access point or locate the antenna so that there is a clear line-of-sight path between the access point's antenna and the client systems. This corrective measure may include installing additional access points to cover the troublesome locations. If you cannot relocate or practically add access points, you may find it easier to work with one or more antenna configurations to get the signal where you need it.

Antenna locations, multiple access points, magical signal problems, and generally expanding the coverage of your wireless system are covered in Chapter 8.

Summary

This is perhaps the most exciting chapter so far—the installation and hopefully first connection of your wireless network components! Yes, it really can be that simple. I cannot imagine a more attractive and successful technology coming to the world of personal and business computing. Problems are relatively few, and for the most part, easy to resolve unless or until the magic starts getting messy.

You may not be able to overcome the effects of bad magic by the brute force application of more access points or the installation of a proper antenna to suit the location and coverage areas. Enhancing and expanding wireless signal coverage is the next topic.

CHAPTER 8

Extending and Maintaining Coverage

Having functional wireless network access a mere 5 to 25 feet from an access point is one thing. Getting a usable signal across the maze of cubicles, around corners, upstairs, downstairs, across the quad on your office campus, or across town is quite another, or two. In Chapter 7, we got you "on-the-air," and now it is time to get wireless everywhere—or to the locations you want it to reach.

In many cases, multiple access points provide the cleanest solution to roaming about your office or campus and staying connected. But more often than not, you may not have access to the parts of a building, or a specific rooftop, to get your wireless signal where it needs to be.

There are many options to extending or improving your coverage or increasing the strength of the signal you have without adding more access points. These options fall into three basic categories—antennas, power amplification, and signal redistribution cabling. Each of these present different design and cost considerations that you must apply to your specific installations.

We normally associate radio signal problems with not having enough signal. But you may encounter a problem—even with a strong signal—that you cannot easily identify. The problem is *multipath*—a condition where too much, or in this case too many—of the same radio signal is not a good thing.

Multipath

If you could hear a voice transmission across your wireless radio signal, you would be able to know and learn to recognize the effects of multipath on your signals—raspy, distorted, and basically ugly sounding audio. The signal could sound worse if you increased the transmitter power or used a better (higher gain, more directional) antenna. You might have heard this effect listening to your favorite FM radio station in your car while driving through a dense urban setting, going into tunnels, passing under freeway overcrossings, or driving in hilly or mountainous areas of countryside—an otherwise loud and clear signals gets fuzzy and distorts at various intervals as you travel along.

Since wireless data transmissions do not provide a way to listen to them as you can an AM or FM radio broadcast signal, you can only try to imagine the effect multipath has on what your wireless adapter receives in these cases.

Multipath is essentially as it says—the radio signal travels multiple paths from transmission point to reception point. Figures 8.1 and 8.2 illustrate the basic effect of a signal with multiple paths. You might think multiple signals are a good thing—more to choose from, and more is better, right? Unfortunately, radio receivers can provide useful information from only one very strong signal at a time. Multiple signals arriving at the receiver after bouncing off various objects tend to distort each other. Distortion is as ugly to a radio receiver as a sour note is from a musical instrument or an off-key singing voice—the desired information is obscured.

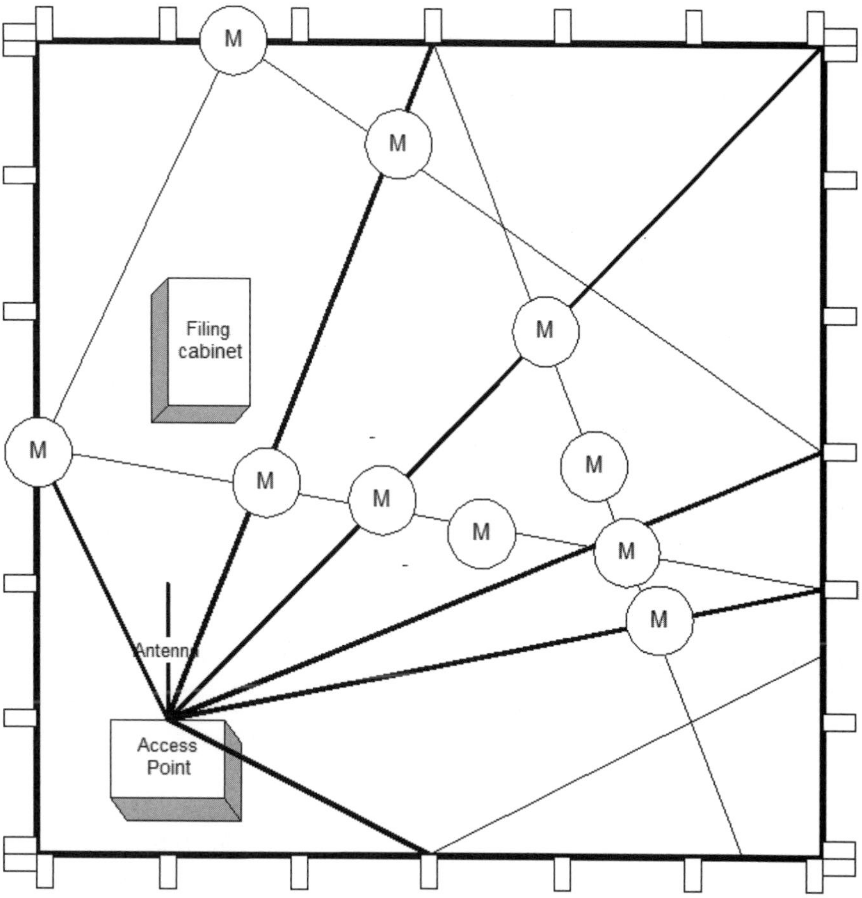

Figure 8.1 A representation of multiple signal paths in an indoor wireless implementation. Where the signal lines intersect are points of likely multipath reception problems and negative effects on wireless network signal integrity.

Figure 8.2
A representation of multiple signal paths in an outdoor wireless implementation.

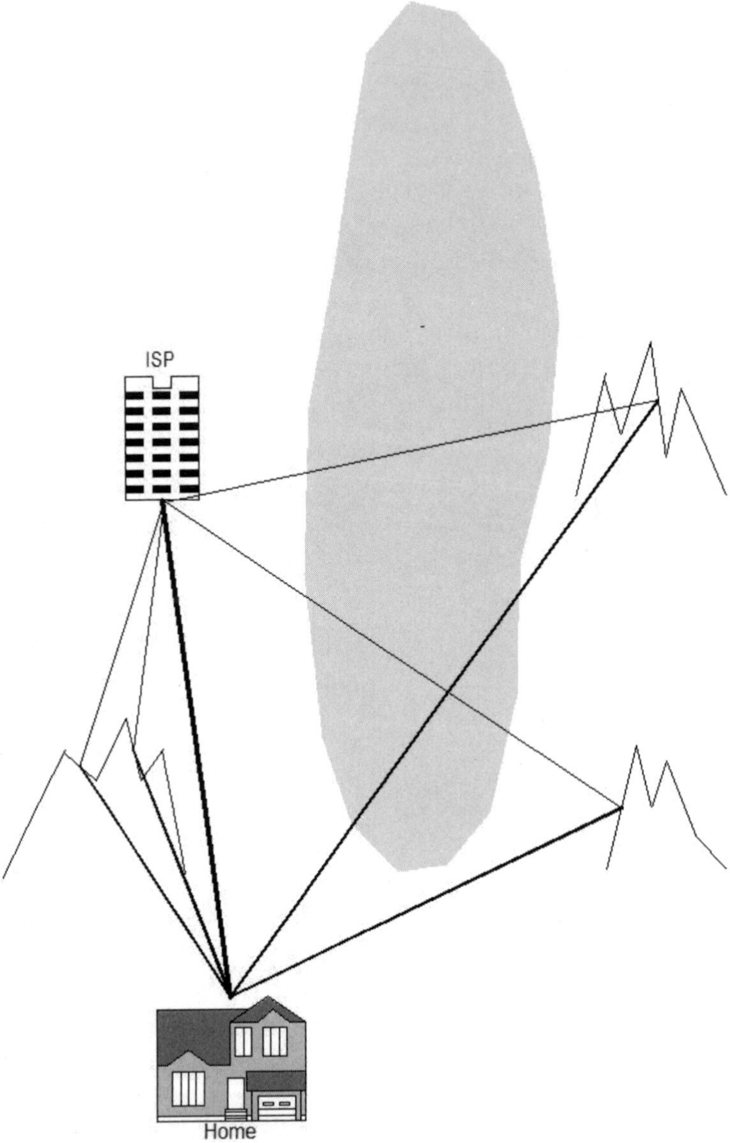

Where the reflections intersect each other or the main direct signal, signal distortion occurs. More often than not, the reflected signal may be as strong or stronger than the direct signal. Occasionally, a receiving station cannot see the transmitting station line-of-sight, and the reflection signal is the only one that the receiving station gets any transmitted signal from. Multipath is not always bidirectional either. It can affect the signal going one way differently than it affects it going the other.

Because signal paths, especially their reflections, can be random and not easily plotted or studied, multipath problems have few and only experimental trial-and-error solutions. The root of the solution is to get one good, strong signal from the transmitter to the receiver. The solution may be different for each signal direction, so the goal is to find a solution that works well in both directions of the signal path.

Solving Multipath Problems

Since multipath problems are most often associated with reflections, the basic answer to the problem is to eliminate or reduce the chances for reflections. Before you start thinking that you have to rip out walls to replace metal frame pieces with wood, redo plumbing with PVC pipe, tear out the electrical wiring and air conditioning systems, replace metal cubicle modules with wooden ones, and generally "go back to nature" to get your wireless network to function well, wait. You do have options.

Your first option is to get an access point closer to the users, so that its signal is much stronger than anything reflected throughout the room. A stronger or more direct primary signal from the access point can overwhelm the reflections and solve the problem.

Another option is to change the type of antenna you are using on your access point—typically from omnidirectional, which will radiate signals in all directions, to a directional antenna. A flat panel style antenna aimed in the direction of most users will work fine. This may also mean using more access points equipped with directional antennas to accomplish omnidirectional coverage, to eliminate many reflections from objects close-in, behind, or directly around the path of the signal from the access point to the users. Going from an omnidirectional to a directional antenna outdoors can also eliminate reflections from objects behind or to the side of the access point.

In cases where there is no direct line-of-sight signal path between the access point and the user's wireless adapter, you may be able to use a directional antenna aimed at a known or suspect reflection point to force a strong but reflected signal to the wireless adapter—more or less bending the signal around a corner or obstacle.

Consider an external and perhaps directional antenna for the wireless adapters on the users' systems—if their wireless adapters have external antenna connections. Remember that a laptop system sits 2 to 4 feet below the top of most cubicle walls, often in a cubicle with a lot of metal in its walls, which will further hide or degrade the direct signal from the access point.

Remember, the goal with wireless is to get a clear line-of-sight or virtual sight line between user interfaces and the access point. If you cannot achieve that, then reflections and blocked signals will always be a problem.

Antennas

Antennas, like wires, are loved or hated depending on who has to install them or have them as a constant visual distraction. They come in many shapes, sizes, and styles, from pure brute force techno-industrial to interior décor-matching varieties. Everything wireless has an antenna. Fortunately, at 802.11a and 802.11b frequencies, the antenna elements are small and easily disguised.

The style of antenna typically suits its purpose. Some style variations are good for all-around signal radiation, and some for partially to highly directional radiation. All-around signal radiation patterns, or omnidirectional antennas, are good for highly mobile environments and centralized locations serving a nonspecific direction. Antennas with directional patterns are best for extending coverage over distances or simply limiting the direction of coverage to a specific area.

Choosing an antenna type and style requires that you understand the area you are trying to cover and which type of antenna is most suitable for the given application.

Line-of-Sight—Placing an Antenna So It Can "See" Clients

From all we have discussed so far, placing an antenna with direct line-of-sight between access points and client wireless adapters is the best scenario. However, this could be interpreted and implemented in many ways.

If the client is another access point or bridge to extend the signal to a still farther away access point or clients, the antenna choice could be a high-gain directional unit. Which directional antenna you use depends on the distance you have to cover, legal power limitations, and physical mounting capabilities for the antenna alone; or you may place the access point at the antenna.

Typical point-to-point installations employ a Yagi type antenna for moderately long distances—exactly how long is not easy to determine. But since a Yagi antenna at 2.4 GHz is still relatively small compared to television or two-way radio antennas, obtaining a 7 to 13 element antenna is not out of reason and more cost-effective than a dish or parabolic antenna. For the highest amount of gain and directivity, using a dish or parabolic-style antenna at either end of a long signal path—even as far as 11 miles or so—is your best investment.

If your access point must be mounted at one end of a coverage area, and the desired area stretches some distance away—a city block or even up to a mile or so—employing a lower gain 5 to 7 element Yagi-style antenna may be most appropriate. The signal radiation pattern will not be so narrow as to exclude coverage to nearby clients and will extend ahead to more distant areas.

If the access point is located in the middle of a cluster of offices and cubicles, with a more or less even distribution of clients spread more or less equally throughout the area, then an omnidirectional antenna is preferred.

If you have a central location for one or more access points, but no ideal place to put a single omnidirectional antenna, such as an elevator/service core that does not provide one line-of-sight location for the surrounding area, using four access points and wide-pattern flat panel directional antennas aimed to cover all four corners of the facility may work best.

Obviously there are limits to how far a wireless signal will travel. An 802.11b signal using full, legal radiated power can be adequately received as far as 11 miles away, using the highest gain antennas at either end. Using antennas, especially highly directional ones, obviously limits the mobility at either end of the circuit.

An 802.11a signal, because it uses much higher frequencies than 802.11a, may be expected to reach 1/7th the distance of an 802.11b signal—call it 1 mile to 1-1/2 mile at the most, even with the best equipment. Sure, some amateur radio operators using much more power and high gain antennas have achieved usable signals across amazing distances at much higher frequencies, but in 802.11-service, we are limited by equipment and legal compliance. Remember, 802.11 was established as a replacement for standard Ethernet cabling, not high-performance metropolitan or regional fiber optic networks. Your mileage will vary, depending on terrain and environmental conditions.

Antennas versus Adding a Bridge and Access Point

Often it does not make sense to push your luck trying to stretch a weak signal across what appears to be a nice clean signal path. Instead, install an additional bridge to span two fixed points or an additional access point for more client coverage, or a combination of both. Antennas are an excellent way to improve signal levels in both directions, since they are not sensitive to signal direction. The effects antennas can have and of how they affect coverage areas were covered in Chapter 3.

Installing an additional access point has obvious advantages—the network is simply closer to the clients. However, getting the network to the access point may be a problem. Installing a bridging system between the location of the network and a distant coverage point, with another access point, is a viable option.

This latter option uses high-gain directional antennas to first bridge the distance, to get the network to a suitable location for an access point. Then the access point provides signal coverage for clients with an omnidirectional or semidirectional antenna.

Signal Amplifiers

Amplifiers, boosters, and kickers are nicknames for almost any active electronic device that provides an increase in output power level, either coming or going. We are referring to a device that connects between a client adapter or an access point and an external antenna to increase the level of at least the transmitted signal, making it stronger at the receive end of the signal path.

A device that amplifies a transmitted signal is often referred to as a *power amplifier*, while a device that amplifies a received signal is most often referred to as a *preamplifier*. Boosting the outbound signal is easy. Doing so and receiving a signal at the same time means a little more sophistication. Boosting the outbound signal and the received signal at the same time requires a lot more sophistication.

In numerous two-way radio systems, used for police, fire, taxicab, and general utility services, the designers and technicians have many options available to them to enhance either the transmit, the receive, or both signals, by separating them electronically or physically. Wireless networking does not provide that flexibility easily, and certainly not at the client side.

The weakest or most variable point in a wireless system is the client—typically roving with a tiny card stuck in the side of a personal computer (PC) or Macintosh portable—leaving the antenna some 24 to 36 inches off the ground and subject to any number of nearby obstructions.

Signal enhancement products that provide signal boosting are available and well used for wireless networking applications. You should decide if, when, and how to use them. If you are enhancing a point-to-point or bridged network path, applying power amplifiers at each end so that the signal is equally enhanced in both directions should be obvious.

If you are trying to enhance the access point's reception and transmission of client signals, it is not practical to add an amplifier at the client side. So the device you use at the access point side, where you presumably have more room and power supply for more equipment, should provide power (transmit) and weak signal (receive) amplification at the same time—also known as a *bidirectional system*. A good source for such a system is HyperLink Technologies and its bidirectional amplifier product line: http://www.hyperlinktech.com/web/amplifiers.html.

It makes almost no sense to provide 3, 6, or even 10 dB more output power from one side of a signal path (the access point) and not provide that same amount of signal gain to the other direction of the path. A client adapter that can hear the access point, but not get its own signal back to that same access point is, well, pointless.

Power amplifier-only units are best suited for point-to-point sections of wireless networking and not the access point-to-client portions. For access point-to-client applications, you and your clients will be better served by bridging in more access points to provide more solid, close-in, client-side access to the network.

Note: *Remember—even though 802.11a and 802.11b are unlicensed radio services, they are not unregulated. Both come with significant power restrictions, depending on the type of antenna you are using.*

Radiating Cable

When I mention radiating cable to just about every radio expert, they are unaware of its existence and skeptical of it. Those who have heard of it get very skeptical and cite one technical limitation or another as to why it would not be effective. Unlike the claims of late night infomercials, this type of product or implementation is not a do-all, cure-all, and you cannot get two or three products for the price of one if you call in the next 10 minutes and mention the secret TV code. But, those who have used it are amazed that it does appear to work.

Generically, I am referring to coaxial cable whose outer shield, typically meant to keep signal contained inside the cable, is opened to allow some of the signal to escape. This is also called *lossy* or *leaky* cable. Almost every radio technician knows that if you violate the outer shielding of a transmission cable, some of the signal will get out. And in most cases, this is a bad thing. But it can be useful if you intend to have it happen.

Making the cable leak signal is fairly simple. Making it leak the right signals to the right places without causing the signal to bounce back to the transmitting device, possibly destroying it, requires some precision and determination.

You could make *radiating cable* yourself, but the process is tedious and imprecise at best. You can buy radiating cable or *continuous distributed antenna*, known commercially as Radiax by Andrew Corporation. It is made in different versions specifically for different wireless frequencies and applications. Details of the technology, case study, and product information are available at the Andrew Web site: http://www.andrew.com/products/trans_line/radiax/default.aspx.

The exciting thing about Radiax is that, if you were considering installing and wiring to another access point, or placing an antenna in the middle of a room, you could string Radiax along your intended coverage area instead of network cabling to another access point. With Radiax, you would benefit from more even distribution of the radio signal along the way. If your building's ceiling structure is such that you should not or cannot lay the cable across the top of a ceiling tile, you can use the decorative style of the cable—essentially a small ribbon of cable that looks like trim—at the edge of the room or across the middle of the ceiling area.

Radiax might appear to be a bit more expensive that other types of cables or multiple access point solutions, but it does save you from managing multiple access points.

Passive Repeaters

A *repeater* is an active electronic system with a receiver that has its output tied to the input of a transmitter. The information within the received signal on one frequency is retransmitted on a different frequency—much like a wireless router might do. Repeaters are an expensive means to move a signal around—necessary in some cases, but overkill in most.

A passive repeater is a means to enhance a signal by conveying it from one place to another without actively receiving and retransmitting the signal or the information within the signal. In effect, radiating coax or a continuous distributed antenna is somewhat like a passive repeater, carrying and leaking signal along its length like a soaker hose for your garden. A typical passive repeater is made up of a pair of antennas connected to each other in a back-to-back fashion—one antenna pointed towards a signal source, the other pointed towards the desired destination.

Passive repeaters, or even an apparatus as seemingly mundane as a large reflector for microwave signals, have been in use for numerous applications. Microwave signals are skillfully reflected off reflecting surfaces, such as parabolic dishes and those massive funnel-looking "horns of plenty" you see on radio towers. If focusing and bouncing signals can work for 7, 10, or 24 GHz microwave applications, it is certainly possible to use them at 2.4 and 5 GHz for wireless networking applications.

To be effective, a passive repeater must have antennas with high gain and high signal captureability. This dictates the use of parabolic or dish antennas, rather than Yagis or omnidirectional antennas. This most obviously leads us to use a passive repeater between two fixed points, where it would be impractical to install bridging or repeater equipment.

Not to be trite or oversimplify (my highly qualified engineering friends will roll their eyes at this and start all sorts of technical rebuttals), but you may be able to accomplish your own passive repeater with two parabolish dish antennas connected to each other with a short piece of high-quality feedline cable. Point each dish at each fixed access point you want to bridge between and see what happens.

A passive repeater can be enhanced further with special bidirectional signal amplifiers that boost received signals from one side to the other. These somewhat expensive devices are built for a specific set of frequencies and designed to avoid feedback in either direction.

If you have tried and like the point-to-point passive repeater, you might think that you could adapt this to a point-to-multipoint application—and you would be correct. Start with a high gain dish antenna aimed at a known access point, run a length of low-loss feedline to a location where mobile users would be, perhaps string a length of Radiax, and end the run with a proper coax termination (a 50 ohm resistor) or omnidirectional antenna. You will have picked up the intended wireless signal and distributed it along or to a desired coverage area—without any electronics.

I have done this to provide signal extension into three levels of basement area from the top of a 14-story hotel building—initially to allow VHF high-band (150 MHz) radio pagers to receive paging signals. It worked surprisingly well, and I found that it worked bidirectionally, allowing two-way radios used in the building to reach out to their base station. It is messy and tricky business running cable down through an elevator shaft to a garage area, but it became the right solution for more than one application. Hopefully your implementation will not be as intense, but will yield equally satisfying results.

Multiple Access Point Networks

Once you have accomplished a single access point network setup, you are either done with your network, or ready to move on to adding access points for more cellular coverage to accommodate your clients. I left this section about multiple access points until after a full discussion about antenna options for a few reasons—to help you consider economical and practical solutions for single access point systems, to keep you from cluttering up the radio spectrum with a lot of radios on different channels, and to save you the confusion of co-channel interference and managing multiple systems.

Adding a second access point or more within the same network is very easy. Simply do what you did for the first one, same service set identifier (SSID) and wired equivalent privacy (WEP) keys, but pick a nonoverlapping channel to use. When your clients move from one access point coverage area to the next, their wireless adapters will automatically scan across the channels, find the same network on the new channel in the new coverage area, and maintain connectivity to the network—amazingly simple. Things may get a little complicated with various virtual private network (VPN) access control and security solutions designed strictly for fixed wired networks, as moving between access points can cause a loss of connection and a need to reconnect, which some VPN systems do not allow. So you will have to consider a new VPN product designed to tolerate wireless or other roving access configurations.

With multiple access points, you must take care not to reuse the same channels, or overlapping channels, if the signal coverage areas of the access points overlap. Using the same channel will result in connection confusion at the client adapters and failed connections.

If the physical layout of your office building or campus setting allows for it, you can reuse the three nonoverlapping channels for access points that have nonoverlapping coverage. Every third floor of a building, or every third building, could reuse the same one of three channels—set the AP in floor/building 1 to use Channel 1, floor/building 2 to use Channel 6, floor/building 3 to use Channel 11, floor/building 4 to use Channel 1 again, and so on.

Along those lines, as you expand your network, you also want to make sure that it is not available to just anyone. You may find it beneficial to use semidirectional antennas to contain your signal within a given desired coverage area, rather than using omnidirectional antennas that broadcast your network everywhere.

Avoiding Channel Overlap and Other Networks

Avoiding channel overlap is a serious consideration for your own network, as well for installing neighborhood and café-area networks—requiring site and coverage area surveys to see who else is using wireless and coordinating with other wireless network providers. Unfortunately, if the other networks in the area are operating in private or stealth mode by not broadcasting their SSIDs, you may never know they are there and could run into serious co-channel interference problems.

If others can operate unseen from you, you can also use stealth mode and be invisible to them by not broadcasting your network's SSID. You can provide the SSID only to those who you want to use your network, as one layer of protection. This puts some of the burden on others who wish to install wireless networks to do the site surveys to find that your network exists and avoid it.

If you need to find other advertised public or open nodes, or even those participating in the Boingo or HereUAre networks, you can find many of them listed at The NodeDB Web site, www.nodedb.com. Thousands of public and other nodes are listed there, but you will want to check the www.boingo.com and www.hereuare.com Web sites to find out where their respective subscriber networks are located also.

The only way you are going to detect the presence of unknown networks is to use a tool like Network Stumbler (from www.netstumbler.com) to show you the presence of identified (SSID is broadcast) and unidentified or stealth (no SSID is broadcast) networks. Obtain the use of a spectrum analyzer, or use the services of an experienced wireless/radio frequency (RF) consultant to do a full RF signal survey of the area to give you a better picture of what is going on around you.

Network Stumbler will identify that an access point is occupying a channel and give you its internal hardware or media access control (MAC) address, telling you specifically that another wireless device is on the air nearby (see Figure 8.3). Once you have found such an access point in operation, a little sleuthing with a directional antenna can help you narrow down its possible location, thus helping you identify the owner of it and hopefully gaining some channel cooperation.

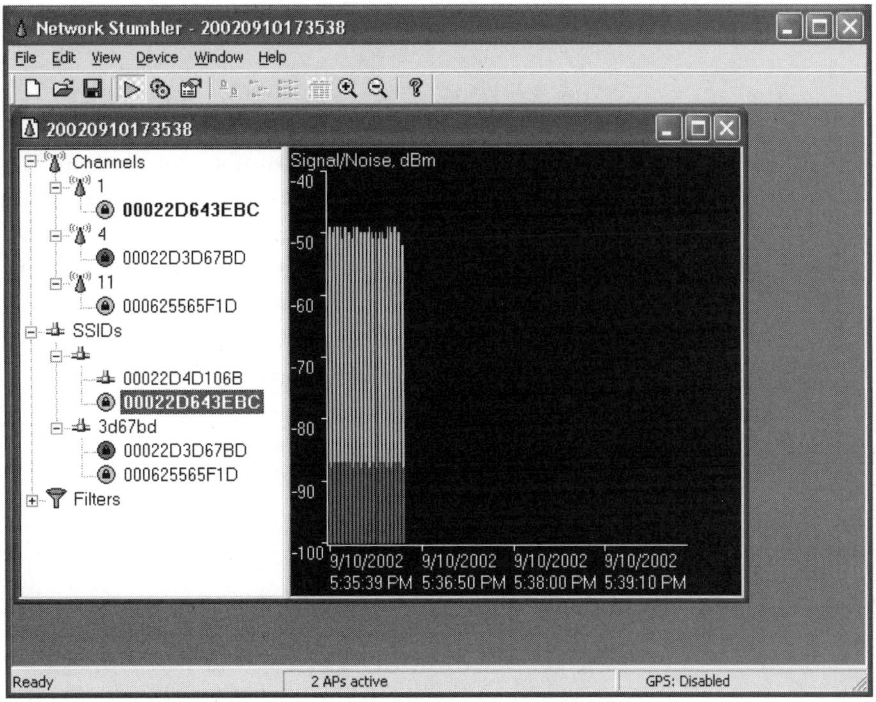

Figure 8.3 Our screenshot of the Network Stumbler program showing signal strength readings and the MAC addresses of active nearby access points, the channels they are using, and their SSID, if available. Note specifically the access point with MAC address 00022D643EBC operating on channel 1 that is not broadcasting its SSID.

Similar signal strength monitoring tools are available and installed with most of the client wireless adapters, and these may be the only test equipment you need—or at your disposal—to determine network performance and interference issues.

If you find other networks operating in your area and need to share channels, hopefully you can work cooperatively with the operators of those systems to determine a cooperative semidirectional antenna scheme for both networks, focusing signals into respective coverage areas to avoid overlap problems.

Channel sharing is more significant to users of wide area, broad coverage community services or commercial wireless Internet service providers (WISPs), who want to be able to enjoy signal coverage in one or more large areas without interference. Users of these networks, like any other, will find they do not have exclusive access to any specific channel or coverage area. So users of a WISP using

Channel 1 for wide coverage may easily find the Channel 1 signal from a local café or home user simply wiping out the signal from the intended WISP. Avoiding this can only be negotiated, as it is certainly not regulated or prohibited.

Remember, too, that you share the radio spectrum with other devices and services, and using a tool like Network Stumbler can help you determine if other devices in the area are wiping out your wireless network signal. The technique for determining interference from devices you own, such as a 2.4 GHz cordless telephone, is very simple. Watch the signal strength of your wireless network with the signal indicator provided with your adapter or Network Stumbler, then use the telephone or other device. If the signal strength indication or network connection drops suddenly, chances are your phone is causing a problem. This technique works for testing the influence of your microwave oven, Bluetooth, and other devices.

If you do not have control over anything you suspect may be causing you interference, and cannot perform the basic tests yourself, you should contact your neighbors to ask them about any cordless devices they have and arrange a test with them. You can also carry your laptop around to various parts of your home or office to determine when and where the interference problem is the most significant. This will at least point you in the direction of the troublesome device. Again, because you are using a shared, unlicensed service, no one is legally obligated to change or fix anything. Your best option is to solicit cooperation—even if it means buying your neighbor another phone—to get your network to work well.

Summary

Wireless networking gets into your system. And the more you get, the more you want—for signal strength and coverage area. Sometimes we have to be reminded, and I try to do this more than a few times, that wireless networking is not as ubiquitous as cellular phone coverage, nor is it meant to do anymore for users than get them off the Ethernet cabling into a less cumbersome environment.

Even cellular phone system coverage is misunderstood, with severe overexpectations of its capabilities and regulatory and environmental limitations that can disappoint us. Sometimes we cannot do anymore to help the coverage of our wireless networks anymore than we can our cellular phone services—but sometimes we can, as this chapter has hopefully illustrated for you.

Be careful what you wish for. Increased coverage means increased exposure of your network to others, and others to your network. Once you get it out there, you want to ensure that only the intended users have access to your system and do not abuse it.

While you expand your wireless network, be wary of not only the regulations of power limitation and tolerance of a shared resource, but also the access control and security risks that come with opening the gate on your once wired-only network to the general public.

CHAPTER 9

Wireless Network Security

Any system connected to the Internet is vulnerable to myriad breeches of security. Any network, connected to the Internet or not, is vulnerable to human hacking or biological bugs; that is, the network users. Every wireless network is vulnerable not only to humans, but to other sources of wireless signals, but especially humans. Vulnerabilities to wireless networks include denial of service by incidental or deliberate radio signal interference, denial of service by deliberate sabotage using known and new transmission control protocol/Internet protocol (TCP/IP) threats, and interception and theft of data by decoding wireless signals. These vulnerabilities can affect the host network (via the access point), interaccess point or bridged systems, and client systems.

A quick review of the material in Chapter 1 tells us that wireless network systems have little or no protection against unintentional radio signals, or those signals from devices in radio services that have priority over wireless networking signals. Intentional interruption or jamming of any radio signal, with the intent to deny services to other users, is strictly prohibited by law, at least in the United States.

Taking or abusing another's data, or tampering with it, falls into an entirely different set of regulations—depending on how the information obtained is used or inserted into someone else's network.

Wireless networks are especially vulnerable because it is nearly impossible to create physical barriers to contain the radiated signals—at least intentional barriers. It is odd that we should have a technology that is so difficult to deploy to where we want it to go amidst a variety of physical obstructions, yet we are unable to create desired obstructions to keep our desired signal in and unwanted signals out.

All of these aspects, and perhaps others not yet imagined or known, create a lot of attention to security issues—a topic that is as timely as it is timeless, as more and more of our daily business and personal lives become digitized, transmitted, stored, shared, and used for myriad purposes. Information security is threatened threefold: denial or lack of information, theft of information, and corruption of information. Covering all three of these in a wired network is a full-time job. Covering them in a wireless network is not only a full-time job, but also an elusive one.

Threats

Physical security of your wireless network traffic is virtually impossible because wireless is an open-air technology, and the spectrum 802.11a and 802.11b uses requires a clear, nearly optical line-of-sight path between two points to be connected. Any physical barrier also creates a barrier to the desired signals, rendering the technology useless—which in itself makes physical barriers threats of their own. You can physically secure most of your equipment much as you would any hub, router, or server, but any external antenna would probably be left exposed—to humans, animals, machinery, and the elements.

Theft of Service or Information

Theft of service is the unauthorized use of someone else's network resources—typically hacking *onto* a neighbor's local campus, café, or business wireless system to gain free Internet access. This is one of the most obvious reasons wireless system operators impose access control restrictions on their wireless networks.

In its simplest form, on an unsecured or loosely controlled network, determining or knowing the service set identifier (SSID) and having or deciphering the network's wired equivalent privacy (WEP) key is enough to gain access. If the wireless network exists simply to provide Internet access, by firewall or router controls, or there is no significant network infrastructure behind the wireless system, Internet access is all you are giving up. If you have more network infrastructure behind the wireless system, it too is very much at risk.

Interception of your network traffic may be done to determine your system's SSID or WEP key. Once through the basic access control, traffic can be sniffed to collect data that are passing across the network. This may sound a bit cloak-and-dagger, and it could be—if you have personal or business information that is worth something to someone else. Mere interception of data was all it took for some crooks to steal and then abuse credit card information obtained from a retail computer store's cash register systems. If all a snoop gets is your credit card data, you may be lucky—if the snoop gets enough personal information, you are at risk of identity theft.

On a business network, all sorts of proprietary data go back and forth. Anything from e-mail to program source code to marketing plans or employee salary information may be available. In such cases, it is not only advisable to implement a very tight access control and encryption plan for the wireless network, but you may want to go as far as setting a policy restricting what type of information people deal with when they are using a wireless connection.

Once someone has access to your network, he may be able to intervene in the traffic between clients and the network. Intervention, or *man-in-the-middle* intrusions, are possible by a bad guy sitting in between a client and the wireless system, setting up a spoofing operation to make the client think it is connected to the wireless LAN and the wireless LAN to think it has a valid client out there. The bad guy will pull out and store valid information and retransmit bogus information. It sounds like "Mission: Impossible" tactics here, but this is quite possible, given enough equipment and skill.

Denial of Service

Denial of service may be accidental or intentional—simply denying clients the ability to connect to a wireless LAN—through deliberate or incidental interference with wireless signals.

An appliance as benign as a wireless LAN-unfriendly 2.4 GHz cordless telephone can be a nuisance or a weapon, depending on who is using it and for what reason. Those wanting to use their own wireless LAN will undoubtedly shelve their cordless phone once they determine it keeps them from using their wireless setup. The little old lady across the street may have no clue or care that her cordless telephone is keeping you from enjoying wireless networking. Someone intent on denying you the use of your wireless system will find some way to use one of these phones to keep you off the Internet.

A cordless phone is not the only weapon capable of denying you wireless network services. A poorly shielded microwave oven, a legal amateur radio station, or government radio service can break your network in milliseconds.

To intentionally deny you service is certainly illegal and also requires that the bad guy knows you have a wireless LAN—by using a tool like NetStumbler to see that you have active wireless gear.

Someone could intentionally or coincidentally create his own wireless network, overpowering yours, which could also deny you services.

Beware that you may also be denying someone, such as a legal amateur radio operator, legitimate use of his radio services by merely operating a wireless LAN, which presents significant apparent noise to amateur radio receivers.

Building and geographical obstructions may also deny you service. These are less likely to be used to intentionally to deny you wireless services from a distant location, but are more coincidental or circumstantial. It would seem that only a handful of very rich people would be able to command the construction of a new building just to block your signals.

No matter the source, if intentional, denial of service could be done to hurt your business by forcing you off-the-air or making your customers patronize a different café—perhaps even one they would have to pay to gain Internet access through. I realize I may have just spawned a few less than ethical ideas by mentioning such techniques, but if they have not become obvious by now, then you are really not equipped to deal with the situation if it arises.

Detection

Detecting threats or problems along the wireless path is a twofold process—differentiating between radio signal-related issues and data issues—and the likely impact on service that each may have. The first level of threat is someone finding out you have a wireless network by passively or actively monitoring the airwaves for 802.11 activity.

Programs such as Ethereal, that puts a wireless interface into *RFMON* (receive only) mode—or uses communications test equipment like a spectrum analyzer—are completely passive and their use is undetectable.

Passive interception of the data along your wireless LAN traffic may go undetected. There is no practical way to determine if some of the radio energy you are transmitting has been lost to another person's receiver, to a leaf on a tree, or to atmospheric conditions. You will not lose data packets, but someone else will have been able to watch and catch them as they pass by.

Discovering you have an active wireless network system does not constitute a theft of service, but it could be, if that service is the distribution of copyright or proprietary material with some associated intellectual or monetary value, and someone receives and records that information. This activity is most likely done to obtain information that could be used in other ways—credit card fraud, identity theft, private investigation, invasion of privacy, detecting illegal activity, etc.

Actively probing your network with NetStumbler or similar software is also not a theft of service or determined threat, but trying to gain entry onto your network through log-on attempts or remote access schemes is wrong. Both can be determined by using robust logging of all network activity at routers, access points, program, and server logging.

A paper titled *Layer 2 Analysis of WLAN Discovery Applications for Intrusion Detection* (http://home.jwu.edu/jwright/papers/l2-wlan-ids.pdf), written by Joshua Wright of Johnson & Wales University, provides specific evidence that wireless network detection and identification programs like NetStumber leave specific, though illusive evidence of their activity on the networks they identify because they actively probe and ask for information from nearby access points, and this probing is a recordable network activity. The study outlined in Joshua's paper can be readily implemented and could be quite useful.

What you do with the information collected is left up to you—since you cannot readily identify who is running NetStumbler nor determine their intent. With hundreds of people "war driving" and otherwise using wireless systems and programs like NetStumbler, the activity is elusive, if not plain harmless, for the most part. I would not like to see dozens of wireless network administrators combing the streets and shaking the bushes around the perimeters of their networks looking for someone who they think might want to take information from their network. At least here, the person is still innocent until damage is done and the person is proven guilty.

That someone can probe your network is a simple call to action to take steps to secure it, at least to the level of equal value of the potential loss you would incur if someone does penetrate your wireless service. This alone should be cause to monitor your network. Using appropriate intrusion detection methods, secure all systems first within with a properly configured firewall; next with adequate access controls, login protections, and file sharing security; then virus protection at servers and workstations. They cannot get you if they cannot get to and adversely affect you.

Identifying Interference

Detecting an interfering signal and discriminating between a legitimate signal source and a possible jammer is nearly impossible without expensive radio test equipment (typically a spectrum analyzer) and a skilled operator that equipment to zero in on signals within the same frequency range as your wireless equipment uses, and determine what type of signal is generating a problem for you.

You can use a tool like NetStumbler to determine if another wireless network is operating nearby. This software will tell you the SSID and channel(s) used, allowing you the opportunity to avoid the preexisting channels, but NetStumbler will not tell you specifically about other sources of interference. If the interference is not another 802.11 network, you may only be able to determine a significant loss of your desired 802.11 signal when the interfering signal comes on the air.

A spectrum analyzer can show that there is another signal within the same radio spectrum. A skilled radio engineer using a spectrum analyzer may recognize and be able to identify the type of signal present and characterize what type of equipment it comes from. With that information, and use of a directional antenna, the location of the interfering signal source may also be determined. This may be a very expensive undertaking, unless you have a friend with the proper equipment and enough time to assess the situation.

Identifying Intervention

Intervention into your LAN traffic may be detectable by staging a known data reliability test between two points, or using packet analyzers to determine irregularities in traffic received at one end of your wireless path or the other. Data transmission reliability is something marginally built into TCP/IP, ensuring delivery of data, but not its integrity. Transmitted data should always get to their destination, but the destination has no idea if the data received are what was actually transmitted.

Creating a robust error-checking routine between two points, to verify that the sent data was not tampered with, is part of what encryption and some data protocols are all about. In fact, wireless networking technology provides encryption, but the encryption scheme is weak and vulnerable to simple deciphering, leading to many forms of wireless network abuse.

Encryption without a cross-check between sender and receiver does not ensure data reliability. Someone "in the middle" knowing the encryption methods used can intercept good data and send bad data to the destination, almost without detection. The destination will not know it is getting bad data unless it has some idea about what is supposed to be sent, which in most cases is impossible. Web sites and e-mail servers do not know or care if you type www.hotmail.com versus www.hotmale.com. Either may be perfectly legitimate pieces of data, but the recipient system has no idea what you *meant* to send. Thus, error-checking only works if you control both ends of the communication and know what data to expect between them. And networks, especially the Internet in general, do not work that way. That is left to specific applications.

Users and operators of corporate or closed network systems are better off than open or community network users because they have control over the user equipment, applications, and data at each end—giving them more control over the end-to-end environments.

Detecting intervention—someone picking up sent data, then corrupting or otherwise replacing what was intended with either garbage or misleading data—requires a detailed look at the data from both ends. Again, this could be implemented as a known data test—sending something that the receiver knows to check against. This may work as a reliable detection if all of the data sent are interrupted and changed before they are received. Smart hackers probably are not going to intervene in every data packet sent. They will look at what is sent, determine if it is of interest and something they want to interfere with, and only then would the data received be different from what was transmitted.

In either case, the intervention process takes some time, even if done programmatically, rather than manually. Thus, a latency or delay-in-transit test may be used as a detection method. If, for instance, data packets normally take less than a typical 1 to 10 milliseconds to be packaged, sent, detected, and unpackaged, and you suddenly find that the data path takes longer than that, perhaps 20 to 50 milliseconds (a guesstimate of the time some program may receive, decipher, alter, recipher, and then retransmit data), you might be able to assume that someone is intervening in the path. Such a test might normally be done with the standard PING or TRACEROUTE network utilities—unless the intervening system ignores user datagram protocol (UDP) packets and only works on TCP packets of data.

You really need a packet analyzer at both the sending and receiving ends of the wireless path to determine if the data received differs from the data sent. This is complicated by the fact that, at some point, both sets of data need to be compared to each other to make the determination of tampering. Packet analysis is perhaps the only way to know for sure if you have data integrity problems or not—but it is not a method you would employ full-time to watch over your network. If the hacker is aware of your detection efforts, the intervention could simply stop for that period of time and resume once he or she has determined the path to be clean.

Preventive Measures

At best, the WEP supported by nearly all wireless network equipment and related software to encrypt wireless data serves as a deterrent to casual network snoopers—*casual* meaning anyone who is not willing to sit around and capture 10 million or more data packets to be able to decipher your WEP encryption key code.

Those intent on sniffing out WEP keys are probably after more valuable data than the occasional e-mail that might pass amid a few bytes of personal web page traffic—and can park equipment near a wireless site and collect the information later, or remotely. Any truly valuable data worth protecting uses methods much stronger than WEP keys to keep it from prying eyes—and of course more expensive in complexity, labor, and cost.

One of the first things you should do before implementing any preventive measures is to perform a security and vulnerability assessment. Internet Security Systems' Wireless Scanner (www.iss.net) and AirDefense's (www.airdefense.net) products are designed to ferret out obvious holes in your wireless system. Performing an assessment is recommended both before and after you have taken steps to secure your network. Otherwise, you may not know if you have really secured the systems or not.

Following an assessment, by all means, plug the leaks. Of course, if your problem is denial of service based on interference or another class of service running equipment legitimately in the 802.11b space, you will have to track down the culprit or move up to 802.11a—which will cause you to re-engineer the radio frequency parts of your system and perhaps add more relay or bridge points to make up for 802.11a's shorter range.

If you experience denial of service due to the presence of another wireless user, identifying the other system operator and employing diplomacy and cooperation are your only legitimate options. If you find another system using noncertified system equipment, exceeding power limits, or employing other unconventional practices, your recourse may take a legal turn, through the Federal Communications Commission.

Access Control Systems and WEP Alternatives

The keys to security are making sure no one else can get onto your network, and if they try, they are held back by the inability to pass the right encrypted data.

Access control systems, similar to those used to log onto e-mail servers or dial-up Internet service providers (ISPs) can help prevent overt theft of services—someone taking advantage of your network access. Software systems such as Sputnik (based on NoCat) provide some level of access protection, as do similar access portal implementations for subscriber networks (T-Mobile, Boingo, etc.).

Almost any virtual private network (VPN)-like implementation will provide tighter encryption as well as access control. Funk Software's Odyssey software combines VPN and RADIUS-based access control for use with Windows 2000 servers and Windows clients—perhaps the only such software available—but support for Mac and UNIX systems is not available.

Mike van Opstal's technique of adding end-to-end dynamic encryption key sharing between Windows clients and a Windows 2000 server through wireless equipment (www.missl.cs.umd.edu/Projects/wireless) appears to be a very sound and practical way to implement wireless security within a completely Windows environment.

Many access points provide media access control (MAC) (network adapter hardware serial number) address restriction/permission capabilities. Although MAC address controls apply across all operating systems, the addresses can be spoofed or faked onto other network devices. The use of MAC address control is limited to the capabilities of your access point and requires less flexibility for clients and system management.

If an access control system does not provide tighter end-to-end encryption methods than WEP, someone can get and abuse your logon information. Access control alone may not prevent interception or intervention. Such a solution must also be applicable to UNIX and Mac users, as well as Windows users.

If you are doing a corporate/enterprise wireless implementation, you are probably looking to implement a solution that integrates with your existing network equipment—such as Cisco—which offers a very complete and robust set of equipment and software.

The Wi-Fi Alliance, a wireless industry trade organization (www.weca.net), recently announced a replacement to the known-vulnerable WEP encryption standard. Wi-Fi Protected Access (WPA) offers stronger encryption and access control between wireless adapters and access points. WPA is due to be available in February 2003 and may appear in firmware upgrades for some existing wireless products. It is expected to be available in new products after release of this new technique. Whether or not WPA will be adopted by all wireless vendors, or the vendors will wait until the more universal 802.11i standard is finalized, is unknown.

Summary

We will not and have not covered exactly what to do in all cases of implementation, troubleshooting, applications, and security—wireless networking is flexible and everchanging. Wireless networking is a relatively young technology being exploited far beyond its original intent and design. New tools, methodologies, and technologies are being introduced regularly to implement, enhance, detect, combat, secure, and add value to this resource.

The most vulnerable part of your network may not be the limitations of technology, and are nontechnical. In addition to the available solutions for the technology at hand, it is important to remember that many security issues are biological or human in nature. Vulnerability includes using simple passwords instead of those that are more difficult to guess or reproduce; using default SSIDs or passwords; sharing passwords with others; leaving passwords on "sticky notes" next to systems; and of course disgruntled employees taking data away from the network on paper, diskettes, CDs, or transmitting by e-mail or file transfer protocol (FTP). The easiest pickings are had when you have direct and obvious access to the information you want. So limiting access to information on a need-to-know basis is also crucial.

Please—take data and network security seriously—not just because of paranoia or cyber-terrorism threats, but because your job and others' depend on it. Networking is part of business, and business is part of everyone's economy. If your data are subject to compromise or tampering, frequent and regular backups of legitimate data can provide a tangible history of the business at hand and is certainly a part of your responsibilities of overseeing any network or data operation.

CHAPTER 10

Software for Wireless Networks

If you want to see how something works, what might be broken inside it, and fix problems or know you have fixed them, you probably need some kind of tool to take it apart. In the wireless world, you have to use somewhat ethereal, indirect tools to see what is happening to the radio signal and the data that hopefully pass between adapter and access point, or directly between adapters in an ad hoc network.

Die-hard techies and serious radio frequency (RF) engineers will drag out expensive test equipment—signal generators, spectrum analyzers, and network packet sniffers/analyzers—to assess the environment of and around a wireless network installation. Unfortunately, most of us do not have $1,000, much less $10,000 or more, to buy a piece or two of highly specialized electronic equipment we will use only once or twice.

Unfortunately, wireless networking is not as logical or measurable as tests you may perform on a hard drive or serial I/O port. You will not find diagnostic programs, but instead, metering software that provides some visualizations of wireless signals.

We have seen a few examples of adapter card–specific signal strength and network availability monitors. These monitors provide a good relative indication of signal strength, but as you get into network design and reliability, you need something a little more absolute than a poor/weak, good, or excellent indication. What you need is something that will tell you in known absolute values which signals exist nearby, and how strong they are.

Fortunately, many programmers took it upon themselves to find out how these new wireless devices work, dug into the inner workings, and pulled out some very valuable data. They found some user-friendly ways of presenting the information to us, so that we could make sense of this invisible connection between computers and networks.

The results are about a dozen programs, most of them for Linux systems, that can help us see, and to some extent understand, what is happening in the wireless networking environment around us—all through the features, functions, and admitted limitations of what a wireless network adapter can reveal to us. Although the world of Linux is a haven and test bed for some of the deepest and most profound network and Internet innovations, Windows and Macintosh users are not left in the dark.

Wireless may be the one thing, next to the Internet, that brings these separate and distinct platforms together for the good of all. It is not about replacing wires with invisible energy fields, it is that all at once, three distinct computing platforms are thrust into working together at the same time. Through wireless and all that it promises for networking and applications outside of pure computing, users of these platforms must configure and exchange a variety of common information in order to establish a common networking ground. It is no longer AppleTalk versus NetBIOS, TCP/IP versus IPX/SPX, or variants and workarounds in between, but purely the same technology and the same terms applicable to all platforms.

User interaction with wireless, wireless security, signal integrity, and failure analysis bring these platforms together. Unfortunately, the tools used to survey and analyze wireless networks and security are not equally available on all platforms. The two most notable applications for hacking or determining wireless network security levels—AirSnort and WEPCrack—are available only for the Linux/UNIX platforms. This forces system administrators of Windows and Mac networks who do not already know it to quickly learn Linux or find someone outside of their environment—usually a high-priced consultant—to help them assess the security of their networks.

Of course AirSnort and WEPCrack could be labeled as tools that have been designed only for the purpose of hacking into someone's wireless network. But in order to assess security, you need something or someone to try to breach it. Better you using these tools on yourself and tightening up security than someone unknown, with motives unknown, trying to breach your network's borders.

UNIX/Linux

I do not profess to be a Linux expert. I can deal with the operating system just so much before becoming frustrated at the lack of concise step-by-step documentation to get you quickly to the point where a new device, feature, or program simply functions. I know I am going to take a lot of flack for saying this, but as cool as Linux is when things are running well, it is not as plug-and-play as the primary consumer operating systems (Microsoft Windows and Apple Macintosh OS 9 and OS X). For Linux to be viable, some degree of detailed technical support must exist with or for the user.

My view includes the commercial distributions of Linux—and especially those for wireless applications. In terms of realizing the user-friendly attributes that make an operating system approachable and practical—and, if not pleasant, at least tolerable to work with—UNIX systems have far to go.

Most of us do not want to GUnzip, untar, compile, link, debug, decipher log files, decipher and edit obscure and esoteric configuration file parameters, learn C and shell scripting to be able to read and extract salient bits of command parameters, and do so over and over again for 12 to 24 hours, only to fail to get a simple wireless network card or two to work. Linux, and UNIX in general, need more user-friendly tools, at least in the context of wireless networking, before it can make a dent in the Windows market.

In reality, it has taken me at least three months on and off, begging for information from various on-line mailing lists and support groups, to get various fragments of information that finally led me to getting a wireless adapter to work with Linux. I think my next book ought to be about 1-2-3 steps through UNIX system configuration for the masses.

These are not religious or philosophical issues, as I have a deep, abiding respect for UNIX experts and the many great things about UNIX-based systems. But this genre of operating system is still about five years behind the DOS-to-Windows, plug-and-play, auto-recovery, goof protection progress that has been made in the WinTel (Windows+Intel) market recently.

There are, however, ways to get Linux to do at least one thing it is good at with wireless devices—routing, firewall, and access control. This can be done without immersing yourself in the struggles of getting this card or that to be recognized and automatically configured at boot time, using external wireless bridges or access points connected to an otherwise ubiquitous Ethernet card in the Linux system. While you avoid the trials and tribulations of configuring Linux for wireless, you will not be able to use AirSnort, WEPCrack, or the other low-level sniffing tools with an external wireless device, but the practical goal is wireless + Linux, leaving the sniffing and packet analysis to those with more time on their hands.

If you have accomplished getting a peripheral component interconnect (PCI) or personal computer (PC) card-based wireless adapter to work with Linux, you are probably familiar with many of the tools and discussion groups available that helped get you through the experience and allowed you to play with wireless all you wanted. For us novices, the next section lists a few must-browse Web sites catering to Linux and wireless hints, tips, and tools.

Resources for Linux and Other Flavors of UNIX

If you scour the Web and hit the usual Linux support sites, you will see listings of some standard tools the Linux community uses to work with various aspects of wireless networking. The first few sites listed can help get you started and provide the files necessary to get wireless networking going on your Linux system. Beware. You will have to know the Linux file system, navigate through the command line, dig around in a lot of readme files, edit a few obscure config files, and compile a few programs to take advantage of many of the following resources.

Jean Tourrilhes: http://www.hpl.hp.com/personal/Jean_Tourrilhes/Linux/Wireless.html
Jean's web pages are chock full of great information and cross-links to help you get wireless going on Linux.

wlan-ng pages: http://prism2.unixguru.raleigh.nc.us
This is a must-visit site to get source code and installable wireless networking files for all that is installable for RedHat Linux and common wireless devices. These files represent some of the best pioneering and growth of wireless networking. Do not miss them.

AbsoluteValue Systems: http://www.linux-wlan.org
This is another must-visit to obtain source code and relevant information to build into your Linux system for wireless networking.

Linux-WLAN List Signup: http://lists.linux-wlan.com/mailman/listinfo/linux-wlan-user

Linux-WLAN List Archive: http://lists.linux-wlan.com/pipermail/linux-wlan-user
The Linux-WLAN list is home to just about everything Linux and wireless. It is more a peer-to-peer discussion medium for those

already familiar with Linux, offering little step-by-step information for novices. But if you want to interact with the two technologies, this is the list for you.

Jason Boxam: http://talk.trekweb.com/~jasonb/articles/linux_wireless1.shtml
This is a small, but information-packed journal of Jason's venture into wireless networking on Linux.

The sites listed above will cross-reference each other and many other sites common to wireless networking, so you cannot go wrong hitting any one of them. Once you have Linux up and running wireless, you may want some of the tools to snoop around wireless networks.

Kismet Packet Sniffer: http://www.kismetwireless.net
Kismet is one of a few tools available to sniff data packets present on a wireless network—valuable stuff if you are into low-level network and data security analysis.

WEP Key Snooper AirSnort: http://airsnort.shmoo.com
AirSnort is the most popular tool for grabbing wired equivalent privacy (WEP) encryption key information from a wireless network. It may be of value as part of a security analysis, but its real purpose is to reveal the keys to other people's wireless LANs. Grabbing someone's WEP key is not for the impatient. It takes at least a million packets to decipher a key. Snooping on a 600-megabyte download gives you few 100,000 packets or so.

WEP Key Snooper WEPCrack: http://sourceforge.net/projects/wepcrack
WEPCrack is designed to prove the ease of breaking the WEP key encryption scheme. It does not sniff for packets. Instead, you must acquire packets using the prismdump program to create a file of captured packets, and then feed that file into WEPCrack.

WAVE Stumbler: http://www.cqure.net/tools08.html
WAVE Stumbler allows you to detect and identify other wireless LANs nearby. It is a good tool for doing site surveys, to see who is on which channel, and perhaps with a directional antenna, find other WLANs.

SSIDSniff: http://www.bastard.net/~kos/wifi/

SSIDSniff falls into the same category as WAVE Stumbler, allowing you to detect and identify other nearby wireless LANs.

Sputnik: www.sputnik.com

Do you want to provide a community network? Get up and running fast with this CD-ROM-bootable instant portal. The software forces users of a Sputnik-backed access point to log into the Sputnik.com server. The service is free, and the Web site maintains a list of affiliated community hot spots.

NoCat Authentication: http://nocat.net

NoCat appears to be the choice of gateway and access control programs for many open/community and closed/commercial wireless network hot spots. It is the foundation for the Sputnik portal program.

Absolute Value Systems: http://www.linux-wlan.com

This site hosts drivers for Linux-based wireless networking.

SOHOWireless LANRoamer: http://www.lanroamer.net

LANRoamer is another option for creating a wireless network hot spot, similar to the Sputnik project. Download the CD-ROM image file, burn a CD, put the CD in a system with a wireless card and access to your network or the Internet, and you have an instant wireless portal site.

Trustix Firewall: http://www.trustix.com

Finally, here is a firewall for the rest of us who are and do not want to be proficient at IPChains and similar scripts to control what goes in and out of our networks. Trustix Firewall is a secure Linux implementation designed to make any x86 system into a firewall appliance, with a graphical interface for configuring it specifically as a firewall to go between your LAN and the Internet or other connections. It also provides IPSec virtual private network (VPN) services between two systems that have static Internet protocol (IP) addresses. While there is no specific wireless component to this product, it treats wireless connections as it would any other Ethernet connection. It is a good tool for any network.

Apple Macintosh

I am similarly concerned by the lack of information and easy, logical accessibility to essential system and feature configuration that would make it about 110 percent easier to do many common, expected things with a Macintosh operating system. By common, expected things in this context, I mean being able to install, troubleshoot, and support Ethernet connections.

I barely maintain about 10 Mac G3s, G4s, and a few iBooks, have become quite familiar with the user interface, control panels, program installations, and the like, but there is a lot missing from the Mac. For all the easy-to-use hype, I would at least expect one complete panel of "idiot lights" to tell me what is happening or not with these systems. I'd even settle for a simple Link LED indicator for the Ethernet connection, but apparently that is asking too much. OS X is the best thing to happen to Apple since it first hit the market. Maybe there is hope, only because OS X offers a full range of UNIX-based network troubleshooting tools—at least PING and TRACEROUTE—without having to scrounge for, download, and install several different third-party tools to provide these features to OS 9.

Resources for Macintosh

Apple OS 9 and OS X, along with its AirPort product series, supports wireless networking just fine. But if you want to dig into wireless with your Mac, you need additional tools—the common wireless local area network (WLAN) presence survey tools and perhaps something to sniff WEP keys off someone's WLAN. Macintosh resources include:

APScanner (for Mac): http://homepage.mac.com/typexi/Personal1.html
APScanner is one of two known tools for detecting the presence of nearby wireless LANs.

MacStumbler: http://homepage.mac.com/macstumbler
And of course MacStumbler is the other wireless LAN survey tool to consider.

AirSnort on Apple iBook: http://www.macunix.net:8000/ibook.html
If you absolutely must sniff out someone's WEP key and do it from a Mac, you will want to know how to get AirSnort running on your iBook.

Microsoft Windows

As popular as Microsoft Windows is for personal and business computing, the number of wireless-specific tools available for Windows falls well behind Linux. This shortfall does not prevent you from using Windows for access control or as a gateway for a wireless network. Windows for desktops provides Internet connection sharing. Windows 2000 can act as a remote access server to a LAN or the Internet, and will host RADIUS and other forms of access control and user authentication.

Resources for Windows

NetStumbler: http://www.netstumbler.com
NetStumbler is one of the most universal tools to use for detecting wireless network activity, providing significant amounts of data about each wireless access point you can receive. It will reveal the media access control (MAC) address of active wireless devices, channels used, signal strength, service set identifiers (SSIDs) or lack thereof, as well as whether or not encryption is used at a particular access point.

ISSWireless Scanner: http://www.iss.net
Internet Security Systems' Wireless Scanner provides automated detection and security analyses of mobile networks utilizing 802.11b to determine system vulnerabilities.

AiroPeek—Packet sniffer: http://www.wildpackets.com/products/airopeek
For the true LAN techie, packet sniffing is everything. Chances are you will need to update your wireless adapter firmware and drivers

to get it to work. If you need to discover an intruder or a new threat to your network, you may have to dig down and look at streams of data packets to determine the cause.

Funk Software Odyssey: http://www.funk.com

Odyssey is an integrated package of the company's Steel-Belted RADIUS remote access authentication software with 802.1x EAP-TLS security for Windows 2000. Odyssey provides a complete access control and security solution for wireless LAN deployments.

WLANExpert: http://www.allaboutjake.com/network/linksys/wlanexpert.html

I really wanted to love WLANExpert until I discovered it does not run on Windows 2000 or XP. If you do not mind running it on Windows 98 or Me, you will be fine, and you may want to, so that you can enjoy its features. It works with most Intersil Prism2-based WLAN cards, covering LinkSys and similar products. Two of the best features are built-in antenna testing and reporting on whether your attached antenna is good or bad. It is most useful for external antenna connections or detecting a broken internal antenna, and it has a module that lets you set the transmit power for your LAN card.

Roger Coudé's Radio Mobile: http://www.cplus.org/rmw/english1.html

If you are planning numerous or complex wireless networks that have to cover long distances or irregular terrain, you simply cannot do without Radio Mobile. Radio Mobile uses standard geological survey maps containing terrain data to show you the signal strength of a signal throughout a selected area. This is a freeware program providing features similar to very expensive commercial radio site planning and coverage software.

Secure Wireless Network How-to: http://www.missl.cs.umd.edu/Projects/wireless

Mike van Opstal provides an excellent how-to guide for configuring a Windows 2000 server and Windows clients for secure, non-WEP authentication and network access. Click on *802.1x Implementation and Setup How-To*. The how-to is a succinct set of documents, rivaling anything Microsoft offers on the topic.

Generic References

The following sites provide a wealth of information and references for wireless networking in general and building community wireless networks.

Personal Telco: http://www.personaltelco.net
This is the Web site for a Portland, Oregon-based grassroots movement to create what it calls alternative communications networks—primarily community wireless LANs to distribute Internet access to more of the public. The site contains how-to documentation and links to several wireless resources.

New York City Wireless: http://www.nycwireless.net

San Francisco Wireless: http://www.sfwireless.net

Seattle Wireless: http://www.seattlewireless.net

FreeNetworks.org: http://www.freenetworks.org

Southern Calif. Wireless Users Group: http://www.socalwug.org
These are more grassroots movements to distribute Internet access to more of the public through wireless networking. These sites contain how-to documentation and links to several wireless resources.

Bay Area Wireless Users Group: http://www.bawug.org
This is not just a grassroots movement, but perhaps the most technically skilled or attended and mentored wireless group in the U.S. BAWUG's site and mailing list enjoy contributions from some of the foremost experts in networking and wireless technologies.

BAWUG List Signup: http://lists.bawug.org/mailman/listinfo/wireless

BAWUG List Archive: http://lists.bawug.org/pipermail/wireless
The BAWUG mailing list is one of, if not the best, general mailing lists to post questions and search for answers on many, many aspects of wireless networks, products, and implementations—heavy on the Linux side, but many list members do speak Mac and Windows too.

Open Wireless Node Database: http://www.nodedb.org

80211HotSpots: http://80211hotspots.com

Will you be traveling with your laptop and wireless adapter? Visit these sites before you head out. NodeDB is a list of typically open/public wireless hotspots around the world; most of them providing free access. 80211HotSpots provides a more targeted list of generally commercial/subscription-based wireless LANs.

Wayport: http://www.wayport.com

T-Mobile: http://www.t-mobile.com

Boingo: http://www.boingo.com

Wi-Fi Metro: http://www.wifimetro.com

Traveling professionals probably will not want to trust the ability to get wireless access from open community networks. For them, there is a proliferation of subscription-based wireless access services. You will see the logos for these national and international services at hotels and coffee shops.

HyperLink Technologies: http://www.hyperlinktech.com

HyperLink Technologies is a full-service wireless equipment and system planning vendor.

802.11 Planet: http://www.80211-planet.com

WECA: http://www.weca.net

Wi-Fi Org: http://www.wi-fi.org

Every market must have its trade associations—commercial advocates of specific technologies and products. None are definitive, but most of them participate in legislation and technical standards organizations that can or will affect the features and functionality of a particular technology or service.

AirDefense: http://www.airdefense.net

AirDefense sells a dedicated appliance to assess and manage wireless network security issues.

RF Connectors: http://www.therfc.com

Demarc Technology Group: http://www.demarctech.com
Parts are parts and we all need them to get and keep our new toys running—from little wires to obscure connectors to full-blown engineered and certified systems. If you need something, one of these vendors probably has it.

Tim Pozar's Site: http://www.lns.com
Tim is one of the foremost qualified authorities on wireless systems, from broadcasting to wireless networks. It's his job to know what works correctly and legally in this domain. He's a busy man, but always glad to help where appropriate.

Summary

With the support for wireless networking provided in the current operating systems, and the software that comes with your network card, you can easily jump in on the basics of the wireless wave. For more intense wireless projects, you will find the software and information links provided here to be invaluable in getting you farther along into a robust and secure wireless infrastructure.

If you need to know more about the signals floating around in the wireless spectrum, no amount of software for any operating system will help you, leaving you to seek out expensive and precise test equipment from Agilent (formerly Hewlett-Packard's test equipment division), Tektronix, Anritsu, IFR, or Motorola.

If you need more specific information about a particular network product, technology, or problem, consult any of the Web sites and list servers listed, or use your favorite Web search engine. You may be amazed at the wealth of specific data available. If you feel the prospect of implementing a wireless network is way over your head, you can probably find a suitable local vendor to help you design and build a network to suit your needs.

CHAPTER 11

Wireless Access and Security Solutions

This chapter will present two distinctly different implementations of integrated wireless security and access control applications and a security analysis program. The first is Funk Software's Odyssey product. Odyssey combines their Steel-Belted RADIUS authentication services and end-to-end data security technologies for Windows environments. By highlighting this type of product you will be able to see the various elements and principles of remote access and wireless security come together in a way that is tremendously simple to use.

The second is WiMetrics WiSentry. WiSentry is not a specific access control product. Rather, it is a security monitoring tool, but it has the functionality of allowing or denying wireless clients and access points to use a wireless network.

If you'd care to follow along with this chapter you can load these programs from our CD-ROM, download a demo version of Odyssey from Funk's web-site: http://www.funk.com/radius/wlan/wlan_radius.asp, get a demo copy of WiSentry from http://www.wimetrics.com, or obtain the ISS Wireless Scanner from http://www.iss.net.

Funk Software: Odyssey Installation

Installing Odyssey starts out by building upon a basic Windows 2000 Server installation with a couple of standard Windows 2000 options installed and enabled before actually installing the Odyssey software. Take each step carefully to ensure that everything works properly with the server first.

You will need a few things to get started:

- An adequate hardware platform to support server software and multiple network cards. At a minimum:
  - Typically a 333 MHz or better Pentium II, III, or IV system
  - 128–256 megabytes of random access memory (RAM)
  - 4 to 6 gigabytes of hard drive space
  - Two 10/100 BaseT network cards installed
- Windows 2000 Server or Advanced Server software. Windows 2000 Professional and XP are also supported for server installations.

- Windows 2000 security certificate server installed, and a locally generated certificate or one imported from a third party, such as Verisign. The certificate is used as part of the security key and authentication processes. If you are installing on 2000 Professional or XP, you will need an external certificate server.
- A wireless access point—Orinoco AP-2000 or equivalent commercial unit is recommended.
- Wireless client PC or laptop running Windows 98, Me, 2000, or XP, and wireless adapter.
- Funk Odyssey client and server software.

Windows 2000 Server Installation

Your best bet is to start with a clean Windows 2000 server installation. The server can be set up as the domain controller or a participant in an Active Directory architecture, but this is not required. To make things easier, and leave the system with adequate performance capacity and security, I recommend that this server not be used for other functions—that is, do not install IIS (Microsoft's Web and file transfer protocol [FTP] server), Microsoft Exchange e-mail server, or similar software. It is convenient to have local domain name system (DNS) services available for faster host lookups, but this is also not necessary. Domain host configuration protocol (DHCP) services to configure clients may be provided by the server, but DHCP can also be provided from most access points.

In addition to the usual server configuration, Certificate Server must also installed, through the selections shown in Figure 11.1.

Once Windows 2000 server is installed, you must configure your network cards for the networks they will be connected to—one to the Internet or local area network (LAN), the other dedicated to tie in the wireless access point. Chances are, you will want to use a private unrouted Internet protocol (IP) address range for the wireless access point and client side—either 10.x.x.x or 192.168.x.x will work.

In my case, the internal LAN subnet uses 10.10.10.x Class C addressing, and because I'm used to typing 10s, I configured the wireless network card's address to be 10.10.0.10, and set the gateway address for this LAN card equal to the static IP address of the wireless access point, 10.10.0.11. These are two completely different subnets, so I would not be confused about addressing or routing.

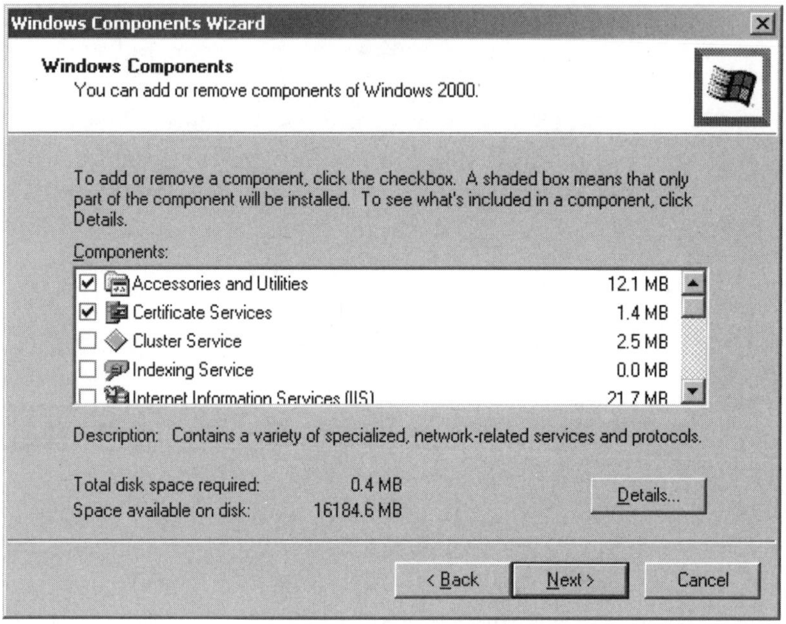

Figure 11.1 Certificate Services selected for installation through Control Panel, Add/Remove Programs, Windows Components Wizard.

After the operating system is installed and set up, the latest server operating system and security patches should be applied, and you should have an up-to-date virus protection package installed and running. We are dealing with secure access control to your internal network or the Internet, and these steps are fundamental to the process.

With the server software installed, configured, and running, you should install the software for your access point hardware, connect the access point, and set up a basic wireless client to connect to the access point, to ensure this hardware and connectivity works. I configure the access point to serve DHCP for clients connecting to it for simplicity, though this can also be done by the Windows 2000 server. You will reconfigure the access point, according to directions for the Odyssey server, after the server software is set up and installed.

Odyssey Installation

Following the updates and patches, the Certificate Server is configured to provide a local security signature for the Odyssey RADIUS server and encryption to use. With this accomplished, Odyssey is ready to be installed.

If you already have a server, be sure to uninstall, not simply disable, Internet Authentication Services, and disable Remote Access Services, as well as any other RADIUS or access authentication services. See Figure 11.2 for the proper dialogs to remove unnecessary services. Failure to do so will conflict with the Odyssey server and its server service will not start. The Event Viewer log will show an error 2147500037 as evidence that there is a conflict with another remote access service.

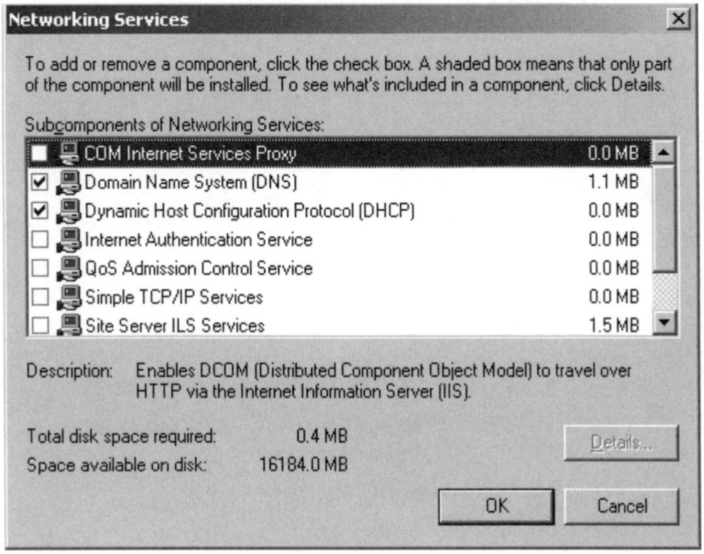

Figure 11.2 Internet Authentication Services must not be installed. It is removed through the Control Panel, Add/Remove Programs, Windows Components, Networking Services selection dialog.

You are reminded of the basics of this requirement during the server installation process. I tried simply stopping and disabling Internet Authentication, but that was not enough. I had to uninstall this through Control Panel, Add/Remove Programs, Windows Components. I also removed, then reinstalled the Odyssey server, to get the server service to start properly.

The client and server software packages come to you as .MSI (Microsoft Installer) file packages, so your server must have the Installer service enabled. The installation process takes only five minutes or less. It automatically starts the server service, or warns you if it has not been started. Opening the Odyssey server administration program gives you access to all of the program features you need.

To ensure the Odyssey server is running, check Windows 2000 Services by selecting Start, Run, Programs, Administrative Tools, then double-click Services. It should show a status of Started and a Startup Type of Automatic, as indicated in Figure 11.3.

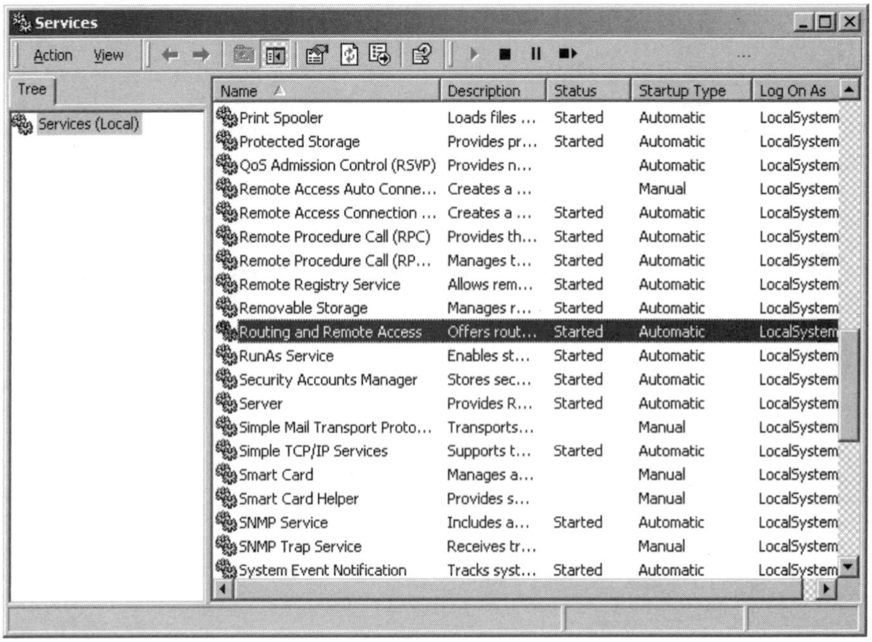

Figure 11.3 The Odyssey server status appears in Windows' Services dialog.

The Odyssey server management program is reached through Start, Programs, Funk Software, Odyssey Server. Configuration is very simple and straightforward. You need only interact with the Authentication Settings, User Trust, and User Identification controls under Settings, to get ready to use the server (see Figure 11.4). The Add Users in the Users dialog is shown in Figure 11.10.

Your first step in the Odyssey configuration is to select the access point(s) to be used. Select Access Point Defaults or right-click Access Points in the left pane to view the access point selections, as shown in Figure 11.5. The server will support more than one, managing access from any access point to the common network it supports. Several common access points are supported and listed.

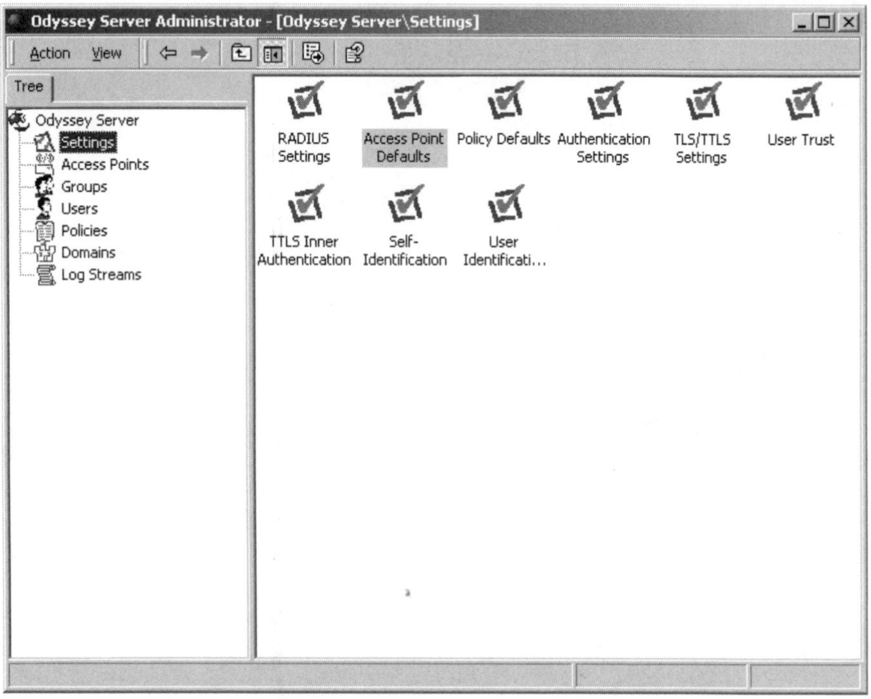

Figure 11.4 The main Odyssey server management dialog shows the very simple interaction to configure and use the server.

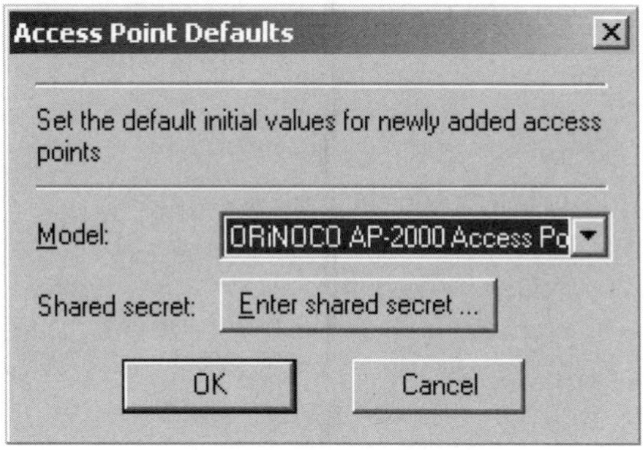

Figure 11.5 Access point selections are made by a simple scrollable menu.

With your access point selected, you must configure its properties by giving it a name and indicating the IP address that identifies it to the server software (see Figure 11.6). Establishing a Shared Secret key, also to be entered in the access point's configuration software, is optional but recommended.

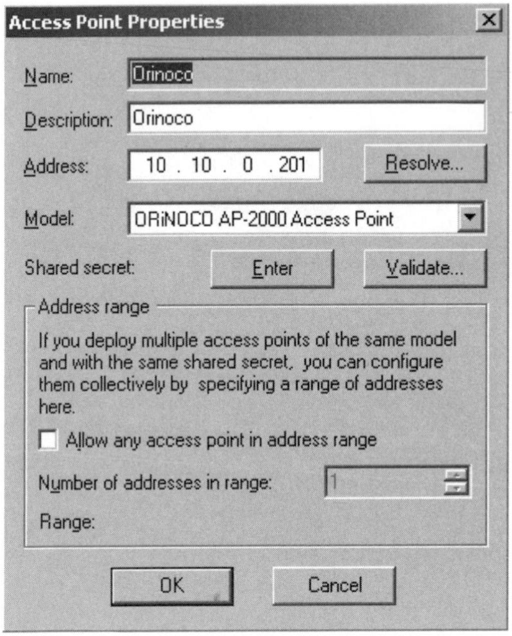

Figure 11.6 The Properties for each access point must be configured for the server to communicate with and support access through it.

Your next step is to select a certificate server to use. Your server must have a Server Certificate installed prior to this step—which an installed and configured Microsoft Windows Certificate Server provides for you. Select the TLS/TTLS Settings icon to present the certificate dialog, as shown in Figure 11.7. Then browse and select a server from which to obtain certificates.

Select the User Trusts icon to add a trust tree for user certificates—Figure 11.8. This server is typically the local server, or may be an external server to sign the certificates used by clients.

One last configuration item before adding users—the User Identification settings—Figure 11.9. This dialog provides several options for how the user is identified by certificate. Only one option is on by default. It is recommended to select them all, unless you know for sure which method to use.

With server configuration completed, you then have to tell the server which users are allowed remote access. Right-click on Users in the menu, then select Add User(s) to access the user selection dialog—Figure 11.10. The top half of the display shows users known to the local server. Highlight one or more user identities, then click Add to include them in wireless access.

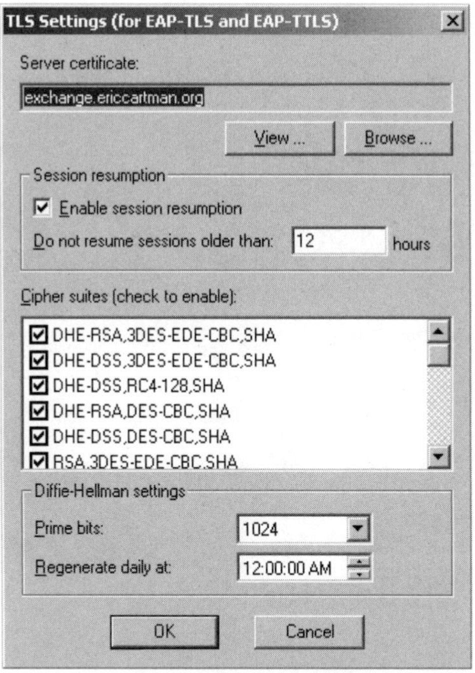

Figure 11.7 The security certificate selection dialog allows you to specify the source of your certificate, the type of certificate to be used, and if you want to allow sessions to resume or expire them after a period of time.

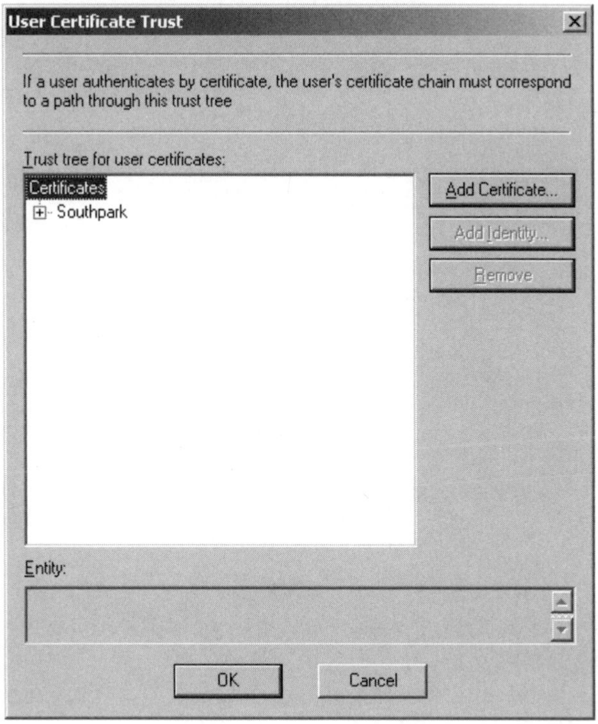

Figure 11.8 The User Trust selection is required for TLS authentication methods.

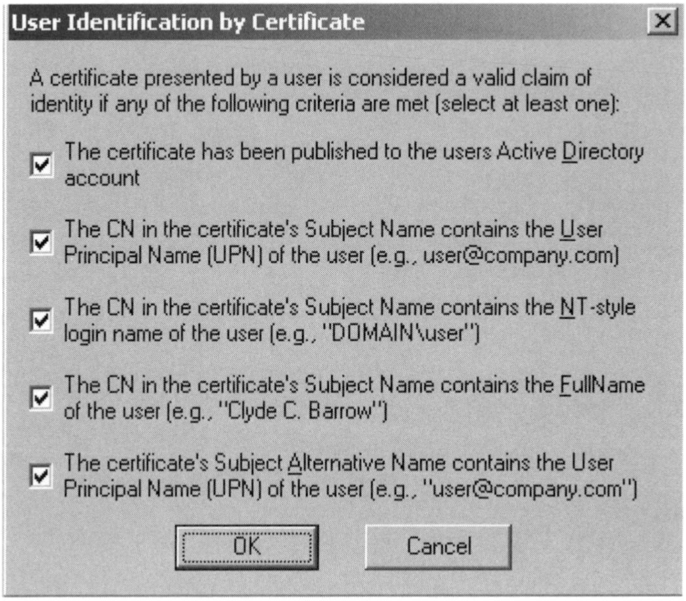

Figure 11.9 The User Identification selections indicate how the user is identified by certificate—selecting all is recommended unless you know which specific method is used.

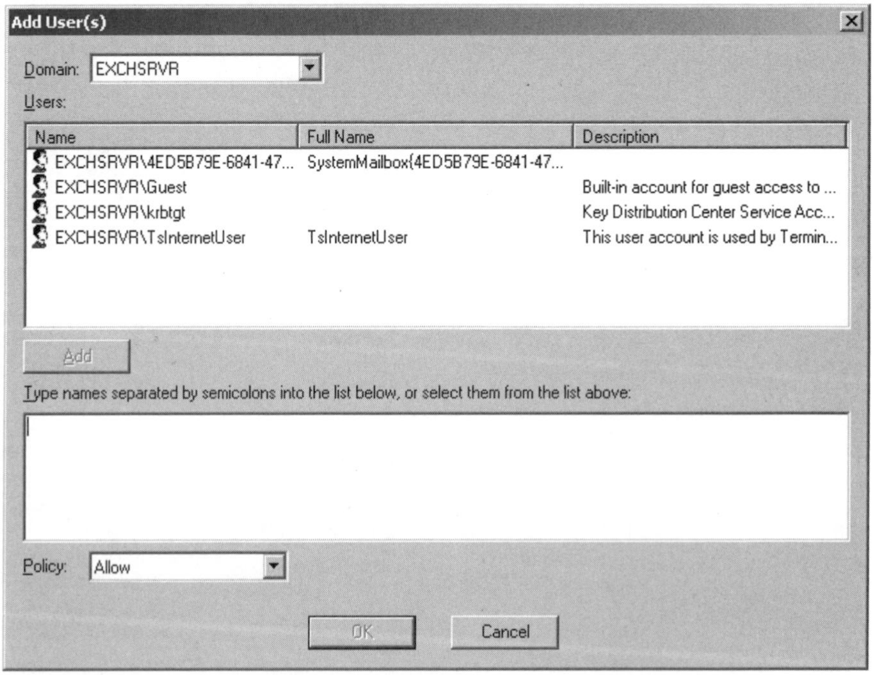

Figure 11.10 The Add User(s) dialog appears very much like the Windows server user dialog. Known users of the server, locally or from an Active Directory structure, appear at the top. Selected users with remote access appear at the bottom.

The above steps conclude the basic setup to establish secured remote access capabilities. You have some final steps to go through to set up your access point to communicate properly and securely with the server, and then a client installation to perform. Then your network is ready.

Access Point Reconfiguration

The Odyssey server requires a very specific configuration for your access point; in this case, an Orinoco AP-2000. The following settings apply for the Orinoco, and there will be comparable settings for the supported access point you choose:

- Disable the RADIUS media access control (MAC) address control.
- Enter a shared secret key to match the secret key entered above for the Odyssey server.
- Enable RADIUS authentication (access point passes authentication requests to specified RADIUS server).
- Tell the access point the IP address of the RADIUS server.
- For encryption type, use 802.1x (or Mixed mode WEP and 802.1x for EAP-MD5 authentication methods).
- Enable encryption for the access point's radio (slot A or B in the Orinoco).
- Set the key length to 128 bits and provide a 26-character key string.
- Set the "Deny non-encrypted data" parameter.
- Enable DHCP settings to suit your network if you want the access point to provide it.
- Reboot the access point.

The server software, and the software installed on the client, does the rest. The access point *knows* to pass on authentication to another authority figure or function, the server accepts the authentication requests and allows or denies access to the network, and the server provides end-to-end data encryption between the client and the server so that the wireless portion of the network is secured.

Client Software Installation

The client software to hook up with the Odyssey server is even easier to deal with than the server side. For XP and 2000, the client software is provided as a Microsoft Installer package file, so the operating system knows exactly how to run it.

Once the program is installed, your next step is to configure it for your wireless network adapter—Figure 11.11. Any card that has its driver properly installed and has been known to work should appear in this listing. The client program supports multiple adapters, including wired LAN cards, to provide support for universal serial bus (USB) and wired-in wireless bridges.

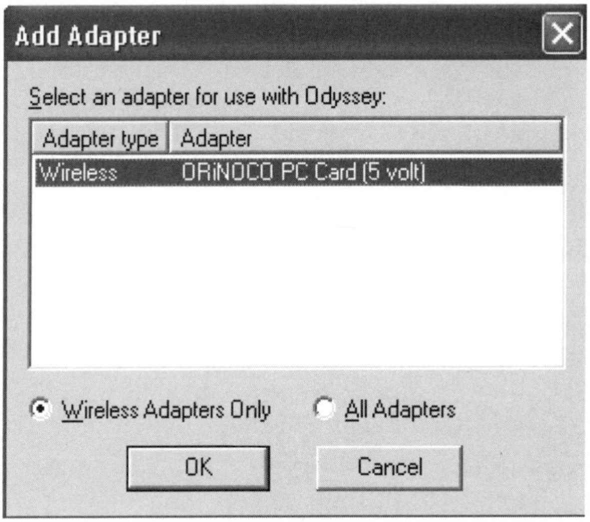

Figure 11.11 The Add Adapter dialog shows all known adapters.

Once the adapter has been selected, you need to instruct the client program to scan for available networks and then configure the one you want to connect to. Figure 11.12 shows the results of clicking the Scan button. Once you choose an available network—preferably the one you know is secure—you will be presented with the Properties dialog for that network—Figure 11.13. Changes to the network configuration can be made from the Networks menu option—Figure 11.14.

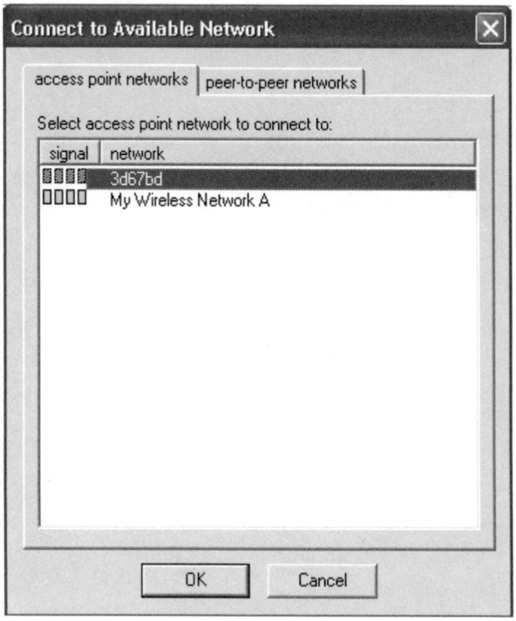

Figure 11.12 Scanning for an available network shows you which wireless network connections are available—however, not which one is secured with Odyssey on the server. You have to know this in advance.

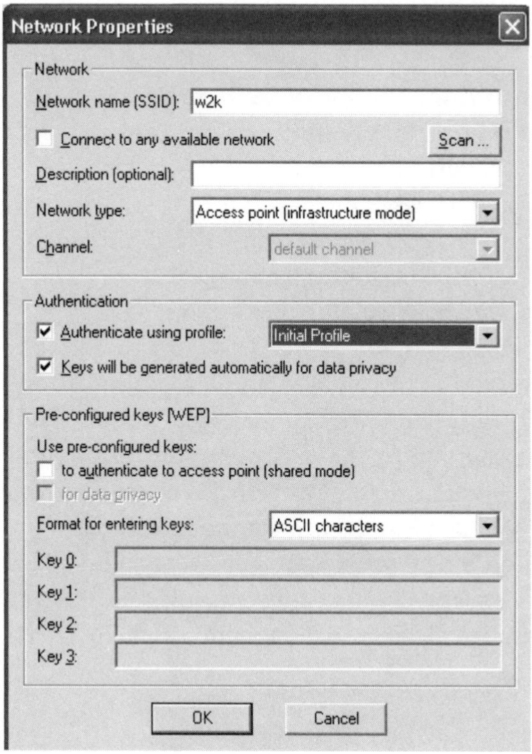

Figure 11.13 After you select a network to connect, you must set up the parameters for the connection. Choosing the default initial profile and automatic keys will get you to your Odyssey server.

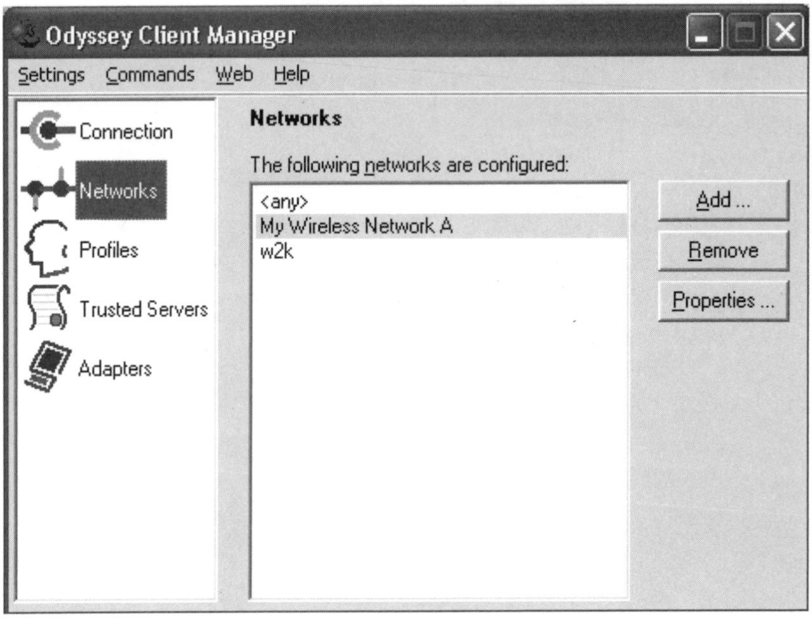

Figure 11.14 You can change the wireless network configuration by selecting the Networks menu option.

Selecting the Profiles option from the Odyssey Client Manager lets you choose from available profiles. Once one of the profiles is selected, as in Figure 11.15, you can determine how it will be used to interact with the Odyssey server.

Finally, you can add or review the servers your client trusts for authentication and connection—it has certificates from—by selecting Networks from the menu (see Figure 11.16).

When you first attempt a connection to your newly secured wireless network, you will see a password dialog pop-up. If you are using Windows server log-on to complete the authentication process, use your Windows network password. Your Windows log-on name is already provided to the program from the username you logged onto your PC from. You will not see the log-in prompt again until your current authentication session has expired, requiring you to validate your log-on again with your password. This is a typical and expected feature—essentially logging you off the network connection if you have been away from your computer for a length of time—to reduce intrusions.

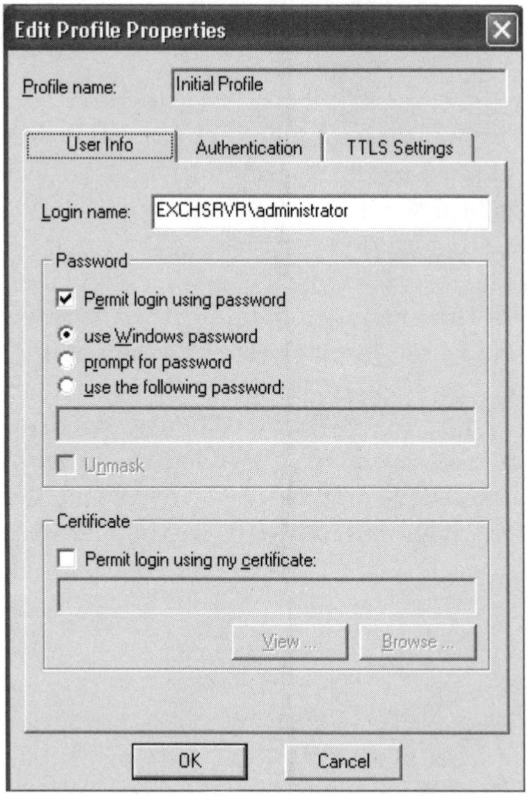

Figure 11.15 The typical profile is to use the Windows server password for authentication.

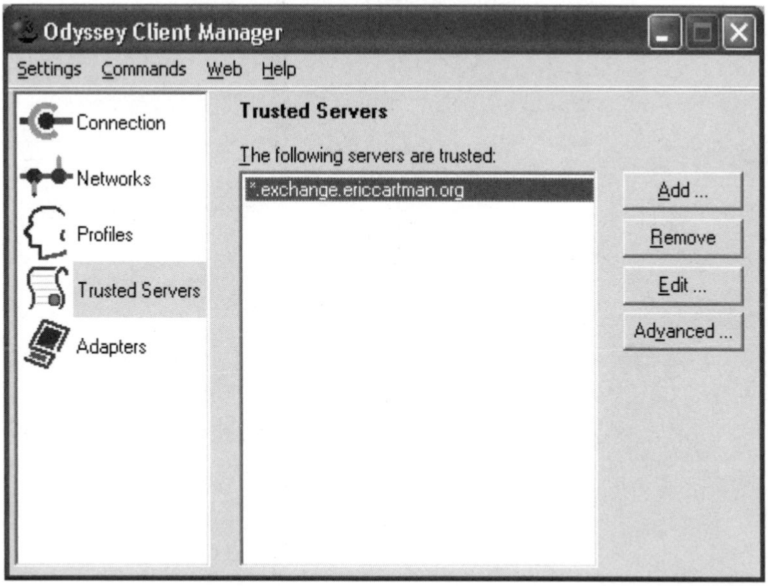

Figure 11.16 Networks your client trusts for wireless connections are shown in the Networks dialog.

WiMetrics: WiSentry Installation

WiSentry is a wireless network security monitoring tool that creates a bridge between your intended wireless LAN setup and your wired LAN. In addition to creating a bridge it provides a *sentry* or access control point on the wireless side of the bridge to either allow or deny specific wireless devices to gain access to the wired LAN on the other side.

It is suggested that you dedicate a Windows 2000 server to this task rather than simply adding another network card to an existing server because any unlikely security gap at the wireless side could expose data on this server. Such a server should not be a Domain Controller in an Active Directory infrastructure, nor should it have any file or resource sharing enabled that might expose data files or access control lists. Figure 11.17 shows the basic configuration for this system integrated into your existing network.

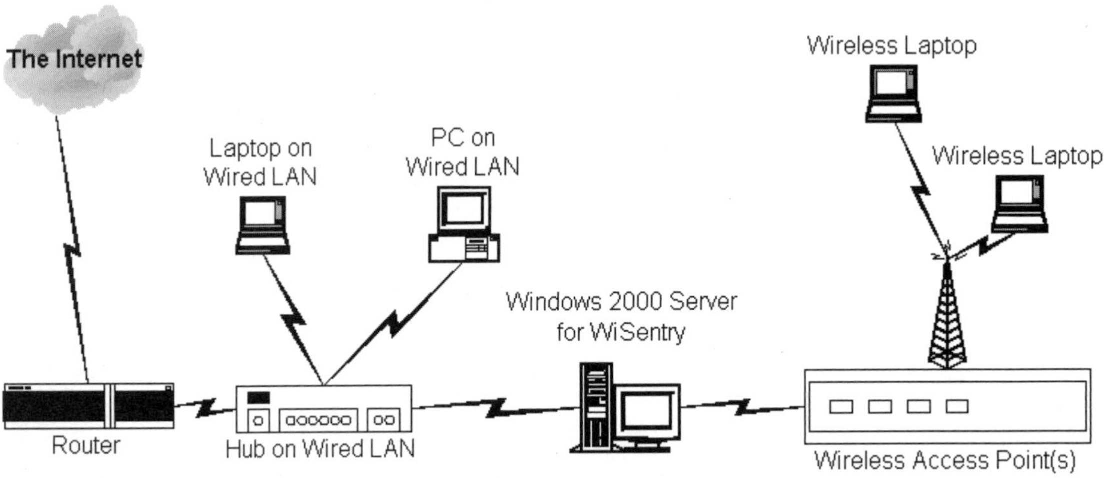

Figure 11.17 How WiSentry integrates onto an existing wireless LAN.

You will need a few things to get started:

- An adequate hardware platform to support Windows 2000 Server software and multiple network cards, at a minimum:
  - Typically a 333 MHz or better Pentium II, III or IV system
  - 128–256 megabytes of RAM
  - 4 to 6 gigabytes of hard drive space

- Two 10/100 BaseT network cards installed
- Windows 2000 Server, or Advanced Server software. Windows 2000 Professional and XP are also supported for WiSentry installations.
- A DHCP server on the wired side of your network—this can be the server on which you are installing WiSentry.
- A wireless access point—Orinoco AP-2000 or equivalent commercial unit is recommended.
- Wireless client PC or laptop running Windows 98, Me, 2000, or XP, and wireless adapter.
- WiSentry software.

Windows 2000 Server Configuration

Start with a basic Windows 2000 Server configuration. Do to install (or disable) Internet Information Server components and Routing and Remote Access, unless you will integrate them into a WLAN portal or provide an underlying login access control. If you do use Routing and Remote Access features, be aware that the server will then contain user access information you probably do not want to expose should the wireless connection be compromised. IIS is fraught with security holes and is simply not an application or service I would want exposed to unforeseen compromises.

As you install Windows 2000 Server, or after the installation is complete, configure the network connections as follows:

- Determine which LAN card will connect to the wired LAN and which will be used for the wireless access points.
- Provide fixed IP addresses within your wired LAN subnet to each of the LAN cards.
- You may wish to configure a specific subnet for wireless services, and configure this into your internal router as well.
- Set the Gateway addresses for each card to the address of your internal router.
- Configure DNS addresses.
- Configure WINS server address as appropriate.
- Configure this server to provide DHCP addresses for the wired LAN subnet. This is optional if you already have a DHCP server on the wired network.

With this basic configuration in place, connect your wireless access point to the LAN card assigned to this purpose, and the wired LAN to the respective LAN card for it. Next, configure your access point, providing the following:

- A fixed IP address
- Gateway address for the wired LAN
- SSID for the access point
- If available, do not enable DHCP from the access point; DHCP will pass through to the server or wired LAN
- Type of security you wish to use—conventional security methods are supported once wireless clients or additional access points are authorized access through the bridge
- WEP keys, if appropriate

WiSentry Installation and Use

The WiSentry installation is straightforward, beginning with a normal Windows installation process, followed by installation of Sun's Java Runtime Environment. A reboot of the server is required to complete the installation and activate the bridge service. Once the server reboot is complete, the installation finishes, and you are ready to run the WiSentry administrative program which serves as the access control point and alerting mechanism for wireless clients.

When run, the WiSentry administrative program (shown in Figure 11.18) begins to sniff the networks for access points. Discovered access points appear on a listing of Active devices. Viewing this list shows you all known wireless devices and what type of device they are, along with the device's MAC address and any IP addresses assigned to them. Color coding indicates if they are unauthorized or authorized. Initially all found devices except the bridge service is color-coded red to indicate it is unauthorized.

Your first action will be to identify which device is your access point, then authorize it so it can be used to pass wireless clients to the wired LAN. This is done by selecting Authorize from the Action item on the top menu bar of the program. Once the access point is authorized you can evaluate all wireless client devices and choose whether or not to authorize them for LAN access.

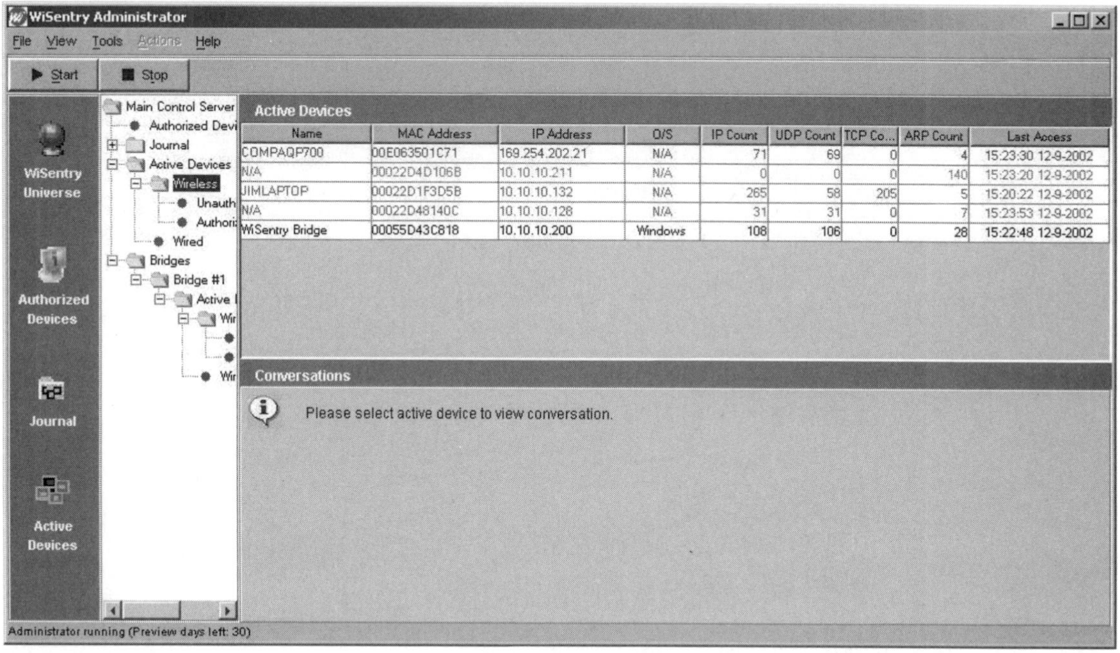

Figure 11.18 The WiSentry administrative program is where active wireless devices are detected, reported, and authorized, or denied access to the wired LAN.

Wireless client devices will be able to associate with an access point but will not be able to obtain an IP address from or access the wired LAN until they are authorized. This enforces that you must know which wireless devices exist and be able to identify them by MAC address or host name before authorizing them for LAN access.

You can leave WiSentry running smoothly by itself, checking every so often for rogue access points and new wireless clients wandering around in range of the WLAN, but you will probably want to set some alarms to pop-up and alert you to any new activity. Figure 11.19 show the alert configuration screen, with the types of possible intrusions that can be detected and how you want to be notified of them.

You can configure the alarms and monitor the system on a separate workstation rather than just the server. As shown in Figure 11.20, when an intruder, an unauthorized access point, or wandering client try to communicate with your network, you will get a pop-up dialog and a list of devices and their classification.

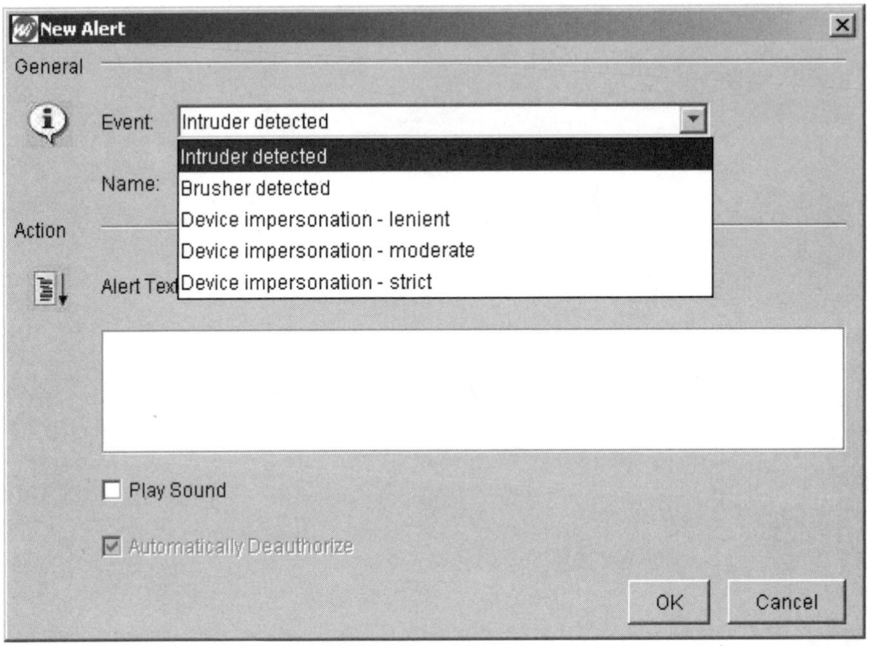

Figure 11.19 Alert configuration in WiSentry provides options for the type of possible intrusion you wish to be notified of and how.

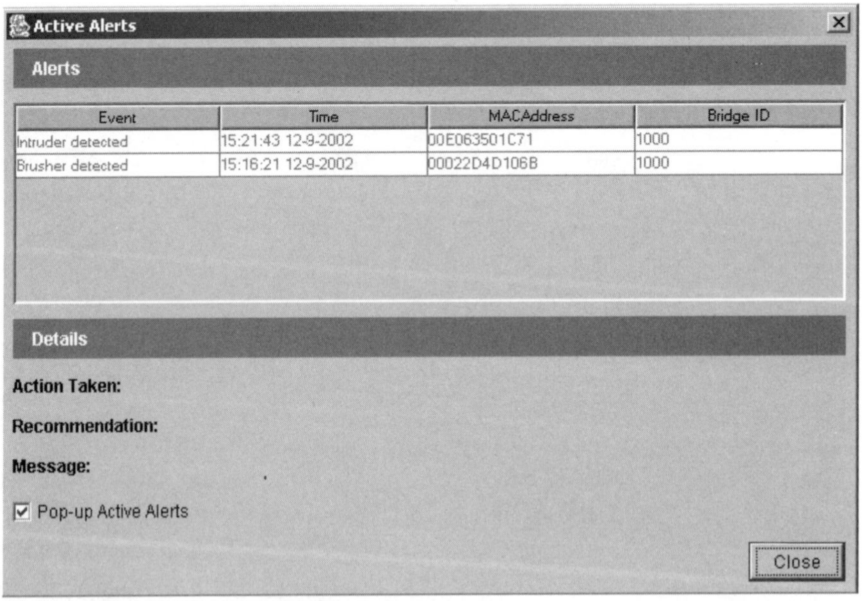

Figure 11.20 The WiSentry alert pop-up tells you what type of device is connecting to your WLAN or if rogue access points have been connected.

Once you receive an alert you will want to review the Unauthorized Devices portion of the administrative screen to get more information about the identity of the intruding device (Figure 11.21) and then authorize it if appropriate.

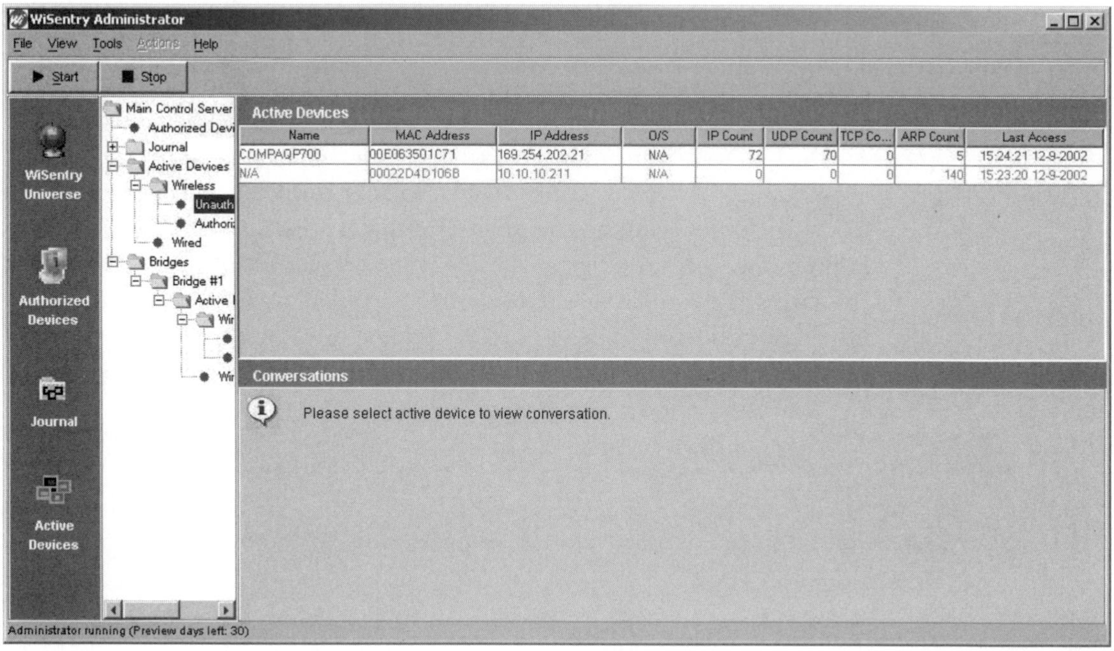

Figure 11.21 WiSentry provides the name, MAC address, and IP address of unauthorized devices so you identify them and determine if you wish to allow them access to your network resources.

As you can see, WiSentry packs a lot of work behind the scenes and makes it easy to deal with WLAN security and access issues.

ISS: Wireless Scanner

While you can control access to and through your WLAN, and you can see which devices are trying to connect to it, it's still a good idea to have an idea of how your WLAN security configuration appears from the inside out. Internet Security Systems has produced a wireless version of their network security scanning software. First, ISS is intended to be installed on a system with a PC Card WLAN adapter—so a laptop or desktop with PC Card adapter is required. Using a laptop allows you to roam about and get close to access points and sniff out unknown or rogue APs. Once installed you should run its driver configuration program to get a driver in place that will allow the scanning software to properly control the WLAN card and take in everything in the air. This driver will likely render the card unable to connect with your present network, and the driver configuration program allows you to switch back to the LAN-functional driver as needed.

Once the sniffing driver is ready to go you can begin taking live scans of the airwaves around you. Data is collected and presented on three different views—the first (Figure 11.22) is of detected access points, the second (Figure 11.23) is of detected vulnerabilities, and the third (Figure 11.24) is of detected wireless clients. The MAC or hardware address for each device makes it somewhat easier to identify the device.

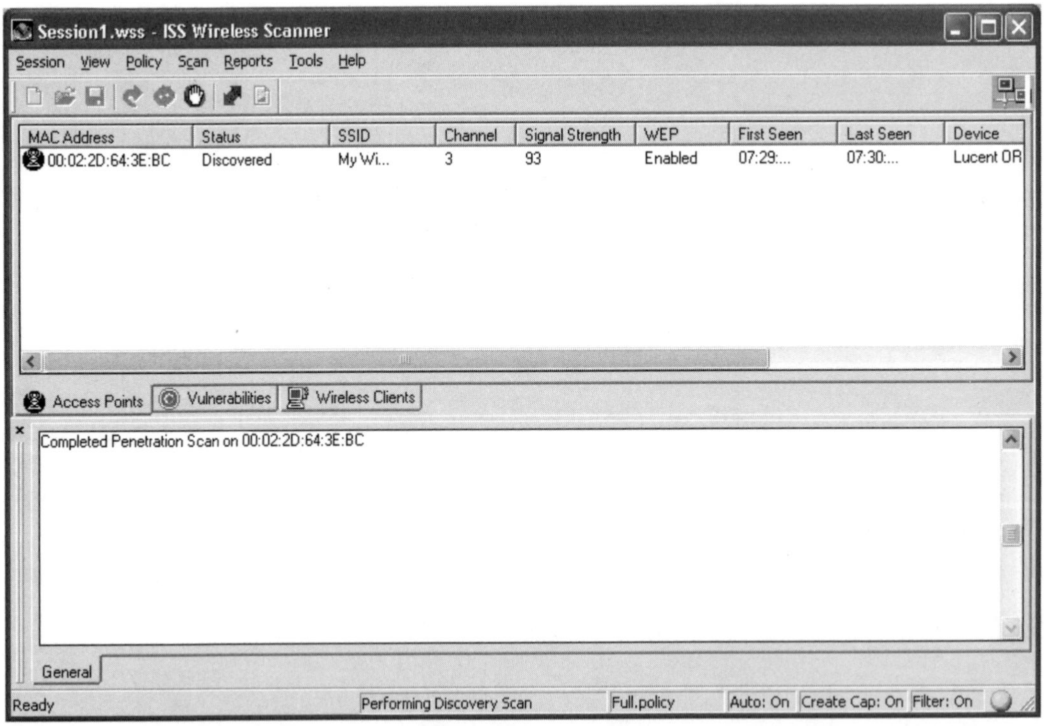

Figure 11.22 The ISS Wireless Scanner summary listing of discovered access points shows MAC address, channel used, signal strength, and time detected.

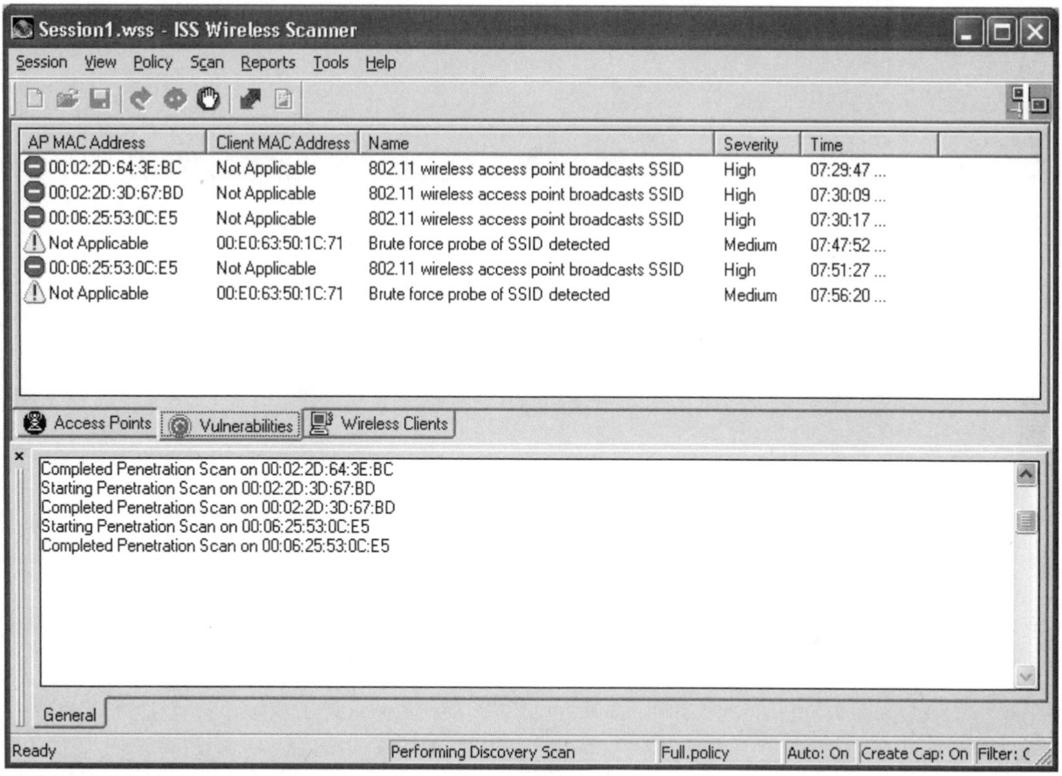

Figure 11.23 The Vulnerabilities view in Wireless Scanner gives a summary listing of potential issues and their severity.

These views are simply summary listings of what has been detected. Once you have collected a data sampling, go to the Reports menu selection and create one of several available reports to understand the WLAN environment, have an inventory of the devices, and an assessment of any vulnerability issues. A sample report of technical details is shown in Figure 11.25.

The Technical Details reports breaks down everything known about detected devices and the vulnerabilities found in them. This report will give you the call-to-action to begin securing your network. The two most common issues you will find in most WLAN setups are either the lack of encryption requirement at an access point and broadcasting the SSID, which can identify the owner or location of a particular access point.

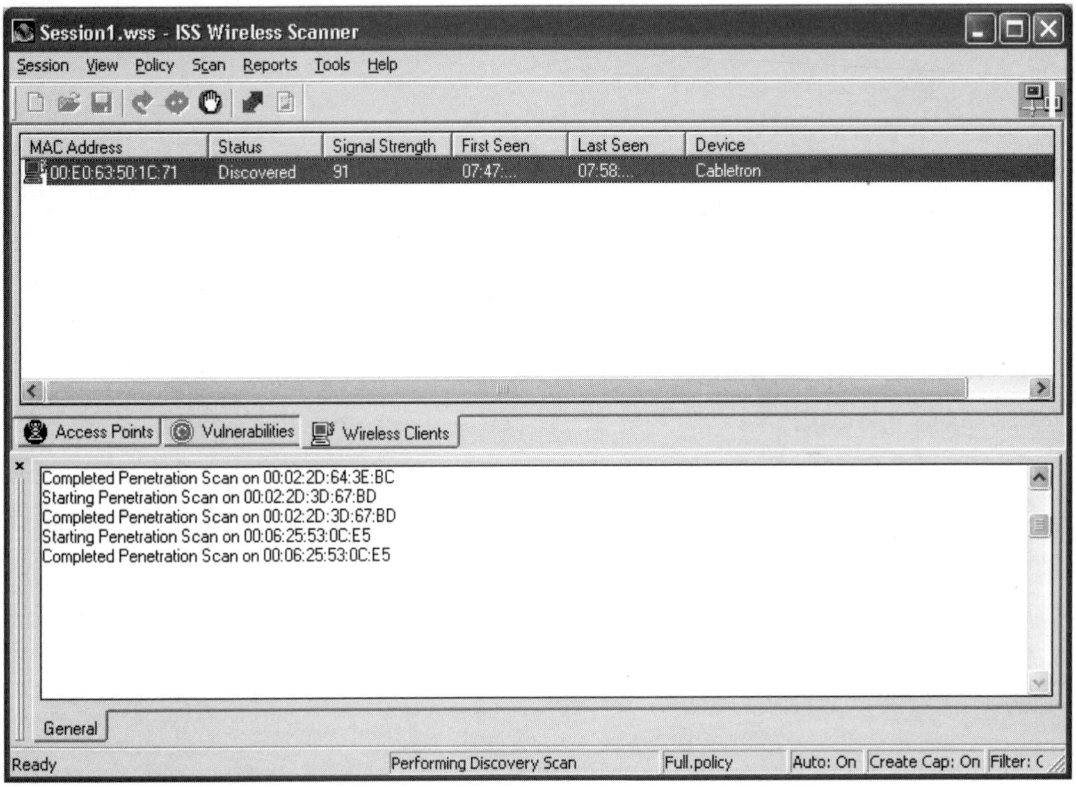

Figure 11.24 The Wireless Clients view shows client adapters that have been detected, their MAC address, and manufacturer.

Summary

There are many ways to approach wireless LAN access, security, and intrusion issues. A product like Odyssey deals with authenticating (or not), specific clients—a front-end positive approach to authorizing access to a network. Odyssey provides end-to-end encryption, but it has no awareness of possible intrusions. WiSentry provides both front- and back-end approaches to access control, and although it is not a specific authentication or encryption solution, it will work with the methods you choose for this purpose. ISS's Wireless Scanner adds another level of detail to knowing what is going on in your wireless LAN environment and will help you tighten up any obvious security gaps.

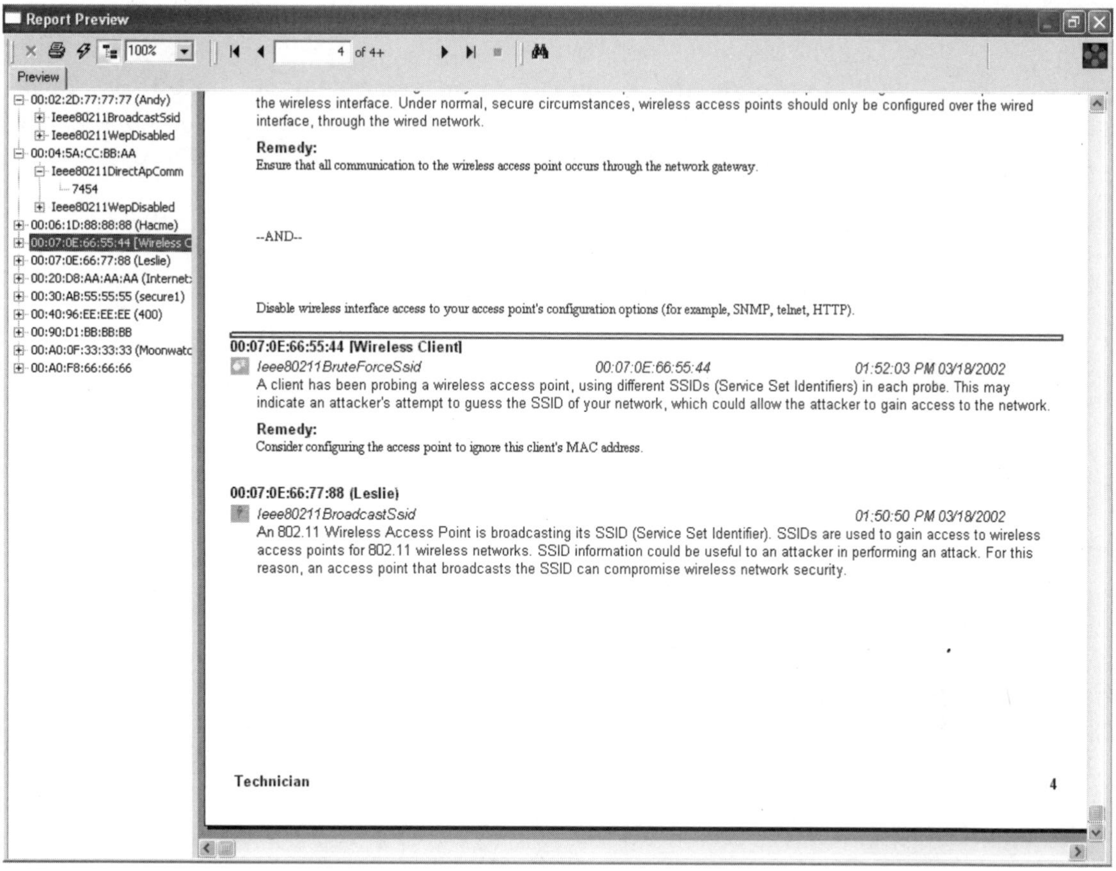

Figure 11.25 The ISS Wireless Scanner detailed report shows the specific problems and solutions for clients and access points with vulnerabilities.

Odyssey and WiSentry are not unlike similar add-on programs that build upon an existing infrastructure and user base to quite simply provide security in the form of access control. Similar features could be implemented using Windows IPSec at the client and server, but managing the process is not as easy, and network options are not as flexible for the client side. Similarly, security alerts about possible intrusions and rogue access points like the ones WiSentry provides, or the vulnerability reports of Wireless Scanner, could be obtained from sniffer products like AirMagnet, but AirMagnet and Wireless Scanner do nothing to stop the intrusions.

Perhaps knowing about these methods and how vulnerabilities can be revealed will get you to tighten up your network as you build it. You might think you can avoid using some of these tools, but as your WLAN grows so will the responsibilities and time to manage all of the components—requiring you to consider something to help give you peace of mind.

CHAPTER 12

System Configuration Data

It seems that every time my friends or co-workers set out to add something new to their personal computer (PC), they run into a conflict with one device or another, or have some piece of misbehaving software that prevents them from doing what they wanted to do or from using their new toy.

My intent with this chapter is to condense years of support work into a quick reference you can use to get yourself out of trouble if you are adding a network card or other adapter to your system, when creating a new wireless or shared network system. This information is not limited or specific to wireless networking. It is also useful for adding any type of peripheral to your system—which you are likely to do when your experience expands and you try to grow your computing interest beyond one simple PC.

Legacy Devices

Legacy devices, if not preset or fixed in their configuration when built into the motherboard or system board, require us to manually set jumpers (tiny connections between two protruding connector pins) or switches on system boards or I/O cards, usually in accordance with a table of possibly dozens of variations of settings, and in comparison to or in contrast with other devices in our PCs. Legacy devices typically do not lend themselves to automatic or software-driven reconfiguration, as may be possible with today's plug-and-play devices.

Several legacy devices that we have no configuration control over are:

- Central processing unit (CPU) and numeric processor using fixed addressing and interrupt request (IRQ) 13
- Clock and timer resources using fixed addresses and IRQs 0 and 8
- Memory and device addressing chips using DMA channels 0 and 2
- Keyboard using fixed addressing and IRQ 1
- Diskette drives using known/expected addressing and IRQ 6
- Video display adapter using known/expected addressing

These listed devices are part of the system board or basic input/output system (BIOS) programming and, as with other devices we will see, must remain as-is for a PC to function as a PC.

Almost all PC devices prior to implementation of the plug-and-play standard are considered legacy devices. These include add-in cards and other accessories, and to some extent, the basic PC system itself. In most cases, legacy devices present the bulk of the configuration and conflict issues we face in dealing with PCs. The next section addresses the most common types of add-in devices with which you could encounter configuration problems.

Logical Devices

Logical devices are those that have obscure abbreviated names associated with a function or a particular device. They are associated to a specific I/O address by program logic that assigns logical names to devices in the order they are found. This is true even for plug-and-play/universal serial bus (USB) devices—although the rules and results of plug-and-play and auto-configuration seem quite out of order, random, and illogical in some cases.

IBM originally provided for a handful of devices its developers believed we might use. These include:

- COM (serial) and LPT (parallel) I/O ports (which are probably the ones we are most often concerned with)
- Disk drives (A:, B:, C:, etc.)
- Keyboard and video output (combined as the CON: or system console)

This is a good list for the most part. Unfortunately, this list of common logical devices has not been expanded, except to add LPT2:, LPT3:, COM3:, COM4:, and the occasional special hardware and software interfaces that give us other unique COM and LPT devices.

In actual use with programs and DOS, these devices must be expressed with their numerical designation followed by a colon (LPT1:, for example, and COM2:), while generically, it is LPT and COM. Specifying only LPT or COM in DOS commands will result in an error message, and the desired command or operation will not occur. For the console and devices of which there is only one of that type, there is no number. You may see CON, but the computer must use CON:.

The logical assignment of parallel I/O (LPT) ports to specific hardware addresses is not as critical for most applications as is the assignment of serial I/O (COM) ports. Most software that uses the COM ports work directly with the hardware, bypassing the features built into the system BIOS (because doing so is much faster than using the BIOS features). Because most communications applications access the hardware directly, but make their own assumptions about logical names and physical addresses, the physical and logical device matching, in the order shown in Table 12.1, is expected and critical. Communications applications also require specific, matching IRQ assignments to function properly.

Consider Table 12.1, a listing of the most common physical and logical devices encountered in a PC system, to be a foundation set of rules for your system configuration.

TABLE 12.1

Logical versus specific physical translations for common PC devices

Logical Address	Physical Address	IRQ	Device Name
COM 1	3F8-3FFh	IRQ 4	1st Serial I/O Port
COM 2	2F8-2FFh	IRQ 3	2nd Serial I/O Port
COM 3	3E8-3EFh	IRQ 4	3rd Serial I/O Port
COM 4	2E8-2EFh	IRQ 3	4th Serial I/O Port
LPT 1	3BC-3BFh	IRQ 7	1st Parallel I/O Port (on monochrome systems)
LPT 1	378-37Fh	IRQ 7	1st Parallel I/O Port (on color systems)
LPT 2	378-37Fh	IRQ 5	2nd Parallel I/O Port if LPT1: is at 3BCh
LPT 2	278-27Fh	IRQ 5	The accepted LPT2 device on color systems
LPT 3	278-27Fh	IRQ 5	3rd Parallel I/O Port

(Note: *h* indicates a hexadecimal number.)

The issue of logical versus physical devices in a PC is not always an easy one to understand, much less explain. Yet this issue is one of the most significant rule-creating and binding aspects of a PC system, and the root of many conflicts. The easiest way to deal with this issue is to simply follow the original rules that IBM defined for all of the devices in your system. In fact, that is what is advocated throughout this book—knowing the configuration rules and complying with them.

Logical assignments occur during the Power-On Self-Test (POST) that runs when you boot up your system. The system BIOS performs a series of equipment checks, looking for specific devices at specific physical addresses in a specific order. As these devices are found, they are assigned sequential, logical port numbers. BIOS uses this information to refer to the I/O ports for any application that happens to rely on the system BIOS to provide access to these ports. Thus, when you are working directly with DOS or its applications, such as PRINT, and you send a file to be printed to LPT1:, DOS passes some control over the printing to the system BIOS, and the BIOS sends the file to the physical device associated with the "name" of LPT1:. The process works similarly in Windows 3.1-Me and changes dramatically with Windows NT, 2000, and XP, avoiding BIOS assignments altogether and replacing them with similar functions within the operating system.

Where problems originate is in the fact that POST bases its naming strictly on a first-come, first-served basis. Although the logical and physical addresses are designed to be matched as shown in the table, and those addresses are what your system and devices will be looking for during operation, the actual order in which these logical devices are assigned may differ.

The apparent confusion and variable assignments for LPT ports (as noted in Table 12.1) begins with IBM providing a parallel port at 3BCh using IRQ7 on monochrome display video adapters. Any parallel port added to a system had to be at either 378h or 278h. When IBM introduced color systems (CGA, EGA, and PGA), it did not provide a parallel port on the card. Any parallel port provided with or added to these systems was configured for address 378h. Quite possibly, this is because you could have both a monochrome display adapter and a color display adapter in the same system, working at the same time. Subsequently, for a color system with an add-in parallel port at 378h, a second port was provided for at 278h.

Always keep in mind that the numeric designation indicates a logical ordering of devices. A good way to remember this is that, in order to have a No. 2 or a second of something, you must have a No. 1 or a first of something. You simply cannot reserve, save, or leave gaps in the logical numbering of the devices, as some people have wanted to do.

Changing Your Configuration

We usually cannot, and probably would not want to alter the extremely low-level internal configurations of our PC system boards (direct memory access [DMA] channels, clock interrupts, etc.). However there are numerous devices we can, and often must, deal with the configuration of throughout the life of any PC system.

Among the frequently added, changed, or removed devices anticipated in the original IBM PC, and subsequently the PC/AT, we typically encounter configuration issues with:

- Serial I/O ports, including internal modems (COM)
- Parallel I/O ports (LPT)
- Video display adapters (MDA, CGA, EGA, PGA, VGA)
- Disk drive interfaces (AT, IDE, SCSI)
- Network interface cards

Developments after the first PC and AT systems provided us with a few new device types to find resources for:

- Pointing device interfaces—bus mouse and PS/2
- Small computer system interface (SCSI) host adapters
- Multimedia/sound cards, with and without CD-ROM interfaces
- Video capture boards
- 3-D video accelerators
- Custom document scanner interfaces
- Internal integrated services digital network (SDN) adapters
- Add-in or built-in infrared I/O ports

All of the devices in our systems require system resources. We can usually take for granted that each device consumes power, creates heat, and must be cooled by one or two meager fans. In addition, all devices in our PC system consume computer-specific resources other than power and space.

Of the devices we can have active simultaneously, not counting the internal system board resources, these are typically:

- Mouse (IRQ 12)
- COM1 (IRQ 4)

- COM2 (IRQ 3)
- LPT1, 2, and/or 3 (usually *not* using IRQ 5 or 7)
- Hard drives (IRQ 14, 15)
- Diskette drive (IRQ 6, DMA 2)
- Sound card (IRQ 5 and/or 7, and DMA 1, 3, or 5)
- CD-ROM (w/ disk drives, sound, or SCSI—IRQ 11, DMA 1, or 3)
- Network interface (likely IRQ 5, 7, or 10)

This list makes a fairly full and typical system nowadays, though I know folks who try to add scanner interfaces, infrared I/O ports, extra COM ports, etc., and simply fail to realize that something must be sacrificed to gain any satisfaction with any one or more of these.

The installation of any new device, or any changes to a device, must be done with the limited availability of these resources in mind, and a knowledge (through the inventory described in Chapter 1) of which resources are being used by other devices.

I/O Addresses

Every hardware device plugged into the I/O slot connectors inside our PCs requires a unique hardware address. During program execution, data and commands are written to or read from these locations.

IBM originally defined that specific devices occupy very specific addresses. Some of these devices are internal to the system board or specific to IBM products and uses. Among these, some addresses are reserved, or are to be avoided, because of other system- or IBM-specific uses, leaving approximately 25 possible addresses for all the possible devices, features, and options we may want to put into our PCs. This is a situation where some devices require 4, 8, or even 32 locations each.

The addresses that are defined, but not specifically reserved, are used for the common I/O devices that IBM planned for and anticipated in its original system developments. These are the devices we are most familiar with—COM ports, disk drives, and so on. In the progression from the original PC to the PC AT, a few new devices were added, or the primary address of a major functional device (the hard drive adapter, for example) was changed to accommodate the growth from 8-bit to 16-bit systems and more options.

Tables 12.2 and 12.3 list the specific I/O addressing for PC-, PC/XT-, and PC/AT-class systems. Many of the technical terms in the tables are beyond our need to define and understand in the context of configuration management, but we do need to know that something is assigned at a given address. This list is compiled from the dozens of I/O devices, specifications, and commonly available PC reference material.

TABLE 12.2
The Original IBM PC and PC/XT Device Addresses

I/O Address	System Use or Device
000-01Fh	DMA Controller—Channels 0–3
020h, 021h	Interrupt Controllers
040-043h	System Timers
060h	Keyboard, Aux.
070h, 071h	Real Time Clock/CMOS, NMI Mask
081-083h and 087h	DMA Page Register (0–3)
0F0-0FFh	Math Coprocessor
108-12Fh	Not Assigned; Reserved by/for IBM Use
130-13Fh	Not Assigned
140-14Fh	Not Assigned
150-1Efh	Not Assigned; Reserved by/for IBM Use
200-207h	Game Port
208-20Bh	Not Assigned
20C-20Dh	Reserved
20E-21Eh	Not Assigned
21Fh	Reserved
220-22xh	Not Assigned
230-23xh	Not Assigned
240-247h	Not Assigned
250-277h	Not Assigned
278-27Fh	LPT 2 or LPT 3—3rd Parallel I/O Port

(continued on next page)

TABLE 12.2

The Original IBM PC and PC/XT Device Addresses (continued)

I/O Address	System Use or Device
280-2Afh	Not Assigned
2B0-2DFh	Alternative EGA Port
2E1h	GPIB 0
2E2h, 2E3h	Data Acq 0
2E4-2E7h	Not Assigned
2E8-2Efh	COM 4—4th Serial I/O Port
2F8-2FFh	COM 2—2nd Serial I/O Port
300-31Fh	IBM Prototype Card
320-323h	Primary PC/XT Hard Disk Adapter
324-327h	Secondary PC/XT Hard Disk Adapter
328-32Fh	Not Assigned
330-33Fh	Not Assigned
340-34Fh	Not Assigned
350-35Fh	Not Assigned
360-363h	PC Network Card—Low I/O Port
364-367h	Reserved
368-36Ah	PC Network Card—High I/O Port
36C-36Fh	Reserved
370-377h	Secondary Diskette Drive Adapter
378-37Fh	LPT 2 or LPT 1—1st or 2nd Parallel I/O Port
380-389h	Not Assigned
380-38Ch	BISYNC_1 or SDLC_2
390-393h	Cluster Adapter
394-3A9h	Not Assigned
3A0-3ACh	BISYNC_2 or SDLC_1
3B0-3BFh	Monochrome Video Adapter
3BC-3BFh	1st Parallel I/O Port—Part of Monochrome Video Card

(continued on next page)

TABLE 12.2

The Original IBM PC and PC/XT Device Addresses (continued)

I/O Address	System Use or Device
3C0-3CFh	EGA Video
3D0-3DFh	CGA Video
3E0-3E7h	Not Assigned
3E8-3EFh	COM3—3rd Serial I/O Port
3F0-3F7h	Primary Diskette Drive Adapter
3F8-3FFh	COM 1—1st Serial I/O Port

TABLE 12.3

The Original IBM PC/AT Device Addresses

I/O Address	System Use or Device
000-01Fh	DMA Controller—Channels 0–3
020h, 021h	Interrupt Controllers
040-043h	System Timers
060h	Keyboard, Aux.
070h, 071h	Real Time Clock/CMOS, NMI Mask
081h, 082h, 083h, and 087h	DMA Page Register (0–3)
089h, 08Ah, 08Bh, and 08Fh	DMA Page Register (4–7)
0A0-0A1h	Interrupt Controller 2
0C0-0DEh	DMA Controller Chs. 4–7
0F0-0FFh	Math Coprocessor
108-12Fh	Not Assigned or Reserved
130-13Fh	Not Assigned
I/O Address	System Use or Device
140-14Fh	Not Assigned
150-1EFh	Not Assigned or Reserved
170-177h	Secondary PC/AT+ Hard Disk Adapter
1F0-1F7h	Primary PC/AT+ Hard Disk Adapter
200-207h	Game Port

(continued on next page)

TABLE 12.3

The Original IBM PC/AT Device Addresses (continued)

I/O Address	System Use or Device
208-20Bh	Not Assigned
20C-20Dh	Reserved
20E-21Eh	Not Assigned
21Fh	Reserved
220-2FFh	Not Assigned
230-23Fh	Not Assigned
240-247h	Not Assigned
250-277h	Not Assigned
278-27Fh	LPT 2 or LPT 3 —3rd Parallel I/O Port
280-2AFh	Not Assigned
2B0-2DFh	Alt. EGA
2E1h	GPIB 0
2E2h & 2E3h	Data Acq 0
2E4-2E7h	Not Assigned
2E8-2EFh	COM 4—4th Serial I/O Port
2F8-2FFh	COM 2—2nd Serial I/O Port
300-31Fh	IBM Prototype Card
320-323h	Not Assigned
324-327h	Not Assigned
328-32Fh	Not Assigned
330-33Fh	Not Assigned
340-34Fh	Not Assigned
350-35Fh	Not Assigned
360-363h	PC Network Card—Low I/O Port
364-367h	Reserved
368-36Ah	PC Network Card—High I/O Port
36C-36Fh	Reserved

(continued on next page)

TABLE 12.3
The Original IBM PC/AT Device Addresses (continued)

I/O Address	System Use or Device
370-377h	Secondary Diskette Drive Adapter
378-37Fh	LPT 2 or LPT 1—1st or 2nd Parallel I/O Port
380-389h	Not Assigned
380-38Ch	BISYNC_1 or SDLC_2
390-393h	Cluster Adapter
394-3A9h	Not Assigned
3A0-3ACh	BISYNC_2 or SDLC_1
3B0-3BFh	Monochrome Video Adapter
3BC-3BFh	1st Parallel I/O Port—Part of Monochrome Video Card
3C0-3CFh	EGA Video
3D0-3DFh	CGA Video
3E0-3E7h	Not Assigned
3E8-3EFh	COM3—3rd Serial I/O Port
3F0-3F7h	Primary Diskette Drive Adapter
3F8-3FFh	COM 1—1st Serial I/O Port

The addresses that were not planned for or assigned by IBM make up the only address locations that are available to be exploited by new devices. IBM did not and could not anticipate the existence of these devices before they existed. New devices not defined by IBM had to squeeze into the few address spaces left. The addresses shown in Table 12.4 are typical of non-IBM add-on devices.

TABLE 12.4
Common Aftermarket or Non-IBM Devices Listed by Addresses Used

I/O Address	System Use or Device
130-14F	SCSI Host Adapter
140-15F	SCSI Host Adapter (as may be found on a sound card)

(continued on next page)

TABLE 12.4 Common Aftermarket or Non-IBM Devices Listed by Addresses Used (continued)

I/O Address	System Use or Device
220-22E	SoundBlaster (SB), SoundBlaster Emulation
-or-	
220-23F	SCSI Host Adapter
-or-	
228, 289	AdLib Enable/Disable Decode (port is active if Sound Blaster emulation is available and active)
238, 239	AdLib Enable/Disable Decode (port is active if Sound Blaster emulation is available and active)
240-24E	Sound Blaster; Sound Cards Emulating Sound Blaster
280-283	Network Interface Card
-or-	
280-288	Aria Synthesizer
-or-	
280-2FF	NE1000/NE2000 Network Adapter
290-298	Aria Synthesizer
2A0-2A8	Aria Synthesizer
2B0-2B8	Aria Synthesizer
300-303	Network Interface Card
-or-	
300-31F	NE1000/NE2000 Network Adapter
320-321	MIDI Port
-or-	
320-33F	NE1000/NE2000 Network Adapter
330-331	MIDI Port
-or-	
330-34F	SCSI Host Adapter
340-35F	SCSI Host Adapter
-or-	
340-35F	NE1000/NE2000 Network Adapter
360-363	Network Interface Card (non-NE-type)
-or-	
360-37F	NE1000/NE2000 Network Adapter
388, 389	AdLib Sound Device (if no Sound Blaster emulation is active)

The addresses listed above may or may not be available on all particular I/O devices of the types listed. For example, not all SCSI host adapters give you the option of selecting either 130h, 140h, 220h,

230h, or 330h. Similarly, these adapters do not use all of these addresses, but may offer them as alternatives.

As you can see, there are at least six aftermarket device types (I/O devices) we will frequently encounter. To accommodate these, there are 14 address locations (possible addresses) available (14 is the number of unique addresses in the table, once repetition is accounted for and eliminated). Since all devices cannot be configured to work in just any or all of the 14 available addresses, there may still be overlap and conflicts, despite the fact that there are more addresses than there are device types. Industry acceptance has limited the addresses that certain devices may use to only a few addresses per device type—such as four predetermined COM port addresses, three predetermined LPT port addresses, etc. Thus, our configuration issues begin.

Interrupts

IRQ (interrupt request) lines are used by hardware devices to signal the central processing unit (CPU) that they need immediate attention and software handling from the CPU. Not all of the devices in your system require an IRQ line, which is good news, because we have only 16 of them in an AT or higher class system. Of those 16, three are dedicated to internal system board functions (the system timer, the keyboard, and a memory parity error signal). The use of the other signals depends on the devices installed in your system and how they should be or are configured.

For industry standard architecture (ISA) or non-extended industry standard architecture (EISA) and non-Micro Channel systems, it is the general rule that IRQ lines cannot be shared by multiple devices, though with some care and well-written software, they can be shared. But since there is no easy way to know which devices and software can share IRQ lines with other devices, this is something we will avoid doing. Table 12.5 shows the predefined interrupts that the PC needs.

TABLE 12.5
IRQ Assignments

IRQ	PC, PC/XT	AT, 386, 486, Pentium
0	System Timer	System Timer
1	Keyboard Controller	Keyboard Controller
2	Not Assigned	Tied to IRQs 8-15
3	COM2: 2F8h-2FFh	COM2: 2F8h-2FFh
4	COM1: 3F8h-3FFh	COM1: 3F8h-3FFh
5	XT HD Controller	LPT2: 378h or 278h
6	Diskette Controller	Diskette Controller
7	LPT1: 3BCh or 378h	LPT1: 3BCh or 378h
8	Not Available on PC or XT	Real Time Clock
9	Not Available on PC or XT	Cascades to and Substitutes for IRQ 2
10	Not Available on PC or XT	Not Assigned
11	Not Available on PC or XT	Not Assigned
12	Not Available on PC or XT	PS/2 Mouse Port
13	Not Available on PC or XT	NPU (numerical processing unit)
14	Not Available on PC or XT	Hard Disk
15	Not Available on PC or XT	2nd Hard Disk Adapter (later systems)

Add-in devices usually provide a number of options for IRQ assignments to avoid conflicting with other devices when installing and configuring them. Some typical IRQ assignment options for add-in devices are shown in Table 12.6.

TABLE 12.6
Add-In Device IRQ Options

Add-In Device Type	IRQ Choices
SCSI Host Adapter	10, 11, 14, or 15
Sound Cards	5, 7, 10, or 11
Network Card	2, 3, 4, 5, 7, 9, 10, 11, or 12

Plug-and-Play

Having come to this point, you may be ready to jump in and say first that "plug-and-play is supposed to handle all of this and make it better...", and then "but, but, but...plug-and-play doesn't set my devices the way you indicate things should be set...." BINGO!

This observation is certainly true, and equally or more troublesome if you are trying to maintain a stable configuration. You must endeavor to get all of your legacy devices into a proper IBM-standard configuration, including all of the subsequent aftermarket items such as SCSI, sound, and network cards, before considering plug-and-play issues. This approach will reduce the variables you will have to deal with and make the entire configuration process easier.

Plug-and-play is a system BIOS-based automatic configuration process. Plug-and-play is an additional set of firmware or BIOS code, working as part of the system POST, to identify and capture device configuration information. Your plug-and-play BIOS does not configure the devices in your system. Instead, it lets devices configure themselves based on remaining and commonly used I/O resources.

Plug-and-play senses the hardware in the system and stores information about it. If there are changes to the system, it starts a reconfiguration process, so a new or changed device can have a chance to get needed system resource assignments. If nothing has changed, the system goes on about its business. If something has changed, it makes that information available to the operating system. You will see the effects of any changes in most current operating systems—certainly Windows, the Apple Macintosh, and later versions of Linux.

Plug-and-play is not just about assigning addresses and IRQs to devices. Sometimes it simply supports a connection method, such as USB or IEEE-1394. Most, if not all, peripheral component interconnect (PCI) and advanced graphics port (AGP)-based and PCMCIA/PC card devices are plug-and-play devices. Plug-and-play must be supported within all USB and IEEE-1394 devices because they are not assigned system resources—just their connection points at the computer are.

Since PC card, USB, and 1394 devices are connected externally and may be added or removed at any time, without plug-and-play support, these devices would have no way to identify themselves to the software that uses them. That software depends on PC card,

USB, and 1394 drivers to tell it what devices are out there. Plug-and-play tells the operating system that PC card, USB, or 1394 capabilities exist.

Plug-and-play can get itself into trouble if the system BIOS or the device is not completely and properly compatible with each other. If the BIOS or the device does not issue or does not respond properly to the "new device needs configuration" processes, a device may either be ignored or may disable another device. There is no manual recourse or method to correct this or force a specific configuration or correct one—other than to ensure you have upgraded BIOS in your PC and firmware in your devices.

This also relates to a common myth about being able to change a device configuration within Windows. Remember that plug-and-play senses, stores, and reports device configuration information. That is key to understanding the processes. Plug-and-play does not provide the ability to change the addresses and IRQs of hardware devices. In most cases, Windows does not have this ability either. Windows does, however, have the ability to reassign the logical names and order of some devices within itself.

The drivers for specific devices may provide an interface *through* Windows to change a particular device, but this is *not* a function or desire of the operating system. It is a feature of a very limited set of hardware and associated drivers. Windows NT, 2000, and XP typically do not support this reconfiguration through driver features, by the rules for these operating systems, as providing such features would render their hardware security useless.

Another issue, not related as much to plug-and-play but to certain devices, is the seemingly random reassignment of logical device assignments, as is common with a variety of USB-to-serial and USB-to-parallel, and perhaps even some USB-Ethernet adapters. For example, a USB-to-serial port adapter is a common accessory for those needing to hook up a personal digital assistant (PDA) or connect to some other serially interfaced device on a PC or laptop that does not have a COM port. The adapter's driver software creates a virtual serial port and program interface so that the PDA or terminal software can communicate to the outside device through USB.

This driver software is typically smart enough to avoid assigning a logical COM port name used by another COM device such as a modem or a single physical COM port. Instead, it will make up or make available a selection of virtual COM port names—COM5 on up to COM15 or so. Fair enough—make up a nonconflicting port assignment and make its use available to whatever software needs to talk to the device—*except* that this assignment can, and quite often does, change from day to day, reboot to reboot. At any given time between reboots or on any given day, your USB-serial port may change assignment from COM5 to COM13 to COM10, etc., making you have to first determine which port name is given at the moment, and then reconfigure the PDA or terminal program to match this "moving target" port assignment. This is something you have to be aware of, if you need to use a USB-to-serial port adapter to connect to an access point or router as you set up your network devices.

There is some good news. Rarely if ever do these problems exist with PC card devices such as wired network interface cards or wireless adapter cards. Typically you will encounter driver-related problems more than hardware problems with these devices.

Summary

This chapter may still seem out of place amid the context of wireless networking, but it addresses an important issue when adding to or changing existing systems to accommodate and maintain your systems.

Unfortunately, the topics of operating systems and device drivers are not as clear-cut and rule-based when it comes to figuring out which driver or piece of software is conflicting with another and crashing the system. It is in most cases easier, however, to change drivers and software than to reconfigure or replace hardware. For these issues, you should be vigilant in contacting the vendors of your system boards, laptops, and wireless devices to obtain and apply patches and updated firmware.

CHAPTER 13

Creating a SOHO Wireless Network

Up to this point, we have discussed the elements of and criteria for wireless network components and their implementation. In this chapter, we will build a typical wireless network to service a small office or home office (SOHO) using off-the-shelf products and personal computers (PCs). Our goal is to provide 100 percent coverage in and around 2,000 to 2,500 square feet of home or office space, using a single access point, with connectivity to the Internet through a digital subscriber line (DSL).

We will describe how to enhance this setup with a small local file server and a personal Web site to show the steps necessary to keep a standard residential dial-up point-to-point protocol over Ethernet (PPPoE) DSL connection "always on" and how to arrange domain name system (DNS) services for the likelihood of a dynamically assigned Internet protocol (IP) address so that the Web site is always accessible.

The components and software we will use during installation and operation of the network include:

- Speedstream DSL router provided by telco/Internet service provider (ISP)
- LinkSys BEFSR41 firewall/router/hub
- LinkSys WAP11 wireless access point
- Laptops and various desktop PCs
- Windows XP client OS
- Dedicated file and web server with Windows 2000 server
- Tardis or similar timekeeping software
- ZoneEdit DNS services
- GLSoft ZEDu ZoneEdit Dynamic Update software to maintain DNS updates
- Norton AntiVirus
- ZoneAlarm security software

Figure 13.1 shows a SOHO wireless network without servers—a configuration for which 802.11 wireless networking was intended. Figure 13.2 shows the same network with a server added to provide web and e-mail services to the local users and the Internet. The components for these configurations are readily available off-the-shelf from most computer stores and on-line sites. They are very easy to install and manage on their own, and fit together to create a modular, easily maintained, almost hands-off network configuration.

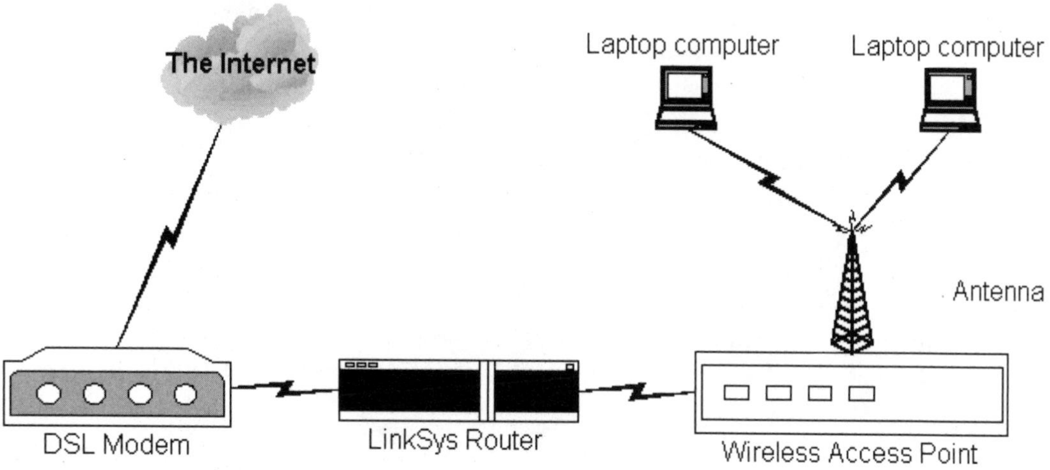

Figure 13.1 The diagram of our desired wireless LAN and Internet connection configuration.

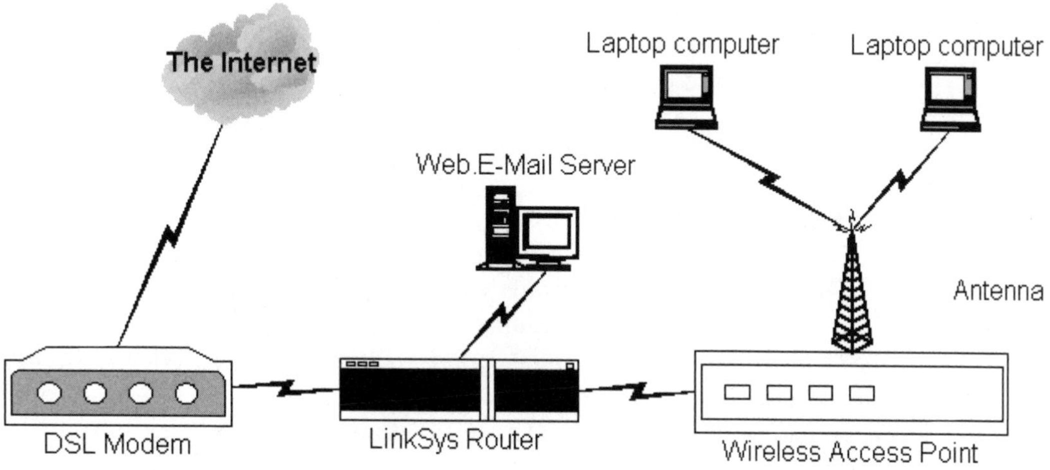

Figure 13.2 The diagram of our desired wireless LAN and Internet connection configuration with the addition of a local file and web server.

The steps we will take to set up and verify the components, individually and together, are:

- Install DSL modem and its connection management software to establish service.

- Install and configure LinkSys router to take over DSL PPPoE connection management.
- Configure LinkSys router to provide domain host configuration protocol (DHCP) services to internal PCs.
- Test PC connections via wired network.
- Configure LinkSys access point for wireless client connections.
- Configure laptops for wireless access to network and Internet.
- Configure LinkSys router to pass through Web server to Internet.
- Configure time services to maintain connectivity.
- Configure DNS update software to maintain DNS records.
- Configure ZoneAlarm to reduce exposure and intrusions to the network.
- Enjoy the system!

I *highly* recommend that you at least glance through this entire chapter *before* you begin any of the installations and setups—especially if your only network adapter connectivity is wireless. There are many interdependent and orderly steps to take, and knowing what they are before you begin will help you understand the choices you have and decisions to make.

DSL Installation

DSL service installations vary by ISP and type of service. Self-installation kits for residential services are quite common and very easy to deal with, almost easier than setting up dial-up accounts. Critical to these installations are making notes of the log-in information—the user log-in specifics and password—as well as DNS and e-mail server information. You need to make a separate note of and retain this information so that you can apply it to the subsequent router configuration.

Allow an hour or two to become familiar with the steps, perform them, and verify that everything is working correctly before making any changes or additions to the basic installation. If you have not done this before and get ahead of yourself and begin adding the router or other elements to the process, you will create mistakes and have to backtrack. If you have done this before or are doing the installation for someone else, let them observe and show them the critical steps and how you do things, so that they can avoid or solve problems for themselves in the future.

Note: *To family, friends, consultants, and experts helping others—this is not rocket-science and most of us are not rocket scientists. Never leave your clients in the dark or so dependent on you that they cannot function on their own with the basic steps. Being an elitist or arrogant is not a service to your client. Yes, they may be less technically sophisticated than you, and may become overwhelmed and not remember everything, but they can and will recall the steps you took and showed them, regardless of whether you have to provide help over the phone or e-mail or they have to figure things out on their own. Besides, the ISPs designed the process for dummies, and most people can do this themselves by following the instructions. Your help will speed the process and make you look good!*

Out here in SBC's PacBell service area, when you order residential DSL with self-installation you are sent a package containing your DSL modem, all of the jacks and cables necessary to interconnect the modem to the phone line and filter the DSL and voice services from each other, probably a new network interface card for your PC—assuming you do not already have one, software installation CD-ROMs, and very clear instructions on paper, as well as within the software installation process. Similar installation kits come with other service providers.

When you receive your installation kit, review the printed materials carefully, take inventory of the contents of the package, then select the installation software CD for your operating system—PC/Windows or Apple Macintosh.

The DSL phone line installation process may be documented on paper or be limited to the CD-ROM setup software. This can be a little complicated if you do not have a network adapter already installed in your PC and have to run the install software first to get instructions, then shut down the PC to install a network adapter, and then restart the software installation process to continue.

The best recommendation is to get the network adapter installed first, however you need to proceed to do that, so that you do not have wireless or other technical complications to deal with. If a server is involved in the process, chances are it will be hard-wired to some portion of your network through a hub or connected to the router.

Use this system as the workstation to install the DSL connection software on, thus avoiding wireless adapter, access point, and router configuration complications. Follow the appropriate instructions to install the network adapter in your PC if necessary, or locate the network connection on the laptop PCs, and then begin the software installation.

If your PC or laptop's network configuration is completely dependent on wireless, follow the instructions to get the wireless adapter installed, and begin to configure your access point so that you at least have a signal available to test this phase. You may have to modify this process and jump into the wireless access point and router configuration sections sooner than later.

Follow the installation instructions to connect the DSL modem to your phone line and connect the filters in-line at all of your telephones. My house has convenient central wiring and duplicate wiring for separate multiple phone lines throughout the house, so I was able to install a single filter/splitter at the main junction box, connect common line to the regular telephones to the filtered side of the splitter, and connect a single straight line to the desired location for placement of the DSL modem—near the rest of my network equipment.

Test your telephones and connect a telephone to the wire intended to go to the DSL modem to ensure you can get a dial-tone and dial-out, to be sure you have everything hooked up correctly. Having confirmed that all is well, you are ready to proceed with the DSL modem installation. If you cannot obtain a dial-tone and make a call on the phone line connection for the DSL modem, stop and retrace your steps and double-check your connections. More than once a poorly inserted RJ-11 plug has foiled even the simplest installations for the most competent technical people (yours truly!).

Connect the DSL modem to its power source, the phone line, and its network cable directly to a PC, laptop, or the server network adapter. You will observe a series of indicator lights on the modem, showing the progress and status of its connection to the phone line and the PC (or Mac). The instructions should tell you how to proceed with respect to what you observe.

Continue to follow the installation instructions, which will take you through the connection software installation, registering your service and establishing your log-in identification and password and probably your e-mail address for this ISP. Make note of every option, log-in information, password, and server information provided as the process goes along. You will need this to configure your router and other devices later on.

Note: *You do not need to use this e-mail address as your primary one, but you will want to check that e-mail account periodically for news and information from the ISP, if it does not have another primary e-mail address on record for you already.*

When the software installation is completed, you will probably end up with a new web browser, or at least your existing browser will be configured with settings specific to this ISP, as well as a Web-based or local e-mail access configuration. Do not panic. You should still be able to use your old/existing applications with your new connection, even AOL and Compuserve. Check your other Internet applications to make sure they still work OK, and resolve any problems before you go any further.

Once you install the router and configure it to handle the "dial-up" processes for the DSL line, you may not need all of this new software. You should leave this software in place and available after you complete the rest of your installations and configuration. It will be handy and perhaps necessary to go back to a known working configuration if you need support for your DSL service later.

Note: *If you have business DSL services, the ISP will likely dispatch an installation technician to wire the dedicated DSL jack, install the DSL modem or a combination DSL modem/router, provide you with the IP addresses assigned to your modem and for your network devices, and configure basic services into a router, if provided.*

At this point, you should have planned any web or e-mail server installations and will need to determine which IP address will be used for which service and host system. If the e-mail services and web host will be running on the same system, you may use the same IP address for both of these services and the server system. The router may or may not have to be instructed as to which services (e-mail, Web, FTP, etc.,) will be passed on to which IP address.

My preference is to assign discrete IP addresses to each type of service so that they can be separated to different server systems as necessary. These multiple IP addresses can be assigned in Windows 2000's Network

Properties on the same server system and network interface initially, to be changed later on if needed. If your business DSL service includes a built-in router, you can skip the next section.

Having verified a working software installation and Internet connection for at least one of your workstations (PC or laptop), you are ready to transition from single-workstation to multi-workstation capabilities with your router.

Router Installation

The LinkSys router is configured entirely through a Web page interface to the internal workings of the unit. The instructions provided are very clear on the initial setup and for some specific details. Beyond the basics, you are on your own—or at least at the mercy of the help provided in this section. We have several things to set up in the router:

- Connect one of your workstation's—PC, laptop, or server—network connection to port 1, 2, 3, or 4 on the router.
- Log onto the router and establish a new security password.
- Configure the internal network IP addresses and DHCP configuration you want to use.
- Select DNS addresses.
- Select PPPoE dialing and log-on parameters to be used with the DSL modem.
- Later, select IP addressing and port configuration for internal Web or e-mail services.

The first thing we will do is connect the router—install the power supply and connect the power cable; move the network cable for your PC, laptop, or server from the back of the DSL modem to one of the workstation ports on the back of the router; then add a new cable from the wireless area network (WAN) port at the back of the router to the network connection on the DSL modem. The router just got inserted in the path between workstations and the DSL modem, and it will perform a variety of useful functions for you—including controlling the modem's dial-up connection to the ISP.

To begin, open a web browser and type in 192.168.1.1 in the browser's address space, then press Enter to log onto the router. You will first see a log-on dialog—Figure 13.3. Leave the User name field blank, then type in the default password of admin, and click the OK button or press Enter.

Figure 13.3 The LinkSys router log in dialog box.

If you successfully log in, you will see the first page of the router's configuration program—Figure 13.4. On this first page, the Setup page, you will see basic information about the router's present configuration. Do not change anything on this screen yet.

Click the Password tab at the top of the LinkSys page to bring up the Password page—Figure 13.5. Determine and type in a new Router Password. It must be entered exactly the same on both lines for validation. Then ensure Disable is selected for UPnP Services, and No for Restore Factory Defaults. Click Apply. The new password will be stored, you will be notified of a five-second pause, and then the log in dialog will appear again. Skip the User name field and type in the new password to make sure you can get into the router configuration.

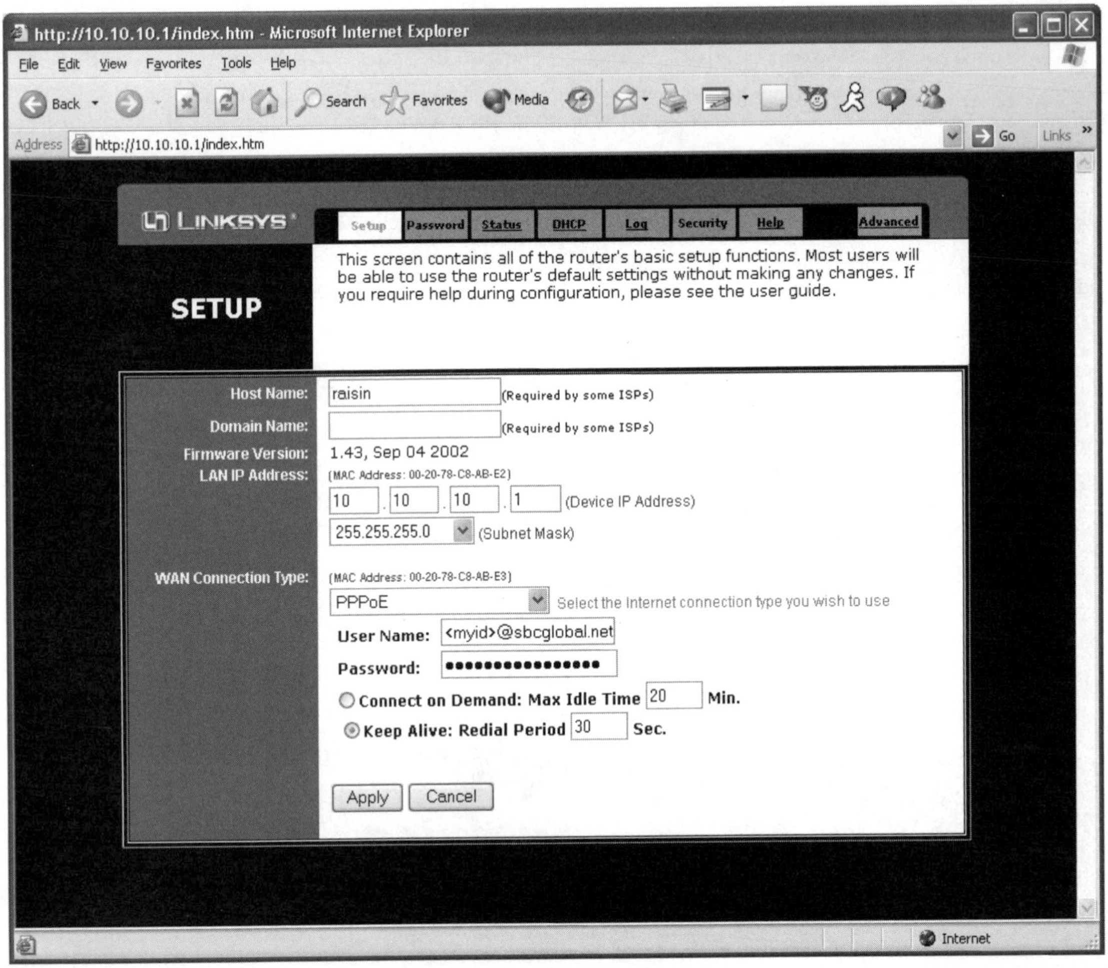

Figure 13.4 The LinkSys router main configuration page.

Note: *If you goof up on the password, you will have to follow the documented steps to reset the router to the default configuration, losing any changes made so far.*

Return to the Setup screen (Figure 13.4) and begin the reconfiguration of the router. I find the default values for the router to be predictable and useful, but with predictability comes vulnerability. In most cases, your internal network client systems will automatically

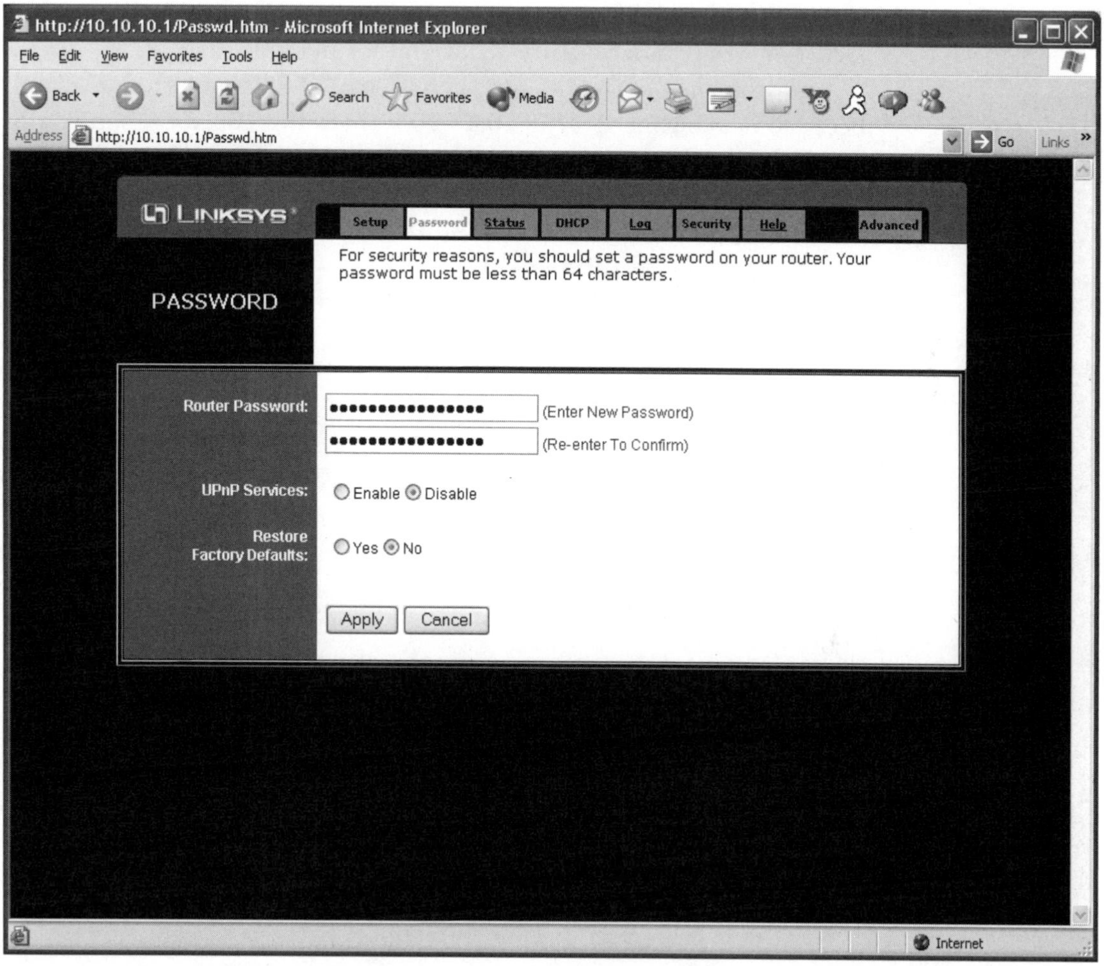

Figure 13.5 The LinkSys router password security configuration page.

be given private, nonroutable IP addresses—from either the 10.x.x.x, 169.254.x.x, or 192.168.x.x address ranges. The LinkSys by default comes configured to use the 192.168.1.x address range, giving us a place to start. Using default settings is OK in a private/home network, but at work, with several other users tinkering about, you probably want to select a different address range and change the default password for the router to reduce the chances of tampering.

The Host Name and Domain Name options are optional and I have never found them, as suggested, to be required by some ISPs, unless you have fixed IP addressing and they are changing their DNS servers to suit your installation (not likely).

I address my network into what I call the 10-net range, if only because it is easier to type 10.10.10.x than 192.168.x.x when configuring fixed addresses into workstations. Thus, 10.10.10.1 becomes the router's new IP address. This IP address is then used as the gateway address on client workstations that do not use DHCP automatic client configuration values.

The subnet mask numbers tell the router if connections between specific hosts' addresses need to go through the router to the WAN port (DSL line), or remain on the LAN side. Since we do not have a big network (over 255 clients), we can use a Class C (or smaller) mask value. If we had multiple 10.10.10.x subnets, we could narrow the last octet of the mask down to typically .224, .192, .128, or other values defining how many host addresses live within each subnet of our address range. The 255.255.255.0 Class C value is the easiest. If we had a situation to support more subnets, we could as easily make them use 10.10.11.x, 10.10.12.x, etc., network ranges.

Next, we have to configure how the router will work with the DSL service—see Figure 13.6—for the WAN connection type values. If you have business DSL service with fixed IP addresses and your DSL equipment does not include a router, you would make the selection of Static IP, and then assign one of your fixed IP addresses to the WAN side of this router. For residential dial-up or PPPoE DSL services, select PPPoE and then enter the log-on name and password you used for the workstation DSL software configuration above.

The next two values determine how your DSL connection is maintained. The Connect on Demand value defines how long the connection will remain active before it is dropped at your end for inactivity and has to be redialed, (because you were not surfing the web or collecting or sending e-mail, etc.), which leads to the perception of slow service. The default value of 20 minutes is fine. This selection is fine for the occasional user and someone who is *not* running a mail, Web, FTP, or game server on his DSL line.

The alternative Keep Alive: Redial Period value sets the router to never allow the modem to disconnect from the ISP side of the connection. The default value of every 30 seconds works OK, defining how often the connection is pulsed or redialed to ensure that it stays alive to prevent disconnection from the ISP. This selection is preferred if

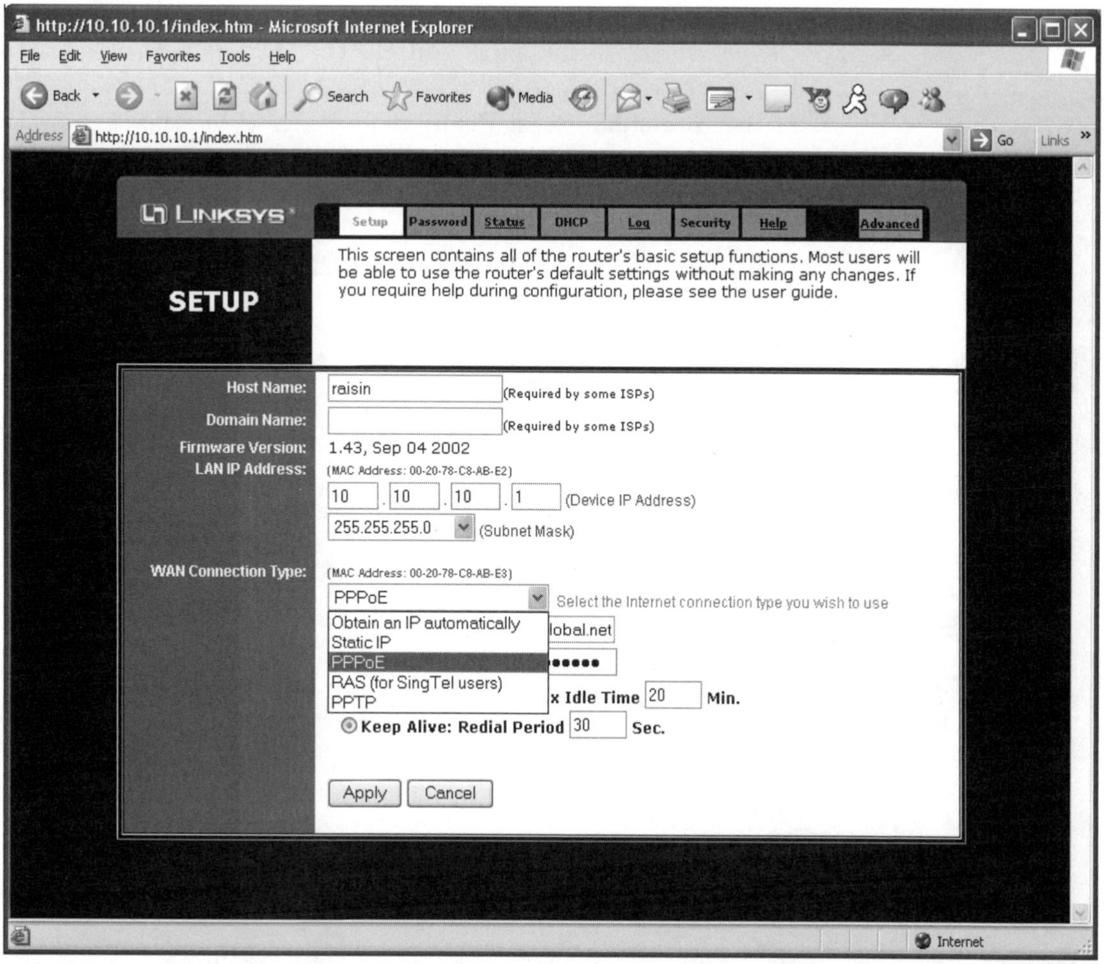

Figure 13.6 *PPPoE selection to use the router to dial-up and log-on to establish your DSL connection.*

you have a server running that needs to be accessible from the Internet, and thus needs to maintain an IP address at a DNS server.

Keeping the connection alive can and will also be assisted by a couple of applications you can run on an always-on workstation or your web/mail/FTP server—the automatic DNS update utility program and the time correction service.

Click the Apply button to save these values in the router. At this point, your browser still thinks the IP address of the router is the original 192.168.1.1 address, but the router will be using the new address you just set it for, and your workstation is using some randomly or previously assigned IP address that has nothing to do with your new router configuration.

After the router has reset itself, you will need to type its new IP address into your web browser to access it, log into the router, and access the remaining configuration items. Select the DHCP tab at the top of the page to get the screen shown in Figure 13.7. This screen is where we define the values for DHCP, allowing client PCs

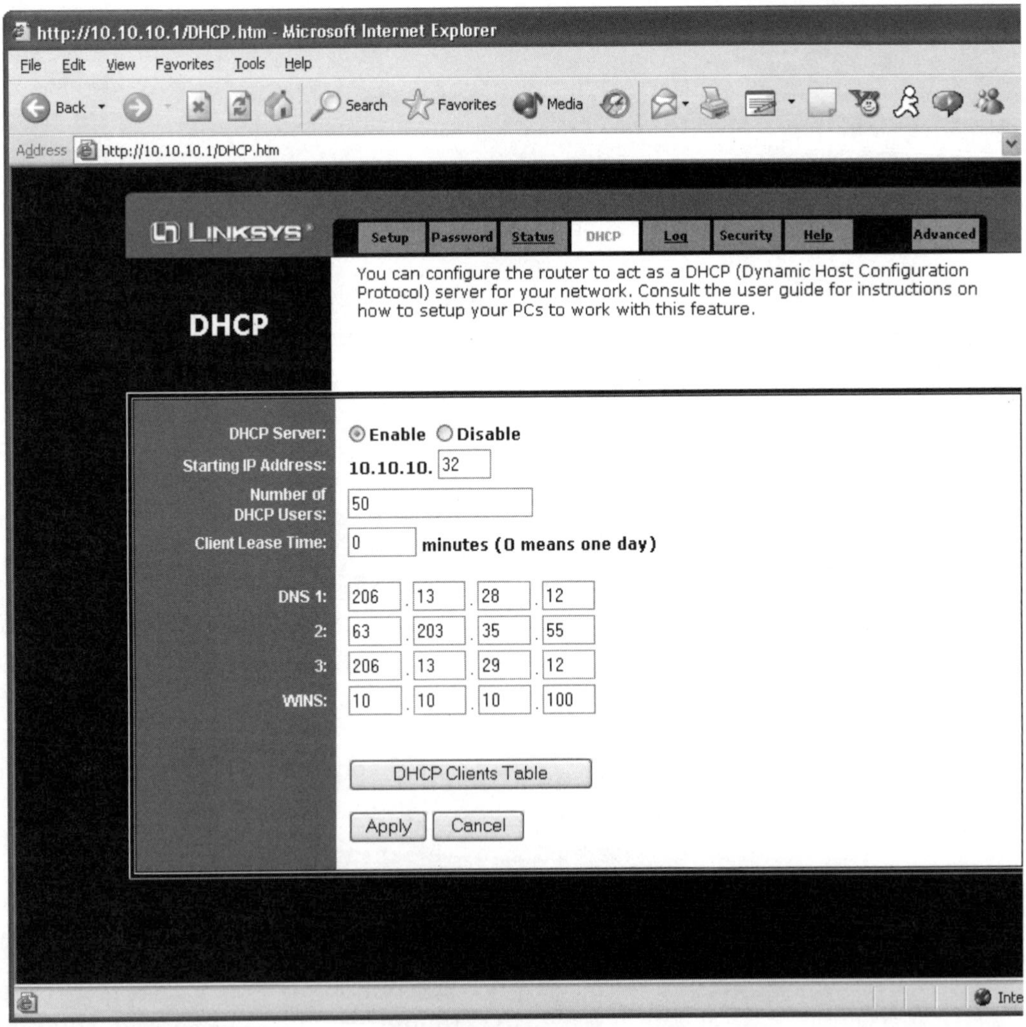

Figure 13.7 The DHCP configuration page of the Linksys router.

and Macs to obtain IP addressing, routing, and DNS information automatically so that you do not have to configure each and every workstation. (Using DHCP is the default value for most PC and Mac network settings.) First, select the Enable button following the DHCP Server label.

The first portion of the address range your workstations will use is determined by the IP address you set for the router in the first page. The range used for the last octet of the IP address is up to you. Determine which address you want the automatic configuration process to assign to the first workstation that requests DHCP configuration. Subsequent workstation requests will get subsequent sequential addresses. Since some devices you put on your network will need to have fixed, preset IP addresses, do not start at 1. A starting address of 16 or 32 seems reasonable under most conditions, allowing plenty of addresses for servers, network printers, etc. How many clients you need to support with DHCP is set next.

Most of us do not have more than a few PCs, some may have a small handful, others may have dozens. The Client Lease Time sets how long a DHCP-assigned IP address stays assigned to a specific system before the address is expired and a new one must be issued. The value of 0 (zero) for an entire day seems adequate in most cases.

Put in the IP addresses for DNS servers given to you by your ISP—these are then dispensed to workstations in response to their DHCP requests. Typically you are given only two addresses, which is adequate; a third is optional. If you are running an internal Windows server and will be using its network naming services, you can also include that server's address for distribution via DHCP. You may now click Apply to make the new settings take effect.

If you want to verify your new DHCP settings using your workstation—to see if it gets a fresh IP address and the various settings from the router—log off your workstation and restart it. Provided the workstation's networking parameters are set to get new IP information automatically (using DHCP), it will get this information from the router, which you can verify easily. For Windows 95, 98, 98SE, and Me users, go to Start, select Run, type-in "winipcfg," then click OK to bring up a dialog box showing your current IP address information. For Windows NT, 2000, and XP users, go to Start, Run, type in "cmd," then click OK to open a Command Prompt box. At the command prompt, type in "ipconfig," then press Enter. In either case, if the address information comes up in the 169.254.x.x range (and that's not the address range you put into the router), then the workstation did not get a new assignment via DHCP from the router. If you get a fresh 10.10.10.x subnet address, it would appear that DHCP works fine.

If you will be running an Internet-accessible mail, web, or FTP server, or using special application services such as pcAnywhere, web-cam services, etc., you will have to select the Advanced tab at the upper right, then the Forwarding tab at the top of the page to reveal Port Range Forwarding values—see Figure 13.8—to define which ports need to pass through to which specific hosts, according to their fixed IP addresses.

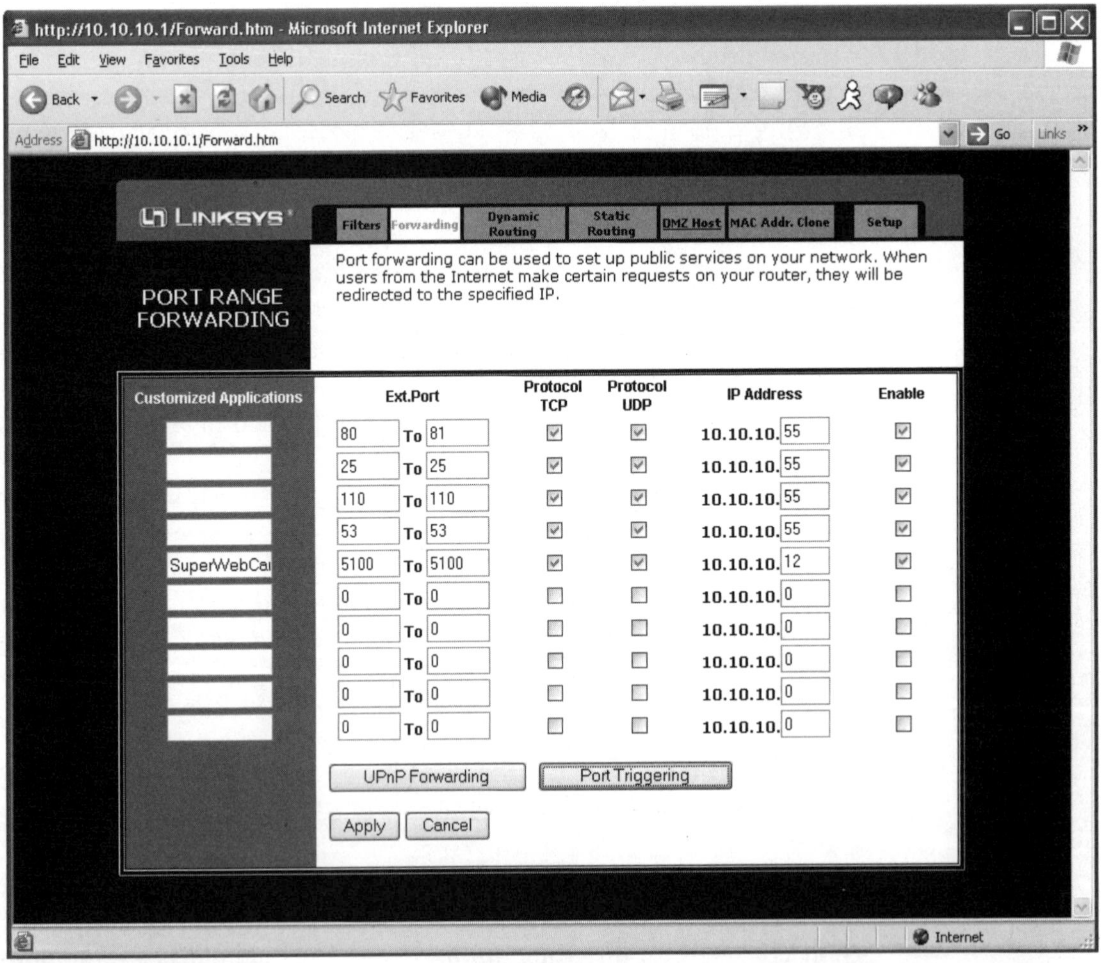

Figure 13.8 Setting up the router to pass web and e-mail services to an internal server.

On this page, you enter the specific transmission control protocol (TCP) and/or user datagram protocol (UDP) port numbers for the services that will pass through, and the specific IP address for the PC, Mac, or server host device to which you want those services to be directed. In this case, we have Web, mail, and DNS services running on a single PC with the internal IP address of 10.10.10.55. Any request for either of these Internet services that comes into the IP address assigned by our ISP will be directed to this server. As mentioned previously, these services could be running on separate PCs, or on the same PC. But that PC could be given multiple IP addresses—one for each service type, for possible separation later. We also allow Port 5100, for a special web camera, to pass through to a PC with the IP address of 10.10.10.12.

Click the Apply button for any changes to take effect, and you should be ready to test your DSL connectivity through the router. To test your new configuration beyond connecting to the router, at your workstation, the one you are using to configure the router, type in the web address for any external Web site you would like—www.yahoo.com or similar. This should cause the router to sense that it needs to find this host somewhere external to your internal network (not a host on your new 10.10.10.x network), out on the Internet, and cause the PPPoE dial-up process to start, activate the DSL or equivalent status light on your DSL, then give you access to the desired web page.

If this process succeeds, you are quite ready to begin adding other fixed/wired workstations and devices as necessary and verify that they work at accessing the Internet, that network printers can be used, servers and file shares can be accessed, etc. Then begin adding your wireless access point and wireless clients to your newly configured network.

Access Point Installation

The LinkSys WAP11 comes in two models—the earliest provides a universal serial bus (USB) port for configuration purposes; the later models have only an Ethernet port that uses simple network management protocol (SNMP) software for configuration. I recommend finding an earlier model unit with the USB port, because it is easier to gain access to configure the unit if you were to lose control of it via SNMP over the Ethernet connection.

Connect the power source for the access point and run a straight-through Ethernet cable from the access point LAN connection to an available port on your router.

To control the WAP11, you must install the configuration utility software that comes on the CD-ROM with the product or is available by download from its Web site—www.linksys.com. Once installed, the software tells you that you must reboot your PC before using the configuration utility software—which is not the case for the SNMP version. Simply cancel the message that pops up and double-click the WAP11 SNMP Configuration Utility icon that appears on the Windows desktop.

The first screen that will appear is the log-on screen for the access point, including the default IP address the unit is programmed for and a password entry area. The default password is "admin." Type it in, then click OK to begin the connection to the access point. If successful, you will see the first screen of the program, as shown in Figure 13.9. This screen will tell you the version number of the access point firmware, the media access control (MAC) or hardware address of its Ethernet port, the mode it is operating in (typically Access Point), the extended service set identifier (ESSID), the current operating channel, and whether or not wired equivalent privacy (WEP) encryption is enabled (it is not by default).

To set up the WAP11 properly to add it to our existing wired network configuration, we need to:

- Set the access point service set identifier (ESSID).
- Predetermine and set a channel to use (optional).
- Set a fixed IP address for the access point to use (optional, but preferred).
- Set the WEP encryption level and encryption key (highly desirable).

These steps take about five minutes to accomplish and then we can move on to installing the wireless clients. First, click the Basic Setting tab to reveal the ESSID and access point name settings—Figure 13.10. Change the ESSID to something familiar to you, but perhaps not identifying your business, family, or location. This name will allow you to (as uniquely as possible) identify your access point from others nearby. Once you remember your ESSID, which you

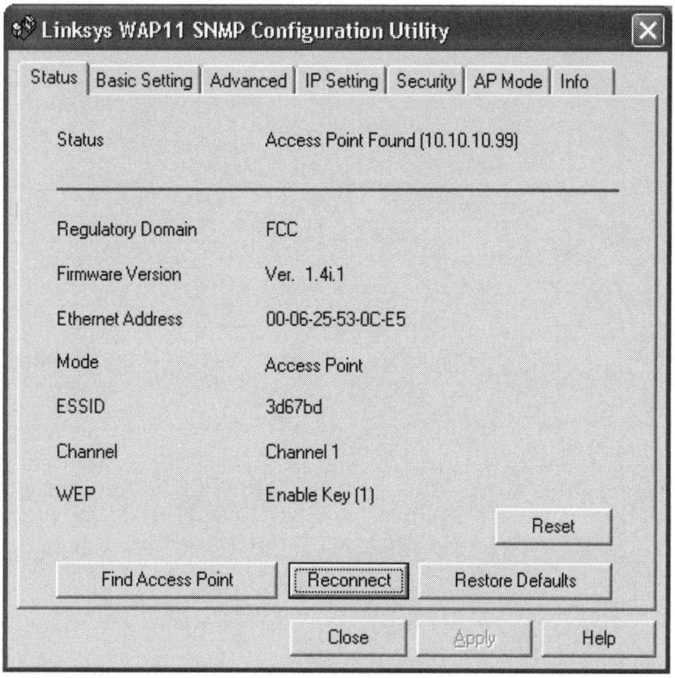

Figure 13.9 The main status page for the Linksys WAP11 wireless access point.

must do or make note of to configure your clients, you can disable broadcasting it in the Advanced setting screen to make it harder (but not impossible) for people to find your wireless network. In my location, I typically choose one of three nonoverlapping channels, 1, 6, or 11. If one or all of those channels turn out to be busy and potentially slow your network because of collisions with others, you may have to choose a channel from other wireless LANs that has less signal strength than the others, and hope you can override their signals close to you with yours. The Access Point Name value is not that critical, but I usually make it the same as the ESSID. I typically click the Apply button after making changes to any one screen to preserve the work I have done so far. After you click Apply, wait for the access point and display to refresh back to the first screen.

The next set of settings you need to change is on the IP Setting screen—Figure 13.11. This is where we will apply a static IP address to the wireless access point—an address outside the DHCP range we set in the router—avoiding 10.10.10.32 to 10.10.10.82. 10.10.10.99 will work, or pick an address lower than 32 if you like to group your network equipment together by address. The IP Mask value should

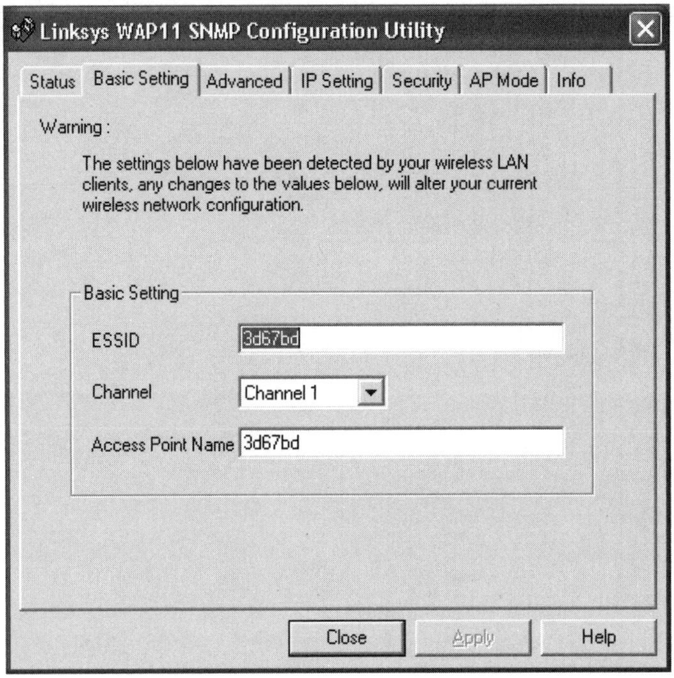

Figure 13.10 The WAP11 Basic Setting dialog with entries and selections for SSID, channel, and access point name values.

reflect that of the local network Class C range we set up earlier in the router—255.255.255.0. You could let the access point obtain an IP address automatically, from the DHCP server in the router, but it is customary to use fixed addresses for all network equipment, to make troubleshooting easier. Click the Apply button and wait for the access point and display to refresh back to the first screen.

Moving along to the Security tab—shown in Figure 13.12—we will set up the encryption level and key value to be used by our clients to connect through this access point. You have the option of using no encryption at all, but why make it easy for your neighbors to tap into your local network and use your services? Select the encryption level—either 40/64-bit or 104/128-bit—you would like to have protecting your network. Be sure that the level you choose is supported by the wireless card you will be using at your client PCs, as many do not support 128-bit WEP keys.

Depending on the encryption level selected, pick a 5 or 13 character word or phrase you would like to use and type it into the Passphrase box; then click the Done button. Clicking Done causes the hexadecimal value of your word/phrase to appear for each key

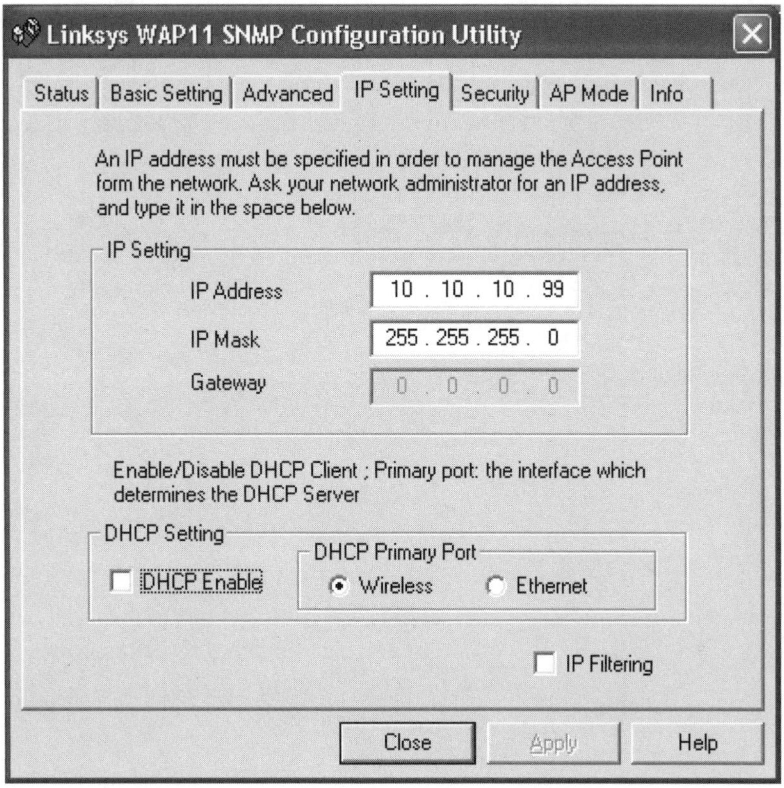

Figure 13.11 The WAP11 IP Setting dialog for specifying the access point's IP address, subnet mask, and if you wish, the access point to use DHCP configuration.

value. Write these values down—the text version and the hex values, or at least the values for Key 1—as you will need to know the hexadecimal values to enter them as the key values for your clients.

Note: *Trying to use text word/phrase instead of the hexadecimal value is the most common cause of failing to connect to a wireless access point—and you do not know this because the client software does not provide an error message telling you the key value is wrong. The lack of error message is partially because you could get the error any time you pass by another wireless local area network (WLAN), and partially to reduce the ease of someone efficiently trying different key values to gain access to your network.*

After you have recorded the values, click the Apply button; wait for the access point to reset with the new values. If you wish, you may change the password used to get into the configuration utility for your access point by selecting the Password Setting button. Enter a new password, then click the OK key. Again click Apply, wait for the access point to reset, then exit the configuration utility. You are now ready to install and test a wireless client.

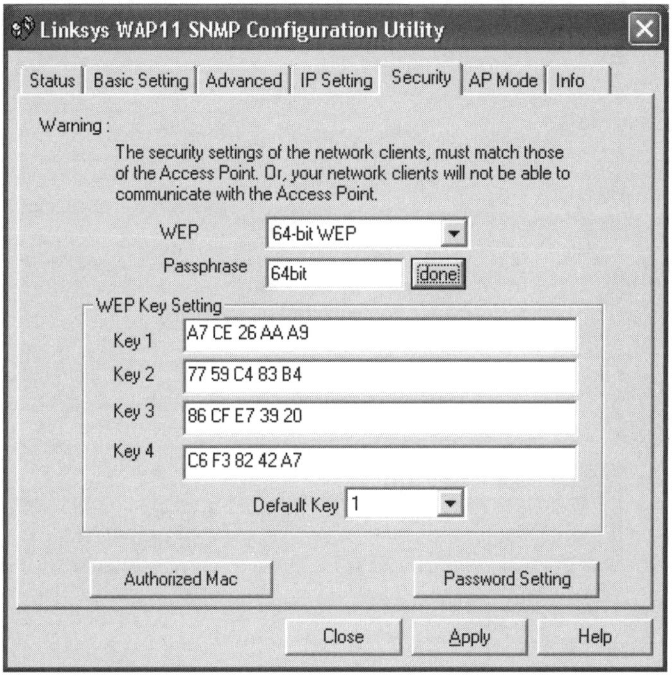

Figure 13.12 The Security dialog for the WAP11, allowing you to set the encryption level and WEP key passphrases.

Installing Wireless Clients

The installation process for your wireless LAN card of course depends on the make, model, and operating platform you are using on the client systems. Existing desktop systems with LAN cards could use the Linksys WPC11 PCI card with built-on wireless adapter, a WMP11 PCI-to-PC card adapter to support adding a PC card adapter, a LinkSys WUSB11 or an Orinoco USB-based wireless adapter, or the LinkSys WET11 wireless bridge unit. Laptops might use either a PC card (most common), a USB-based wireless adapter, or a wireless bridge.

Once the adapter is installed, you will have to configure it—providing the same SSID and WEP key information is used at the access point. Windows XP provides built-in wireless support and will immediately notify you if one or more wireless network connections is available through a pop-up bubble from a new icon in the task bar's tool tray. Right-click the wireless network adapter icon and select "View available wireless networks" to get the wireless LAN selection dialog shown in Figure 13.13 to appear. Type in the proper WEP key information, remembering that you may have to use the hexadecimal value instead of the text value to make the connection work.

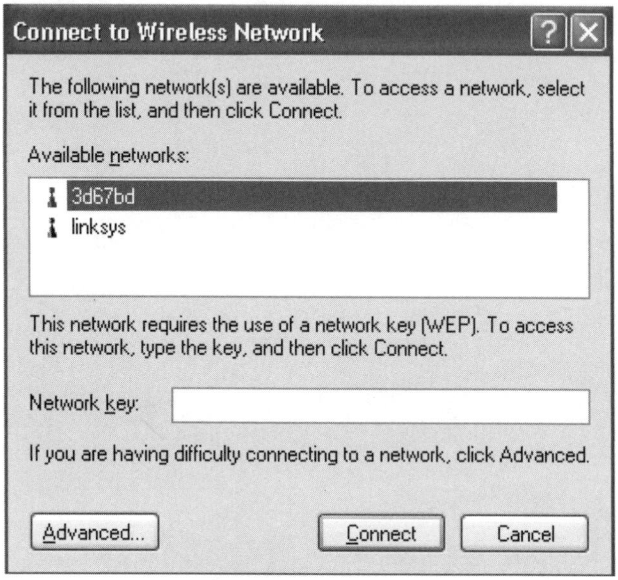

Figure 13.13 Windows XP's wireless LAN selection dialog allows you to select which WLAN to use and provide the WEP passphrase.

To verify that you have a connection to the network and the Internet, you can perform a few simple tests. The most obvious is to open a web browser program and try to connect to a known Web site. If making a web connection fails, you have to troubleshoot your wireless configuration and connection. To get a status under Windows XP, start with a right-click on the wireless network icon and select Status to access the details about your wireless connection—Figure 13.14. What you see is an indication of wireless signal strength and if packets have been passed back and forth. Your first clue to a wireless problem is the signal strength level. If you see any color at all in the ascending scale, your wireless card is receiving an access point signal. If not, move the workstation closer to your access point and try again.

Your second clue that a problem exists is that either the Sent or Received packet counter remains at 0 (zero)—see Figure 13.15. This is your first indication that you are not connected properly to a wireless access point. Your wireless card software may give you similar signal and packet traffic indications.

Figure 13.14 Windows XP's WLAN status dialog, indicating signal strength and data traffic.

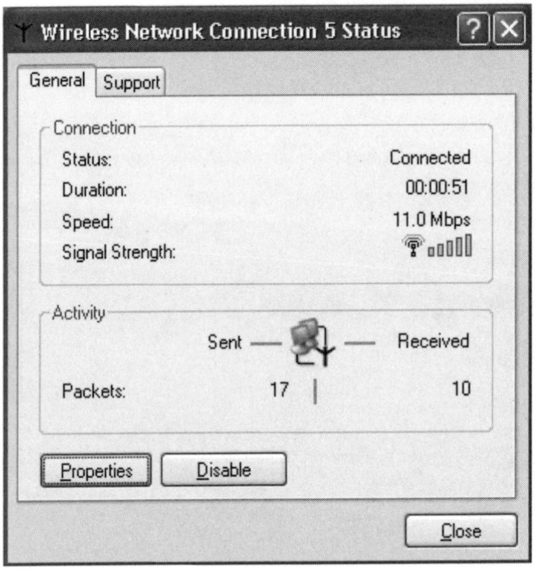

Figure 13.15 Windows XP's WLAN status dialog showing good signal strength, but no received data, indicating a problem in connecting with the access point.

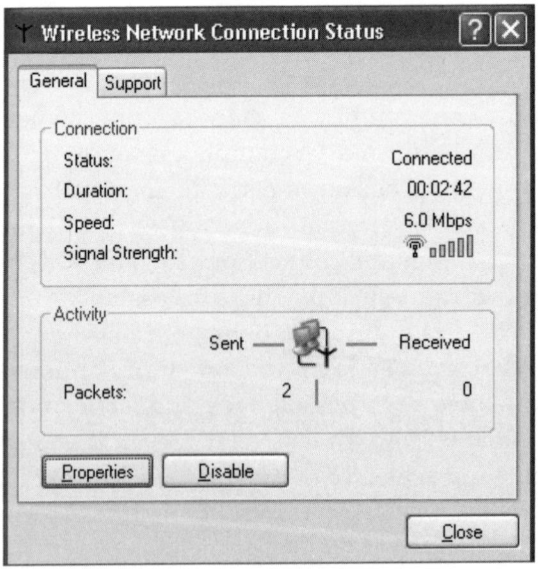

Your third clue comes after selecting the Support tab to get the IP address details—Figure 13.16. This dialog should show the Address Type as "Assigned by DHCP" and IP parameters within the range configured in one of your DHCP servers.

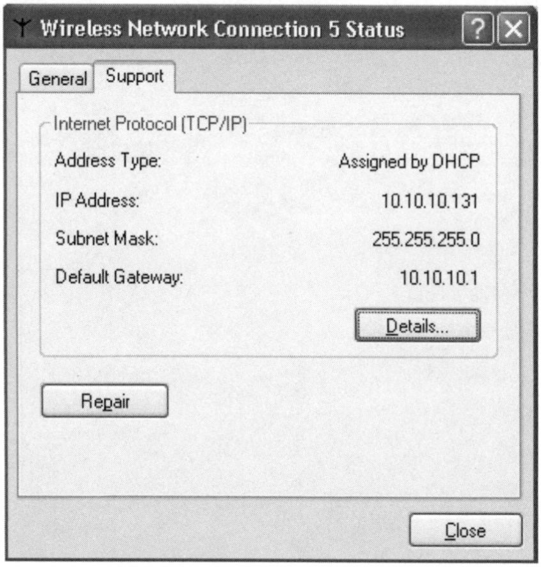

Figure 13.16 Windows XP's Wireless Network Connection status showing a good DHCP-issued address, indicating a successful connection to a local access point.

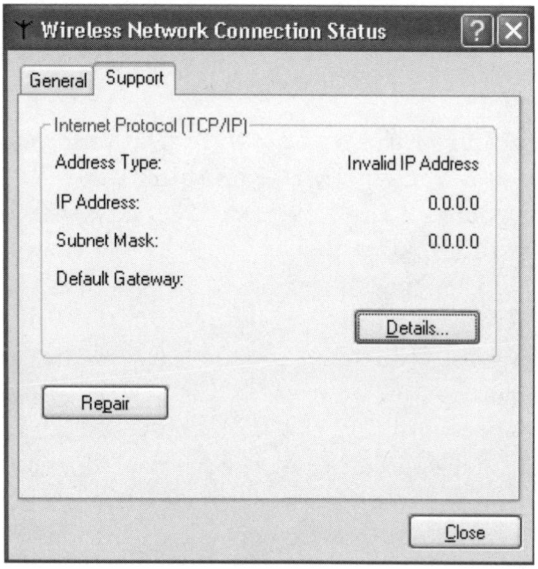

Figure 13.17 Windows XP's Wireless Network Connection status showing Invalid IP Address, indicating a failed connection to a local access point.

If the dialog shows an Address Type of either Invalid IP Address, as seen in Figure 13.17, or Automatic Private Address, as seen in Figure 13.18, your wireless client did not authenticate properly at the access point and could not reach a DHCP server to get a proper address. You can use the WINIPCFG program (Windows 95-Me) or IPCONFIG program (NT, 2000, XP) to get similar information on the IP settings for your WLAN device.

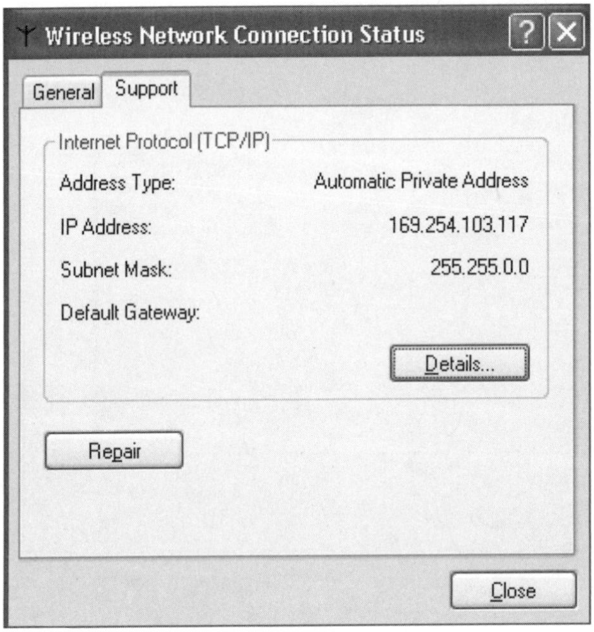

Figure 13.18 Windows XP's Wireless Network Connection status showing Automatic Private IP Address, indicating a failed connection to a local access point.

With either of these last two indications, your possible solutions are limited to retrying what you think the WEP key is at the access point, or going all the way back to ensure that you have the correct WEP key information at both ends of the connection. Since you are focused on this specific situation, it is a good time to go back to the access point configuration program and reset the WEP key values to what you want them to be, and do the same for the client.

Once you get an address in the proper range assigned by your DHCP server and you see both Sent and Received packet counts incrementing, you can then check your connections to LAN servers and the Internet. If they work fine, you can move on to configuring your other workstations for wireless operation.

Configure Dynamic DNS Updates and Always-On KeepAlives

If you are using an always-on business DSL service with static IP addresses, you may skip this section, except perhaps for the information about Tardis time-synchronization software. For services that use PPPoE and provide only dynamic IP addresses, you want to keep that connection on as much as possible. And, in order for people to find your server on across the Internet, you have to keep your primary DNS server updated with your connection's current IP address.

I use the free ZoneEdit, www.zoneedit.com, service to manage the DNS chores for my domains. I discovered that it supports dynamic DNS updates for those of us with dynamic IP addresses. Thus, the ZoneEdit Dynamic Update program, or ZEDu, is the perfect choice to keep the DNS server up-to-date on my current IP address. To use this service, you need to sign up with and configure your domains with ZoneEdit. With that accomplished, you install ZEDu (http://glsoft.glewis.com) on your web and e-mail server(s), in the ZEDu dialog (Figure 13.19), supply your ZoneEdit log in and domain information, tell ZEDu how often you want it to update the ZoneEdit DNS servers, then step away and forget about it. Because ZEDu updates the DNS servers on a regular basis, it also acts as a reasonable keep-alive utility so that your connection rarely, if ever, disconnects and requires a DNS update with a new IP address to be done.

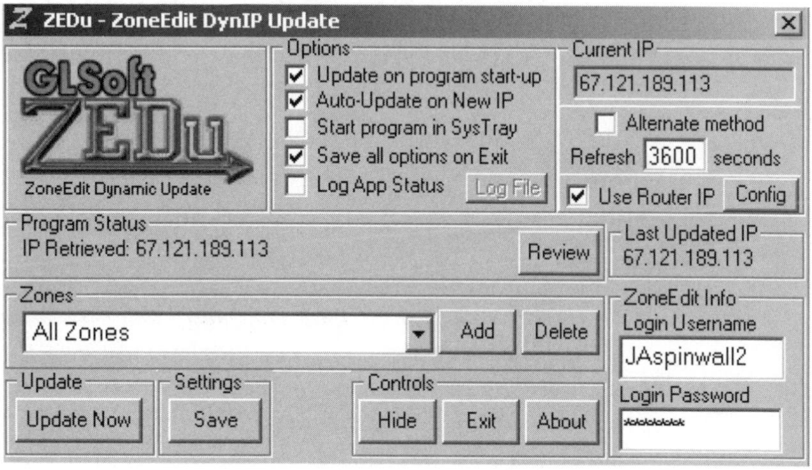

Figure 13.19 The ZoneEdit Dynamic Update program configured to send current IP address information to the ZoneEdit DNS servers.

Because I am a nut about time accuracy, and want an extra measure of DSL connection keep-alive assurance, I also run the Tardis (http://www.kaska.demon.co.uk), time-synchronization software and configure it, as shown in Figure 13.20. This frequently downloads the correct time from the former National Bureau of Standards—now National Institute of Standards and Testing (NIST)—atomic clocks and time servers in Boulder, Colorado. The result is my servers, and any workstation also running Tardis, have their clocks set with the correct time every few minutes. My DSL line is rarely, if ever, disconnected and reconnected, so DNS updates are infrequently needed.

Figure 13.20 The Tardis time synchronization program set up to receive periodic correct time updates from the NIST server in Boulder, Colorado.

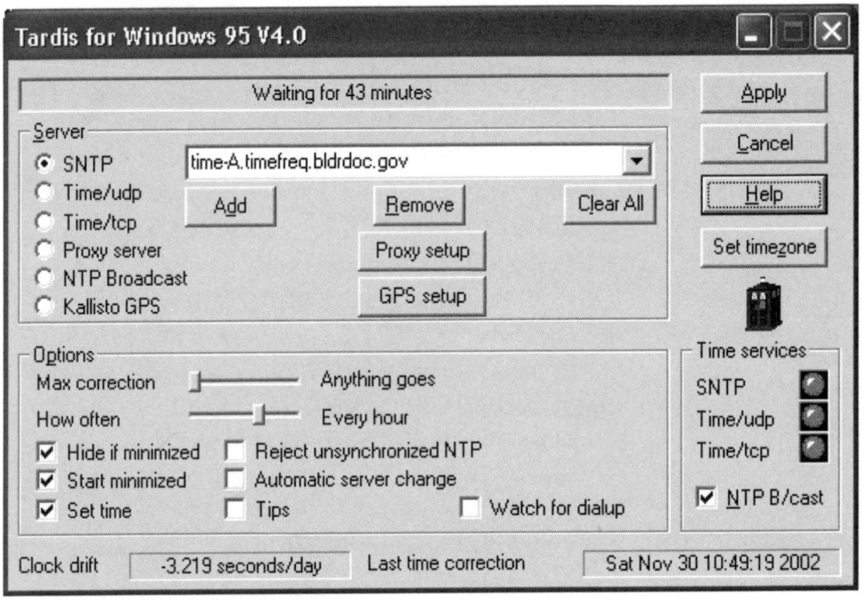

The combination of these three solutions allows you to run one or more servers available over the Internet, but yet behind your dynamic DSL connection and firewall/router.

Note: *Even though you update your domain's external DNS server frequently with the current IP address, there is no guarantee that the update will be picked up by the several thousand other possible DNS servers out on the Internet. While your DNS server could be configured with short duration update and time-to-live settings, the other DNS* *servers that get their information from your server can choose to ignore the timing values from other DNS servers and keep stale IP address information in their databases for several hours or days. If your address changes due to a dropped PPPoE DSL connection, and even if a program like ZEDu updates your server, many DNS servers may retain your old address for a day or more,. Then, people wanting to access your site may end up trying to connect to the old address, or perhaps someone else's site, if they are running a server on their connection.*

Now that we have shown you how to work with dynamic IP addressing, we will try to explain why ISPs make us use PPPoE dial-up ISP services and dynamic IP addresses. The generic answer to these issues is that ISPs do not want you to run servers at home on their budget cable or DSL services. They prefer to sell you fixed IP address services for more money.

One specific answer to these issues is that Internet bandwidth and DSL resources are shared among several hundred different users, and since most users do not use the connection 24 hours a day, or have web or mail servers at home, it seems more efficient to disconnect when not being used.

An advantage to this type of always-on, or more to the point quickly on connection, is that your home systems are not left exposed to Internet-based cyber-attacks—a very important concern since many home-users do not know about or have hardware or software firewalls to protect them. If the connection is down and your IP address changes frequently, it is difficult, if not impossible, to abuse your system. A distinct disadvantage to dynamic IP addressing and the use of routers that combine many users onto one address is that many corporate virtual private network (VPN) secured connections will not work—something to ask your corporate network administrator about if you work from home and need to connect to your company's LAN.

Local Firewall Security and Virus Protection

Considering the wild frontier attitude some people have about the Internet this vast worldwide cyber-expanse is full of "gypsies, tramps, and thieves" to quote Cher. The challenge to find or create the ultimate irresistible marketing tool or cyberweapon is perpetual. Traversing the wired network is bad enough, but the relative unbounded territory of wireless gives the bad guys a lot more anonymity when it comes to trying to steal your data, deny you network services, or trash your systems. We have yet to see a wireless-specific virus, but you can bet someone is out there trying to create one—that could alter your wireless settings to intercept, redirect, or deny your data the path you want it to follow.

Going wireless gives you even more reason to lock your systems up as tight as possible, to reduce the chances of hacking and viruses. Fortunately, the same tools that can help protect your wired systems also serve wireless very well—remembering that basically wireless replaces wires. Unfortunately, so far, the tools we use for wired networks provide no added features or benefits for wireless systems—yet.

Two basic tools in your personal computing protection arsenal should be a reliable software-based firewall to monitor inbound and outbound traffic, as well as program access to and from the Internet, and up-to-date virus protection. My personal choices are ZoneLab's ZoneAlarm Pro and Norton AntiVirus, but there are comparable products on the market you may prefer.

Some of you are wondering why if I already have firewall protection built into my router, I would also use a software-based firewall on my workstations. First, because when you roam about with a wireless system, you cannot be sure that there is an adequate firewall on the wireless system I am using. Second, because a hardware firewall knows only about the network in general and some inbound hacking attempts, and nothing about specific applications. Low-cost hardware firewalls do not know about specific Trojan Horse or remote sniffing applications that may have gotten onto my system and attempted to make outbound connections. ZoneAlarm Pro appeals to the techie in me, as it allows me detailed control and monitoring of every program and host that tries to use my network or Internet resources. Sometimes you want the hardware firewall opened up just a little bit, to apply very specific controls at a specific workstation. ZoneAlarm protects both my workstations and my servers and has saved my web and e-mail servers from attacks and traffic overloads that typical hardware could not.

The use of basic virus protection is obvious—even though I rarely, if ever, use Microsoft Internet Explorer, Outlook Express, or Outlook for web work, e-mail, or newsgroups, I do use Word, Excel, and other products that have considerable vulnerabilities that come with their respective features. Norton AntiVirus has never failed me, whereas other products have cost me several hours, due to their false protection against some of the most annoying bugs on the Internet. Unfortunately, I am at a loss to find a reasonably priced virus protection product for use on personal servers. It seems that protecting a server has a market value of 10 times or more than products for workstations, though they are basically, marginally, the same software doing the same tasks. One way around this is to find a virus protection product that will let you scan the files on mapped network drives from your workstation.

The emphasis here is on protecting anything and everything you can—within reason—and similar protections must be applied to workstations and servers. That same level of caution applies to choosing more secure applications—especially for use on servers that have more direct connections to the Internet than workstations. I do not use Microsoft IIS, FTP, or e-mail server applications—no need for them when I can do the same basic things with freeware, shareware, or lower cost products. For workstations, I am *very* careful to avoid or quickly uninstall applications that plant ad-ware and Trojan Horse programs within the system. Ad-Aware and Pest Patrol are two reasonably trusted tools for detecting and eliminating these kinds of programs, which are not considered viruses.

Summary

Setting up a wireless network in real life can be as straightforward as it appears here. Your equipment make and model may be different, but the basic settings, functions, and symptoms are essentially the same. The most confusing part is probably the translation between text- and hex-based WEP keys.

Router configurations for sharing a DSL or cable modem connection to the Internet are likewise similar and straightforward, especially if you do not get yourself wrapped up in different terminologies used in different products. I find that the tech support for these products from the various vendors is pretty helpful if you get into trouble.

Setting up and running a web or e-mail server under these conditions is well beyond the scope of this book, but it is good to know it can be done and supported in this type of configuration, and with a few easy-to-use software tools.

I cannot emphasize security and virus protection enough. I have foolishly placed unprotected Windows 2000 and Linux systems directly on the Internet and had them scanned for, found, attacked, and rendered almost useless within two hours of first appearing on the Internet. Simply, if you are operating systems on the Internet, there is no mercy. Get protection, install it, configure it, and use it—no exceptions!

CHAPTER 14

Neighborhood and Community Wireless Networks

So you have your home network up and running and you want to share it with the neighbors or have the confidence to build another one and set it up in the local coffee shop or bookstore. You have a lot of options, not the least of which is deciding if you are going to run an open or closed network, operate it free for users, or recover some of the costs.

You can reproduce the small office, home office (SOHO) system shown in Chapter 13 as the foundation for your system, run it open or closed, or sign up with an affiliate partner program like Boingo or Sputnik. If you do not want to start from scratch, you could make an initial investment, order a wireless local area network (WLAN) in a box from Hotspotzz, sign up and share paying subscribers, and maybe bring in a little extra cash.

Every option has some catches and some benefits, depending on how involved you want to be, how reliable you think you can make the services, and how much you can afford in time and money to run a network with more users. You also need to consider the equipment you will use, its certification and compliance with the Federal Communications Commission (FCC) and any local regulations, and your technical ability to install a more significant system than popping an access point atop your bookshelf at home.

Sharing Your SOHO WLAN

Unless you live in an apartment or condo complex with neighbors immediately surrounding you and the location of your access point, you will need to get the antenna out of the den, basement, or garage and up where your neighbors have a reasonable chance of getting adequate signal levels to make things work. This means using an omnidirectional antenna to distribute signal around you for more than one neighbor, unless you are at the end of the block and can focus a directional antenna toward all of them.

Your first concern here is that you are not allowed by FCC regulations to connect an external antenna to your access point—that is you are prohibited from installing an antenna on your roof and running coax to the access point in the den or wherever it is located. You must either buy an access point and antenna system designed and certified for this purpose or move your access point, complete with its attached antenna, to a higher location and supply it with power and Ethernet resources.

If the neighbors are just going to take in wireless signals to supply transmission control protocol/Internet protocol (TCP/IP) to a single specific personal computer (PC) or their entire LAN, they may be able to pick up the signal from your external antenna directly to a wireless device on their end—and what that wireless device is depends on what they will be doing (see Figure 14.1).

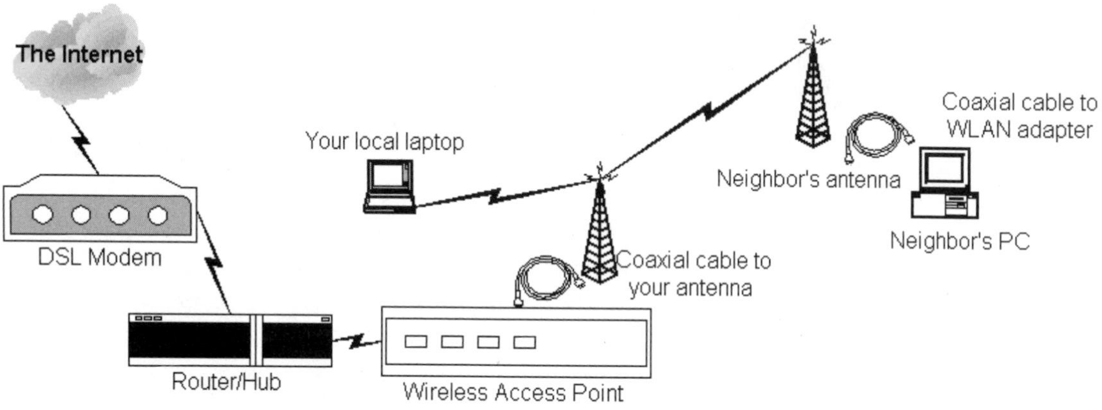

Figure 14.1 A typical neighborhood wireless sharing arrangement involves being able to provide enough signal to the area and the ability for the recipients to acquire your signal. External antennas are typically necessary to accomplish this—one at your end to get signal out to your neighbor and another at your neighbor's end to pick up your signal.

Your neighbors may also need to use an external antenna, likely directional, and they are likewise bound to the restrictions of using equipment that is also certified, end-to-end, as a system. For someone immediately next door, connecting a universal serial bus (USB), PC card or peripheral component interconnect (PCI) card to their PC for wireless access could work fine, but not great. If they have an Ethernet adapter in their PC already, or they are a bit farther away, they may be best served using a small wireless bridge mounted high in a room or in the attic, and connecting it with CAT 5 cable with power-over-Ethernet to their PC to be able to capture signal from your access point.

Many people are doing this type of installation—typically using one off-the-shelf access point wired with a pigtail connection and a few feet of coax to an omnidirectional or a directional antenna on their roof or in an attic as the Internet source, to extend service to a friend nearby who is using a directional antenna wired with a few feet of coax to a pigtail connected to a PC card or other off-the-shelf WLAN client adapter.

Your neighbors should not expect the ability to walk freely about their homes with laptops or personal digital assistants (PDAs) and enjoy your wireless signal from 100 to 1000 feet away. That is just not going to happen. If they want to be able to do this, they will need to install their own local access point and bridge it into your network as if it were another client, as illustrated in Figure 14.2. Here your neighbor's network is bridged to your wireless LAN, feeding the WAN side of a router/firewall/hub device as if it were a digital subscriber line (DSL) or cable modem connection. The router redistributes the wide area network (WAN) connection to a LAN, including wired PCs, and may use a local access point for laptops.

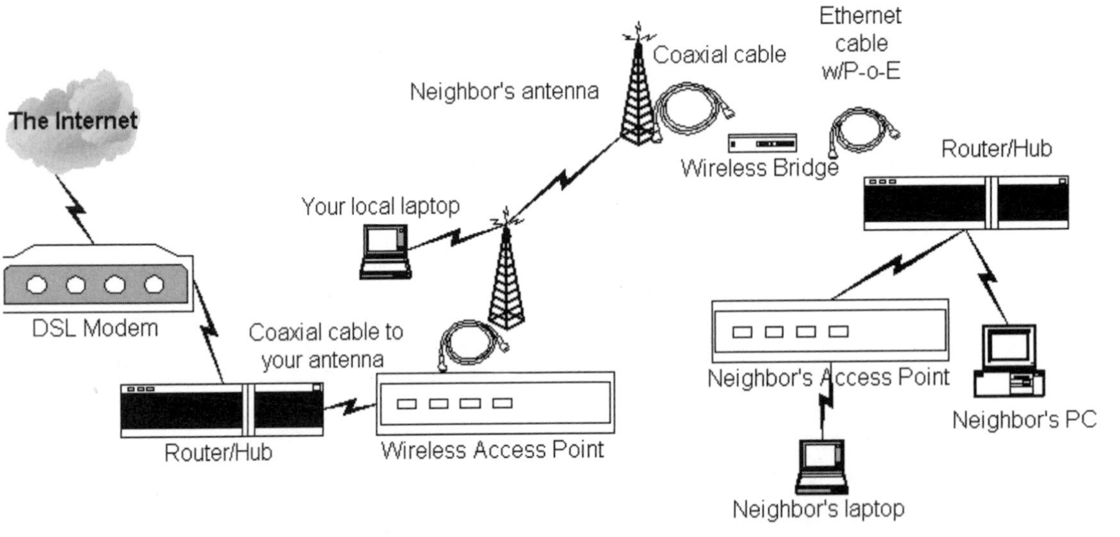

Figure 14.2 A neighbor's LAN and local WLAN bridged into an existing wireless network.

Adding an external detached antenna to a wireless bridge, access point, or WLAN card will typically cover an area from a few hundred feet up to a few blocks. Generally, these systems do no harm to the environment or other 802.11 users nearby, but they are illegal if the wireless device, cable, and antenna is not bought as a certified system.

A nosey or unfriendly neighbor could make a complaint to the FCC, which may or may not prompt an investigation or inspection. And if a violation is determined, the equipment can be seized on the spot, charges filed, and a court date with a Federal judge set. This is not known to have happened yet—*yet*—but it is certainly feasible and something to be aware of.

Indeed, the rules leave a lot to be desired, but they exist, and there are some good reasons for them—not the least of which is that if you make serious mistakes with the antenna connections or do not use the right equipment, connectors, coaxial cable, etc., you could destroy your WLAN equipment or, worse, cause interference to others.

If you work out the costs, buying a properly certified system or two costs less over a year than the expensive monthly DSL service charges for each household. Combined certified systems will cost less than buying separate access points, coax, and antennas. If you get enough people together to purchase at the same time, you may be able to get a quantity discount. You will also get more reliable equipment than off-the-shelf retail products, keeping you and your neighbors sane and happy with the service. A properly installed certified system may not keep you out of suspicion with the neighbors, but it will keep you out of trouble with the FCC. So, think in terms of overall cost, not to mention the convenience of not having to piece together an uncertified system.

If it sounds like this is getting a little complicated, you are correct. The focus of most off-the-shelf products and the intended use for them is to extend an existing network, not build mini-wireless-Ethernet-empires or wireless Internet service providers (ISPs). There is nothing wrong with building a neighborhood or community wireless network. The point is that certified equipment should be purchased and deployed properly for your technical and legal benefit. Certainly, if you expect others to share in the costs, it is to their benefit and should be one of their caveats to ensure they are spending their money on good, clean, reliable equipment that is not liable to be confiscated or fail because of inadequate funds, skill, or careless installation.

Open Community Wireless Networks

If you want to add wireless Internet access to your local business—the classic Internet café, bookstore, cocktail lounge, library, or youth center venues—you will probably start out with a setup similar to the one shown for a SOHO in Chapter 13. That is fine if you are not using your Internet connection or a shared part of your LAN to also run your business systems. If you are running your business on a LAN and intend to share one Internet connection with your public wireless network, the setup gets slightly more complicated in order to protect your LAN from the general public.

Sharing a single Internet connection between two separate LANs—one wired, the other wireless—may be as simple as that shown in Figure 14.3, by adding another router to the setup. The additional router is placed between the existing router and your internal business LAN to block common wireless LAN traffic from entering your business server and computers.

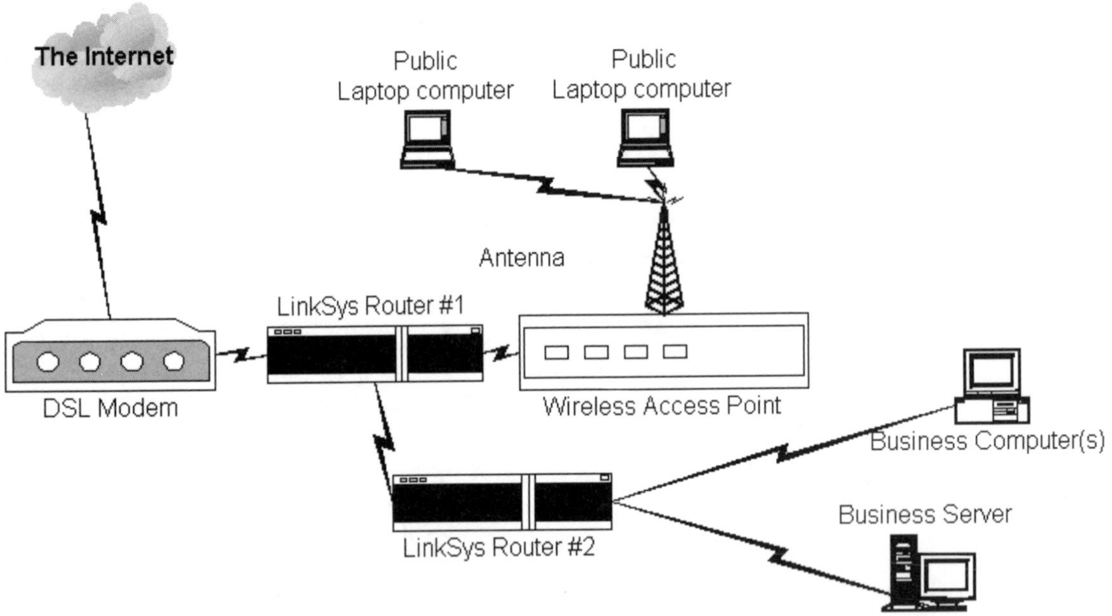

Figure 14.3 Combining a public access wireless LAN into an existing business network with two routers to share dynamic IP DSL Internet services.

For smaller home-brew systems, the LinkSys router products are quite adequate. In fact, the first router could be a combination router and wireless access point like the LinkSys BEFW11S4, and the second a LinkSys BEFSR41. The configuration could be swapped around, using the BEFRS41 with its WAN connection point for the DSL circuit. A WAP11 access point is then run for the public and another BEFRS41 or GEFW11S4 for protection of the business LAN, both connected to the LAN ports of the first router.

Note: *Keep numerous straight* ***and*** *crossover Ethernet cables handy! When you work with a lot of different types of small network gear and changing configurations, you find yourself connecting things in all sorts of ways that may require different cabling to go between Ethernet ports, be it hub-to-hub or PC-to-PC, etc.*

Chances are, your business will be using higher-priced business DSL service, giving you a set of fixed IP addresses. And, the DSL modem may provide some router and firewall features to resist intrusions, but you still need network address translation (NAT) and domain host configuration protocol (DHCP) services to configure the wireless clients as they come onto the network.

If you are using a T-1 circuit for Internet access, I suggest using either a Cisco model 1720 router with a single T-1 and two Ethernet interfaces, as shown in Figure 14.4, or three Ethernet interfaces for DSL service, to provide a more costly, but also higher performance, more reliable solution. With the Cisco 1720, the primary Ethernet port connects to the Internet connection, the wireless access point to one of the internal Ethernet ports, and the business LAN to the other, with the router configured to allow either LAN Internet access, but no inter-LAN or subnet-to-subnet traffic flow.

In these configurations, you will have to be very aware of the network address translations, gateway addresses, subnets, and domain name system (DNS) server allocations you have to set up in each network device. Let's configure a typical setup by the numbers, using Figure 14.5 as our example configuration, so that you can see the different address parameters you will have to use, starting with some assumptions about either a static or dynamic IP address given by your DSL provider or ISP:

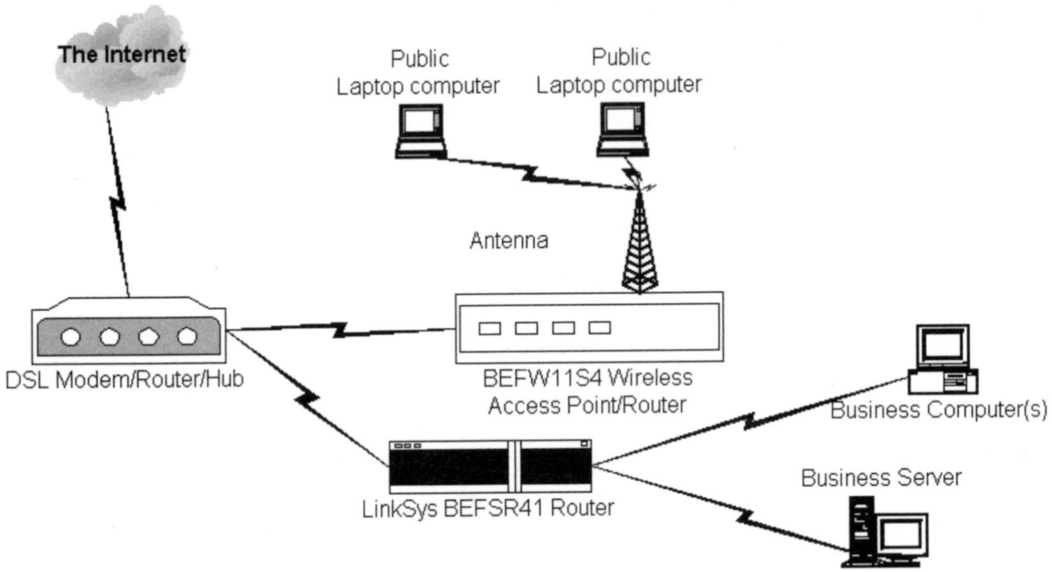

Figure 14.4 Combining a public access WLAN into an existing business network, with the wireless side using a combined access point/router unit, instead of separate pieces of equipment.

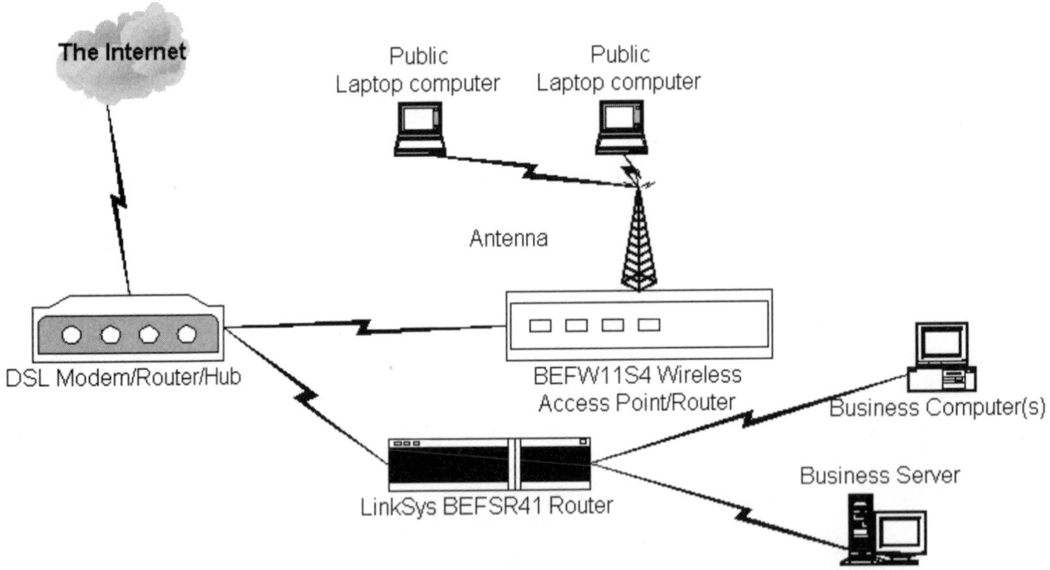

Figure 14.5 Diagram showing a public access WLAN, with a separate business LAN using a Cisco 1720 router.

- DSL provider assigns your DSL modem IP address 24.66.13.13.
  - This address is then used as the gateway address for clients and other devices connected to it.
- DSL provider gateway address is 24.66.13.1 (preset into their DSL modem/router).
- DSL provider assigns you DNS addresses 24.66.13.5 and 24.66.12.6.
- DSL provider assigns you eight fixed IP addresses from 24.66.13.21 to 24.66.13.28.

With this information in hand, you can begin to configure your wireless access point and the router for your LAN. We will configure each of these devices so they have no awareness of each other. To keep WLAN users from seeing LAN users and vice versa, assign different subnets for the clients that use them. First, the WLAN access point:

- Set the WAN gateway address for the WLAN access point to 24.66.13.13—the address of the LAN side of the DSL modem/router.
- Set the WAN IP address for the WLAN access point to 24.66.13.21—one of the fixed IP addresses assigned by the ISP.
  - This address also uniquely identifies the WLAN access point to the Internet at large, helping you trace traffic or abuse if necessary.
- Set the WLAN access point DNS server IP addresses to 24.66.13.5 and 24.66.12.6.
- Configure the WLAN access point's router LAN address to 10.10.10.1.
- Configure the WLAN access point's router LAN subnet mask to 255.255.255.0.
- Configure the WLAN access point's router to provide NAT and act as a DHCP server.
- Configure the WLAN access point's router to issue as many DHCP addresses as needed—a default of 50 is adequate.
- If the WLAN access point is separate from the router, set its gateway address to 10.10.10.1, the address of the LAN side of the router.
- If the WLAN access point is separate from the router, give it a fixed IP address within the same 10.10.10.x subnet as the router's

DHCP configuration, but not within the range of DHCP addresses to be leased out.

- Set the WLAN access point service set identifier (SSID) to a name you like.
- Configure a suitable wired equivalent privacy (WEP) encryption key if you prefer to have some use restrictions on this WLAN.

With this configuration complete, your WLAN clients should be able to associate with the access point by SSID name, obtain a 10.10.10.x-subnet address, acquire gateway and DNS information from DHCP, and access the Internet. Perform a similar configuration of the LAN router as follows:

- Set the WAN gateway address for the LAN router to 24.66.13.13—the address of the LAN side of the DSL modem/router.
- Set the WAN IP address for the LAN router to 24.66.13.22—one of the fixed IP addresses assigned by the ISP.
  - This address also uniquely identifies the LAN router and its users to the Internet at large, helping you trace traffic or abuse if necessary.
- Set the LAN router DNS server IP addresses to 24.66.13.5 and 24.66.12.6.
- Configure the LAN router LAN address to 192.168.10.1 (quite different from the WLAN equipment and users).
- Configure the LAN router subnet mask to 255.255.255.0.
- Configure the LAN router to provide NAT and act as a DHCP server.
- Configure the LAN router to issue as many DHCP addresses as needed—a default of 50 is adequate.

With this configuration complete, your LAN clients should be able to obtain a 192.168.10.x-subnet address, acquire gateway and DNS information from DHCP, and be able to access the Internet. By default, the routers should not pass Microsoft Windows Networking or NetBIOS traffic at all (on ports 137, 138, or 139), so Windows clients on either the LAN or the WLAN should not be able to see across or through the routers to each other, though they should be able to perform Windows peer, workgroup, or server/domain-based networking among themselves. Similarly, neither router provides inbound access to file transfer protocol (FTP), mail, or web servers from the Internet to the LAN or WLAN sides.

If you intend to run an Internet-accessible web, mail, or FTP server on your business LAN, you will have to configure that separately in both the DSL modem/router and the LAN's router. I do not recommend that you run these servers connected directly to the DSL modem/router in this scenario, because even though the DSL modem/router provides some protection from intrusion via the Internet, users of the WLAN may be able to intrude upon it from the inside.

If you do connect your web, e-mail, or FTP server(s) to the DSL modem/router, remember there are thousands of "gypsies, tramps, and thieves" on the Internet. It takes less than two hours for an unprotected server on the Internet to be discovered and exploited with the Nimda virus (Microsoft IIS) or other bugs that can easily destroy all of your hard work. Cleaning up an undetected Nimda mess can take several hours of valuable time. Ensure that you have properly secured the server with current security patches and virus protection, and consider using ZoneAlarm Pro to add an additional layer of protection to it.

Wireless ISPs

You, too, can become a wireless ISP in a matter of an hour or so. With either of two wireless Internet service provider (WISP)-in-a-box kits—from Boingo or Hotspotzz—you get preconfigured WISP equipment, marketing materials, and international awareness that your local hot spot is up and running, ready for business. The Boingo kit costs about $700; the Hotspotzz kit is around $1,000. HereUAre offers a partner program without a WISP-in-a-box kit so that you can roll-your-own system. In each case, you become an affiliate, entitling you to revenue share/royalty for every user that uses your hot spot, and a commission/royalty on every user that subscribes to the service through your location.

If you prefer that the process takes longer, you can roll your own WISP presence. The easiest options are using NoCat or Sputnik portal software running on a Linux system to act as the log-in front end for a more open WISP presence. You may also check into portal software from Portal.com, or access control hardware systems and software from vendors such as Nomadix (http://www.nomadix.com), Colubris (http://www.colubris.com), or Vernier Networks (http://www.verniernetworks.com).

Becoming a WISP makes sense for many businesses that have a lot of visibility to and lingering nearby presence of the general public—once again coffee shops, cafés, and bookstores. I am not convinced that creating and running a WISP intended to cover a wide area is as yet cost- or performance-effective. Because of the costs involved, you need to purchase and install more rugged equipment, arrange for and rent suitable space for the area to be covered, provide that location with Internet access, and deal with the sheer hassle of managing equipment installed at some distance from where you might normally set up and run your business.

There are various methods for providing Internet access to distant locations—depending on where they are, the ability to install more than one antenna, and your cost limitations. Two of these are illustrated below. Figure 14.6 shows a remote location getting Internet access from a satellite Internet provider, and Figure 14.7 shows a high-elevation bridge point to distribute Internet access from one fixed, wired point to other remote access point locations.

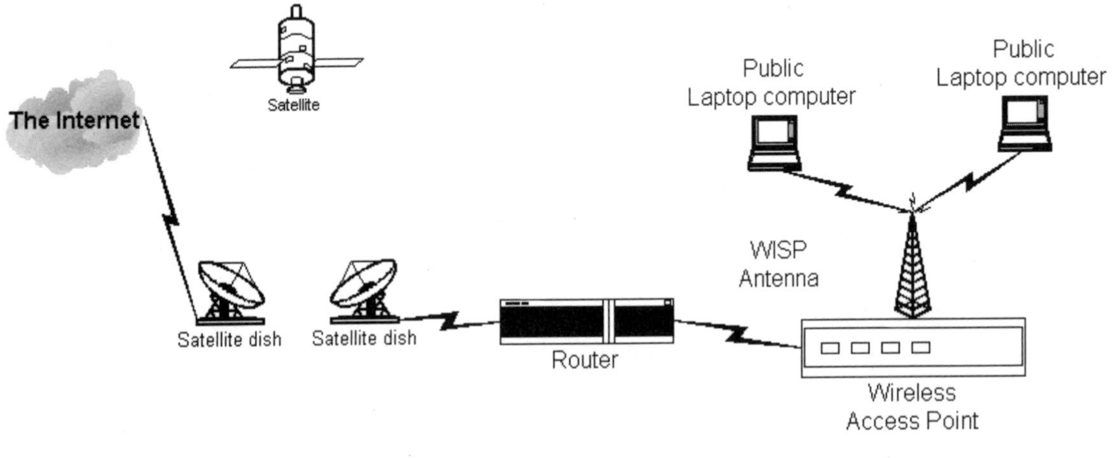

Figure 14.6 Delivering the Internet to a WISP access point by using satellite services for remote locations.

Either of these configurations should work for the preconfigured Boingo or Hotspotzz services or HereUAre—the "satellite solution" for locations without sufficient wired-Internet services—rural areas, campgrounds, truck stops, etc., and the bridge or relay point for urban and suburban areas with suitable building or hilltop locations.

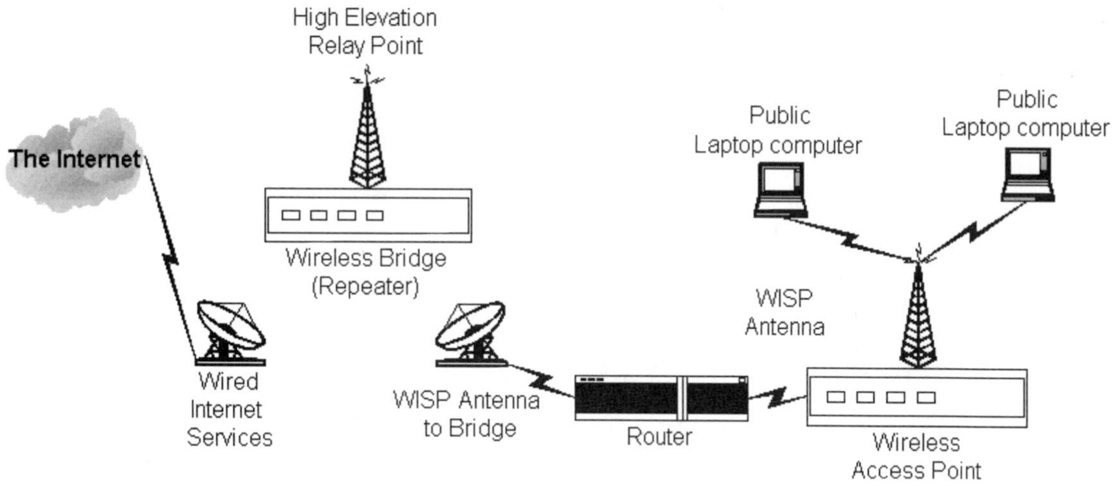

Figure 14.7 Delivering the Internet to a WISP access point by a high-elevation bridge point for urban and suburban locations.

Portal Software

If it can be done with bolted-down kiosks and desktop systems at Internet cafés, why not with wireless? There are about a dozen vendors of fixed location kiosk systems for Internet cafés. And within and running those kiosk systems is software that provides access control based on various payment methods and back end billing systems for cash or credit card transactions. While few if any of these vendors appear to have applied their products to hot spots and wireless access points, there seems to be no reason someone crafty at putting together systems and software could not perhaps apply a front end server application like NoCat on Linux to a Windows server running kiosk software and create his or her own billable WISP.

Some of the products and vendors I have considered using for this are:

- The Web Kiosk Commander by Rocky Mountain Multimedia, Inc. at http://www.rockmedia.com/multimedia/webkiosk.html
- Central Payment Kiosk System at http://flyby.net/byok/cyber_cafe.htm
- Secure Web Portal by Entrust at http://www.entrust.com
- NNU Runtime Engine by NetNearU at http://www.nnu.com, which is newly designed for wireless service

If you wish to work out an external payment system and manually control WLAN access you could use Funk Software's Odyssey software as the secure access control mechanism for your WISP WLAN without a portal front end. In this case, the users would have to perform a client software installation and you would have additional server administration tasks for every subscriber as they sign up for services, but you would be providing a secure WLAN implementation that would offer some additional value to serious users.

Summary

Neighborhoods, communities, parks, cafés, a library—anyplace Internet users spend enough time to boot-up, log-on, and check their e-mail or look up something on the Web is a good place to set up a WLAN hot spot or creates your own WISP presence.

I urge you to do a lot of your own research into your options as far as equipment and software to operate and manage a WLAN system others can depend on. If you can work out the economics of equipping, locating, and running a reliable WISP, or just want to set up a hot spot for friends and fun, you can see that it's very easy to do.

"Big players" AT&T, IBM, and Intel have just recently begun to focus on the WISP market to create a nationwide wireless system available for resale through WISPs and other dealers. So there will eventually be some well-funded and well-equipped competition. And if you publicize your system, the competition, your neighbors, and the FCC will know better where to find you, so pay attention to the details.

CHAPTER 15

Upcoming Standards and Trends

Once a standard is set, technologists begin to move on to find a place to set a new one beyond it. Good enough never is. Such is the case with our first, and if you will, second generation wireless networking devices.

In the United States, and to some extent the entire world, we rely on the Institute of Electrical and Electronics Engineers (IEEE), American National Standards Institute (ANSI), and the Electronic Industries Alliance (EIA), among other such policy, trade, and technology groups, to help define interoperability and compatibility among types of equipment, services, and procedures. Almost every country or certainly every region of the world has similar standards-setting organizations that work along with (most of the time) other such organizations so that manufacturers and consumers can have products that are better, more consistent, and work with each other. Each country or region must also recognize and work with a variety of government agencies concerned with technology, safety, economics, politics, and traditions.

All combined, there are probably a few thousand people involved in deciding how and what can work safely, effectively, and functionally. Those thousands of people are involved in theory, research, development, economic, political, and environment issues that can take years to come to some marketable conclusion before a type of technology or product ever reaches us as consumers.

From that you have to wonder about the staggering rate at which new technologies, capabilities, products, and benefits appear before us. We have all seen a lot of potentially good ideas, or at least companies with them, bite the dust years before they should have, for being ahead of the curve, and certainly ahead of standards and our ability to absorb some products or the cost of them into our lives and work. The Sony Beta video standard, integrated services digital network (ISDN), and X2 modem technology are notable advances that lost out to economy, feasibility, market timing, and standards—de facto or official.

Using Radios and Resources for Networking

The technologies discussed in this book—the integration, or convergence if you will, of computer networking and radio—have clashed and cooperated within international standards bodies and government agencies and come out ahead with almost a new economy, or at least a new segment of a larger economy to our benefit. The clashes and cooperation took years to work their way through, and they are no where near completion. There is always something new on the horizon.

It makes sense having these two technologies together—it always has, at least for the more than 30 years I have been an amateur radio operator, and years beyond that just dabbling and working with radios and gadgets. Radio is a communications medium, just like wires. Getting something communicated over radio means cutting the wires and inserting something new in between the two ends—something that can take the wired signal, make it somehow go over radio waves, and convert the radio waves back into something that can go back onto wires. When I got into working with radio for a living, the wired signals of interest were either voice communications, or an annunciation or alert to someone—your basic walkie-talkie for voice or a pager for some type of tone alert—quite easily done. Of course there was, and still is, broadcast radio and television for news and entertainment.

I can imagine my grandkids in a few years asking me if we had mobile phones or computers without wires when I was growing up. Well, I did have a car phone (the size of a briefcase) when I was 18 years old because my job was to repair them and the control systems that made them work. But the closest I came to any computer at the time was either racks of relays similar to systems used in dial-telephone days, or a cabinet full of digital chips that could finally replace a room full of relays. Oh, there were computers, but only special people got to touch and use them. Mention a wireless computer to someone and they would look at you funny, with a vision of a semitrailer full of computer gear pulling a trailer with a generator and another with a transmitter station and huge antenna trailing behind—no way!

Now all of us are special, unwired, and with more mobile voice and computing power connected to the world by radio than we know what to do with. And still, we don't have quite the resources we need to do everything we want or could do with them together. From want, need, or desire come innovations that need to be optimized and standardized; resources that need to be shared, acquired, or created and in some cases regulated or at least moderated; and products made, sold, and supported to work to our advantage.

We have seen wireless networking, mostly in the context of IEEE 802.11b, but also functionally similar and applicable with and to IEEE 802.11a. Those two implementations, while requiring technologists to come together and eventually agree on how and what to make, also required U.S. and international regulators to arrange the sharing or exclusive use of portions of an elusive, fixed, nonrenewable resource called radio spectrum.

Perhaps, unless you are in the radio business or have studied it, you may not realize just how many and what types of uses *consume* this resource that extends in frequency or wavelength almost literally from DC (the type of energy from batteries) to light. Usable radio spectrum is totally invisible and almost intangible, until you use it. One could say that we have usable, practical radio spectrum from 1 MHz to perhaps 100,000 MHz (100 GHz). If we were to give radio spectrum a unit count, we might think we have 100,000 million pieces of radio spectrum to parse out to its users. And if we did, without considering how we use those units, or if they can be used uniformly for all things, we would have a *lot* of radio space available.

In reality, we don't have 100,000 million pieces of radio things to parse out to anyone or everyone for anything. We do not have that many because of the way we have traditionally applied radio, how we apply it now, and how we will want to use it in the future. Those applications include everything from Morse code or CW (transmitting dots and dashes) to voice communications to radar to television, satellites, and now high-speed data links.

If all radio spectrum behaved uniformly—that is, the same amount of power with the same size antenna carried all signals the same distance—things might be easier. We could theoretically have 100,000 million unique radio resources or channels to use if we were all willing to learn Morse code and accept relatively slow-speed communications—for conversational speechlike interaction, fine; for talk or music radio or certainly television, not possible; for large amounts of complex data, intolerable by even yesterday's standards. Under theoretical circumstances, every person could have several "radio units" to do with what they pleased, with enough left over for every government and business to have a few to play with.

One of the first and most obvious limitations to this theory is that we all cannot, or will not, accept Morse code communications. Thus, the transmission of complex signals works over radio, more than one and typically a few thousand (and in some cases a few million), of those radio units must be consumed to get desired information from point A to point B.

For example, it is known that it takes 3,000 Hz of bandwidth (3,000 of our theoretical radio units) to get an intelligible voice signal from sender to receiver over the radio waves. If a theoretical uniform radio system were possible and used only for voice, our 100,000 million radio units would support only 33,333,333 simultaneous voice transmissions. I suspect the combined phone systems in New York; Washington, DC; Texas; and California handle that many simultaneous phone calls most of every day. Add in a few million modems and fax machines for good measure while you are at it.

We also know that we want more than voice communications, music, and perhaps color television. As used today, FM radio broadcasting uses 100 KHz of bandwidth or 100,000 radio units. We have room for a million of these if that is all we do with our radio spectrum, but we do not need a million of them. An acceptable color television signal consumes six million radio units, and our 100,000 million radio units give us only enough room for 16,666 television broadcasts. Hmmmm—now we have to compromise. How many radio units used for voice signals do we give up to make room for how many FM radio and television signals?

Now we want 10 Mb (or preferably 100 Mb) data communications speeds to use some of our radio resources too. 100 Mb of data transmission could theoretically consume 100 million radio units. That means we have only enough radio units for 1,000 100-Mb data transmissions or 10,000 10-Mb data transmissions. How many voice and FM radio and television signals do we give up to accommodate how many data transmissions?

I think you can see that there may be no easy way to decide how many of the different application types we need or to decide what is fair to distribute across the usable radio spectrum. I left out the parts about allowing some "guard space" adjacent to each radio unit, so that stronger nearby signals would not cross over and interfere with each other. Using every available radio unit leads to other dilemmas—mostly interference—because many of these signals will mix together in predictable mathematical patterns, creating havoc to other seemingly unrelated signals.

Another limitation, quite limited by physics, is that all radio frequencies (RF) do not work uniformly at the same power levels with the same equipment or antennas. Lower frequencies dictate larger antennas and, to some extent, larger equipment. Higher frequencies dictate smaller antennas and equipment, but need more power (which also means more equipment) to go the same distance. Still another is that the distance covered by radio communications varies with the frequency and with the weather—at least the weather at very high altitudes, in the ionosphere—something for which we have no known control. These factors and the general behavior of radio signals at different frequencies are fairly well known and tend to dictate classes of service or application-specific uses for the different frequencies.

We use some of these limitations to our advantage when figuring out what part of the radio spectrum is best for what we intend to communicate over it, and where we want it to go. For this reason, for example, cellular telephones that have to communicate only a mile or two to a nearby base station can use low power at very high frequencies. And those frequencies can be reused over and over again in other cells nearby. You would not use a huge radio with a 200-foot long antenna for something like cellular telephones, as the idea of portable brings up visions of trucks and trailers full of equipment. We know a cellular telephone-type radio cannot broadcast very far, but a larger radio and antenna at a lower frequency can, so we use those parts of the radio spectrum for long-distance communication around the Earth.

What we find is that some of the radio spectrum is reusable or can be used simultaneously by separating its reuse to areas beyond the range it normally covers, and we have been doing that for years. For instance, 1070 KHz AM broadcast radio stations exist in only three to four different parts of the United States, while perhaps 20 or 30 101.3 MHz FM broadcast stations exist in the same larger geographical area. And there are more in other countries, just as there are multiple Channel 3 or Channel 21 television stations around the world.

Because we know how most radio waves and the ionosphere behave, and we have learned how to design antennas to shape the pattern the radio signals emit, we can tailor the amount and direction of signal. An FM radio signal, for instance, does not need to broadcast to the stars, so the antenna pattern is tailored to place that signal down into the local listening area and not much beyond the primary area of interest. Doing this, we can cellularize radio frequency reuse around the country and around the world. One of the enemies of reuse is boosting the power to get a signal just a little bit farther out there. And in doing so, we may infringe on someone else's territory, causing interference. The frequency, direction (or focus), and the power of various radio signals is, thus, regulated to avoid interference between users. This is a very important consideration when dealing with a limited resource, and a resource that has limitations on how much of it we can practically use.

Cellular phone service has or had only 40 to 50 800 to 900 MHz-range channels to share, and there are thousands of phone cells throughout the country. We also anticipate there are, or will be, millions of wireless networking devices using 2.4 or 5 GHz spectrum all over the world—and perhaps several hundred of them within 1 to 5 miles of yours. Above 2 to 5 GHz, it becomes impractical to use radio spectrum for anything but point-to-point communications—satellites, microwave hops from place to place, etc. We do not want to use 10 GHz radios for personal communications devices because water resonates near that frequency and a high percentage of human mass is water…make a phone call, boil your brain.…

So, we have learned since the days of Marconi how different parts of the radio spectrum behave, how the atmosphere behaves, how to control the power levels, and how to tailor antennas to focus and optimize signals to specific areas, with the result that we can reuse the same frequencies many times in different areas of the world. What we have not been able to do is make more radio units available; thus we have to work out better sharing arrangements or repurpose some of the spectrum we have been using if we want to do more or different things.

As good (as in beneficial to us and financially successful to the manufacturers) as the implementation of 802.11a and 802.11b have been so far in the limited amount of spectrum allotted to them, there is speculation and investigation into broadening the use of radio for data networks, and, of course, which part of the greater radio spectrum can be used for it. While computer networking looks to expand, so do police and fire and other types of communications, causing a major reshuffling of television channels and a push to digital TV broadcasts that can use more of the original television space for more and different radio uses. The U.S. Federal Communications Commission (FCC) has begun a new Spectrum Policy Task Force (http://www.fcc.gov/sptf) to study the current and future demands of radio spectrum use, leading to possible changes in who uses what parts of the spectrum and how.

Data networking over wireless is in competition with cellular phones, your local police department, domestic and international broadcast stations, other governments, and probably dozens of other nonobvious uses—weather tracking, military purposes, etc. It is an ongoing struggle to obtain and use more and more of what there is no more of—practical, usable radio spectrum. Like land, they just are not making any more these days. Obviously, a lot of uses and reuses and reallocations have to be thought up, fought over, negotiated, and perhaps even bought and sold somehow (the U.S. government's latest way to make money).

If you are into lobbying or just enjoy the intrigue of geopolitical and economic issues, keep an eye on what the FCC and similar government agencies around the world are thinking about doing with radio spectrum. While you are monitoring the action there, keep in mind the points made throughout this book about certification of wireless devices and complying with radio regulations. Become familiar with the rules and regulations presented in Chapter 1. If we think we deserve more radio spectrum for wireless networking, the chances of getting it will be in our favor if we can and will stay within the laws that govern what we do now.

Going Beyond Current Wireless Networking Standards

IEEE 802.11b and then 802.11a differed in both RF spectrum and modulation technologies, but share wired equivalent privacy (WEP) encryption and support for other security or privacy methods. Pending IEEE standard 802.11g brings some of 11a's technology to 2.4 GHz devices for higher throughput. Pending IEEE standard 802.11i will bring a greater level of security to all three technologies—11a, 11b, and 11g.

Typically, for business reasons—that is, the vendors, dealers, and retailers need to make money—many manufacturers are not waiting for the standards committees to release the standards before making and selling new devices with the capabilities they expect to be approved.

Could there be problems for users of equipment with the new technologies before the standards are final? Yes!

Fortunately, the problems of prestandards wireless technology may not be as significant as the differences and incompatibilities between the separate, competing v.90 and X2 modem technologies of a few years ago, when Internet service providers (ISPs) and users had to pick and choose which one they were going to support, if not both.

At best, we can hope that there will be no technology changes between prestandards-release products and poststandards-release products. Vendors feel that the pending issues are informalities, not technologies. The worst case is that you end up buying a set of prestandards gear that will only work with itself and current equipment, but not with poststandards-release equipment, which is fine if you are not building or expanding a huge network with a lot of users.

Between best- and worst-case scenarios may be that equipment vendors release new firmware for you to upload into your equipment to bring it up to date. This is not an unusual circumstance, as the release of nearly every piece of computer equipment sold is followed by at least two to three updates of firmware or driver software to fix a bug or add an incremental feature. Certainly corporations looking to invest in several pieces of wireless equipment may wish to wait until some technologies have stabilized before purchasing and deploying, to avoid the expense and hassle of updating dozens of access points and client adapters.

802.11g—Higher Speed at 2.4 GHz

The IEEE 802.11g standard will not be released until the spring or summer of 2003 at the earliest. When adopted and released, it will provide for new interradio operating modes and bit-rate (transfer speed) throughput improvements, while integrating four different wireless standards. Though the standard is not yet released, many chip manufacturers feel that the technical issues are solid enough to have made and sold new chips implementing the technologies, and WLAN equipment makers are already making enhanced client and access point products with those chips.

You will see these devices advertising 22 Mbps and possibly 54 Mbps, but none can legitimately claim compatibility with 802.11g until the standard is approved. These throughput levels would be meaningless and a waste unless the wired network behind them is

100BaseT or Gigabit Ethernet, so now wireless portable computers can begin to feel more like their hard-wired counterparts on current local area networks (LANs).

802.11g will remain compatible with 802.11b by keeping functional support for 802.11b's complementary code keying (CCK) for bit transfer rates of 5.5 and 11 Mbps. 802.11g adds orthogonal frequency division multiplexing (OFDM), as used in 802.11a devices, to deliver 54 Mbps speeds in the 2.4 GHz range.

802.11g also comes with two new modes that can provide throughput up to 22 Mbps. Intersil's 802.11g chipset will use a combined CCK-OFDM mode for throughput of 33 Mbps. Texas Instrument's chipset uses a packet binary convolutional coding (PBCC-22) mode for a variable throughput from 6 to 54 Mbps. Other chip vendors may have one or the other or both technologies in them.

While 802.11g is not expected to provide any improvements to range of coverage, testing has shown it to maintain connectivity at the same range or slightly better range than 802.11b; however, 802.11b may still transfer data faster than 802.11g at the far end of the signal range.

For those of us with smaller wireless local area networks (WLANs), say 10 or fewer users in a modest office space, 802.11g's higher throughput will probably be very beneficial. If your WLAN initiative needs to support a lot of users, it is important to consider that 802.11b and g can only negotiate between three available noninterfering channels to minimize interference and maximize throughput. 802.11a (5 GHz) chips, which use ODFM techniques, can handle more available carriers within a channel, which means more users can use the WLAN with less chance of interference. As the WLAN environment gets busier, and especially in enterprises with several WLAN users in a small area, 802.11g devices would not be able to maintain as much effective throughput as 802.11a devices, even though they both use ODFM.

802.11i—Enhanced Security

The IEEE 802.11i standard defines enhancements for the current wired equivalent privacy (WEP), a relatively weak, static encryption key form of data security for wireless devices. Robust security is one thing current wireless LAN products lack. Numerous articles have revealed the results of research into the weakness of the WEP method currently available in most wireless products, and how to crack the 64- and 128-bit encryption keys. Given enough data over time, it is possible for hackers to decipher encrypted data over wireless networks.

Regardless of WEP, many corporations have chosen to deploy third-party security products to tighten up their networks, rather than use one or the other more readily available security features of their network operating systems. For home users, wireless Internet service providers (WISPs), coffee shops, and other "mere mortals" who may not have servers or want to manage them, there is no economical or built-in alternative to weak WEP. The 802.11i standard and its implementation in upcoming wireless products will help solve this problem.

IEEE 802.11i implementations will use IEEE 802.1x standards and stronger encryption. One such technique is advanced encryption standard (AES; http://csrc.nist.gov/encryption/aes), a Federal Information Processing Standard (FIPS) that specifies a cryptographic algorithm for use by U.S. government organizations to protect unclassified information.

Fortunately, taking advantage of 802.11i itself should not require equipment changes. Upgrades to existing access points may be available from your equipment vendor. However, using AES may require new equipment. Some vendors are set to begin implementing 802.11x-like security through an industry-initiated WiFi protected access (WPA) method in early 2003. WPA is essentially 802.11x with a new temporal key integrity protocol (TKIP), but without the AES.

TKIP starts with a 128-bit temporal (temporary) key value that is shared between clients and access points. The key is combined with the device's media access control (MAC) address. Then a large 16-octet value is added, creating a unique encryption key for each device to be used for further communications. TKIP uses the same RC4 method as WEP to provide the encryption.

For home users, or in situations that do not provide a security server as the back-end provider for 802.11i methods, WPA provides a pre-shared key (PSK) mode that uses a single master key that may be manually entered into the access point and client systems. Check your wireless equipment vendor's Web site for information about firmware or driver updates.

802.1x—A Security Standard for All Networks

The use of IEEE 802.1x is a pending industry standard that specifies an access point-based means to communicate dynamic encryption keys to clients, and can be used whether or not WEP is used. The IEEE has given 802.1x the title of "Port Based Network Access Control," meaning that transmission control protocol (TCP) and user datagram protocol (UDP) ports are not open to pass data until the authentication process has succeeded. While 802.1x is not part of the 802.11 standard, the 802.1x is suggested to be part of 802.11i and the 802.11 standard. It is already implemented in Windows XP and many access points. A variety of vendors offer dynamic key management using 802.1x.

802.1x does not provide the authentication methods. You still need to implement an extensible authentication protocol (EAP) such as transport layer security (EAP-TLS) or EAP tunneled transport layer security (EAP-TTLS), which defines the authentication. Since the access point is a medium to pass 802.1x traffic, you can choose the EAP at the operating system, server, and client level of your choice without having to change equipment. The authentication may then be RADIUS or whichever method is used by your network's operating system(s).

Security is further increased with 802.1x because the client has the ability to change encryption keys periodically, thus reducing the time available for hackers to decipher the keys and reducing the vulnerability of the communications.

Summary

Why the emphasis on radio amid the discussion of new and emerging technologies for computer networking? Because, even though we have made tremendous advances in data compression and in application development to limit the amount of data this needs to move between systems, those new and emerging technologies will want more and more of the limited radio resources.

Until we have super-fast multigigabit data transfer capabilities and *huge* disk drives on which we could store "the whole Internet" for ourselves, and smart algorithms to transfer to us only the parts that change—the billions of parts of it that change daily—we will continue to want to move incredible amounts of data around. The Internet is just one segment of all of the data in the world so far. Businesses and governments transfer and use probably 2 to 10 times more data than is on the whole Internet.

It is very important to understand that, as ubiquitous as wired networking is, as the concept of networking itself is, wireless networking thrusts us into a new realm of resources and considerations, along with thousands of others interested in sharing a resource we are newcomers to—radio. Fortunately for us—the consumer at least, but manufacturers and service providers as well—it is in the interest of governments, emergency responders, and even more, consumers as yet untouched by computing and networking, to find and deliver ways to get more data to more people faster by wireless means. Still, we cannot be arrogant about our new-found value and the desire for the things we have. We are not unique or alone. We must cooperate with everyone else who uses the radio spectrum.

Of course with more users, more uses, and more data, the issue of exposure, vulnerability, and who gets to see and use which data becomes more important. Enhanced data security is an obvious, existing, and parallel concern. While most of the world enjoys freedom of speech and the sanctity of individuals, some parts of the world do not. Questions of who is allowed to communicate, and what they are allowed to communicate, are crucial in some corners of the world.

It seems we only want security for the things we evaluate as good or benign, and want no security at all for the things that are perceived as bad. But technology does not know the difference. It does not have a value system, a context, or set of rules to go by. So far, most of us seem to be reasonable people, and we will work these things out. Meanwhile, it is good to know that we can communicate, and can or will be able to do so securely, with relative ease.

CHAPTER 16

Installing Antennas

If you are going to install an antenna outdoors, at home, on a commercial building, or at a commercial tower site, you will probably want it to stay up there for awhile, not rust, be presentable and acceptable to the landlord or site owner, keep water out of the electrical connections and coaxial cable, keep water out of the building and equipment, protect it from lightning, and generally work well for you.

Those are what most people think about first when putting up antennas. They're wrong! We will cover those and more in this chapter, but first, the number one concern when working with antennas is *safety* (see Figure 16.1).

Figure 16.1 The author and fellow climber Steve work together to install a new multiantenna bracket. Cooperation and teamwork is a must on the tower and between tower and ground crews. Safety equipment and procedures are the highest priority.

Be Safe!

None of what we do with wireless networking or other personal or work projects is worth dying, getting injured, or damaging or losing equipment. OK, you are sitting at home comfortably thinking you are going to "whip up" a quick antenna mount to share your wireless local area network (WLAN) with a neighbor and this death thing comes up suddenly. You wonder, why is he telling me this?

Stupid things happen!

Working above ground level, and sometimes at ground level, can be hazardous—so hazardous that the Occupational Safety and Health Administration (OSHA) requires specific awareness, training, and in some cases, safety measures for anyone working on elevated platforms—from ladders up to 2,000 foot radio towers. The rules are not as applicable or stringent at home. After all, we routinely grab a ladder to clean out the rain gutters, paint, change light bulbs, etc.

Going up on the roof gets a little more serious. Most are not flat; it could be slimy or slippery from moss, algae, or moisture; the grit in composition shingles does come loose; wood shake is brittle and crumbles; clay and tile offer no slip protection; and asphalt and gravel flat-tops are sticky and flammable. So, working at heights is not something anyone should take lightly.

Below are some things you can do while you work on your project to really mess yourself or your equipment up and do a really lousy job of looking after your safety and that of others:

- Wear baggy, loose-fitting clothing.
- Wear lots of metal jewelry, especially dangling chains around your neck, metal bracelets, and lots of metal rings on your fingers.
- Wear loose-fitting open-toe sandals.
- Work alone.
- Use a wobbly, old broken ladder.
- Always stand on the top step of a ladder.
- Keep the ladder as vertical as possible so that it can tilt back and fall over easily.
- Work near and grab power lines and other wires.
- Work in the wind and rain at night.
- Ignore and throw away all safety information.

Please *do not* do any of those things—you know better, you should, or you will. Let's review a list of some of the proper considerations and practices you should follow:

- Wear clothing that is tucked in, rolled up, and otherwise not going to get caught on anything.
- Wear *no* jewelry—at least not around your neck, arms, hands, or fingers.
- Wear hard-soled, closed-toe shoes with some grip or grid on the bottom.

- Work with at least one friend.
- Use a ladder in good condition.
- Never stand on the top step of a ladder.
- Keep the ladder at a proper tilt. Stand up straight at the base of the ladder, stretch your arm out, and grab a rung at shoulder level. The angle should match this posture and arm position. Adjust the angle so your arm stays straight and you can easily reach the rung. This is the most comfortable, balanced, and safe climbing position.
- Stay away from and never grab power lines and other wires.
- Work only on calm, dry days.
- Read and heed all safety information.

These are the basics of common sense, with safety in mind. Think of anything and everything you will have to do to get where you are going, stay there and work for awhile, have adequate room away from hazards like power lines, be patient, and consider what could go wrong first—then avoid it.

Once you are mentally prepared to work safely, you have to consider the safety aspects of your installation and wiring. Chances are, you will be mounting your antenna on a metal pipe, er, mast and that mast will be attached to the side of a wooden part of the structure, on a newly installed tripod attached to the roof, or the side arm or leg of an existing radio tower.

The key element here is that you will be using a metal pipe, typically 5, 10, or 20 feet long. Where you place that pipe and as you move it around should be at *least* the full length of the pipe away from any and all electrical wires. It should also not be in a location where it could fall onto any electrical wires below. That little yellow and red warning sticker on many mast pipes and antennas is there for a reason. Even skilled and experienced antenna installers have suffered electrical shock, falls, or death from coming in contact directly or indirectly with electrical lines. Power lines are obvious, but even phone, TV cable, and other lines are susceptible to static and lightning, and should be avoided. Yes, you are going to be putting in new electrical cabling of your own, but you will of course dress that properly, out of the way of other objects and wires.

If you are working on a rooftop or at an existing communications tower, you will probably be near other antennas and cabling. Those antennas will have radio frequency (RF) energy applied to them. You should be aware of and heed any RF safety restrictions posted at the site, or find out about them from the site owner or communications company that services the equipment.

The most common use for antennas at most commercial radio sites and atop urban buildings is high-powered radio paging. It is not uncommon for these antennas to be fed with 250 to 330 watts of RF power at 900 MHz. This is definitely an unsafe power level to work near, even if you are just passing by to get to another spot on the roof or tower. Rooftop owners should ensure that these antennas are placed far enough away or high enough from where workers will be. When working at a commercial tower site and climbing near or past these antennas you are within your legal rights to reduce or disable the transmitter to provide for safety—but do so only after contacting the transmitter's owner or service shop.

Certainly do not grab onto antennas, or climb in front of microwave dishes. Unless you know for certain that a particular transmitter is off-line and the antenna is not radiating, consider the RF signal hot. Many of these systems are running on high power and provide high gain, so there is a serious concentration of unsafe RF around most antennas and in front of microwave dishes (see Figure 16.2).

If there is a television or FM broadcast station transmitting nearby, those stations may be required to reduce power while work is being done. In one case, I was working on a tower at a TV station running only 500,000 watts to an antenna mounted some distance away. Its radiation pattern was directed away from where I was working and the RF level was safe. The next time I returned to the site, a new tower and antenna had been installed a further distance away, placing the tower I was to work on in the radiation pattern of a new 1.5 megawatt transmitter. It was no longer safe to climb this tower without having the new transmitter's power reduced. Pay attention and know where you are, what has changed, and the rules!

Anywhere you are working, and especially when you are climbing towers, special safety gear is *required*. On rooftops, the parapet or ledge must be at least 42 inches tall to provide a barrier to falling over; otherwise a fall protection harness must be used. When climbing anything that places your feet over 6 feet above ground (or floor or roof) level, you are required to wear and use fall protection devices.

Figure 16.2 A typical large microwave relay and communications tower with high RF fields radiating in all directions.

For most of us, this means an OSHA-approved full body harness, a fall protection strap with shock absorber, and positioning lanyards to secure us while working in one place, and of course a safety helmet, with chin strap to keep it with you. This precludes the use of recreational climbing equipment from the local sporting goods store, including those wonderful colored aluminum carabiners. Recreational equipment is not OSHA or American National Standards Institute (ANSI) certified and does not have adequate load ratings. Save the carabiners and nylon straps for equipment bags, but not for use as personal protection (see Figure 16.3).

Figure 16.3
The author strapped in while climbing a tower at 200 feet above the Sierra Nevada. Many towers do not have ladders to climb, so you have to have excellent "monkey bars" skills to get anywhere. (Do not try this at home. Hard hat removed for clarity. Professional climber on closed tower.)

Do not free climb! OSHA and common sense dictates that you must have two points of secure protection at all times—except while transitioning from one secure point to another—and then one protection point must always be attached. Your safety equipment and those attachment points must be rated for 5,000 pounds of load. If you are going to rescue an injured or trapped climber, the protection points and rescue gear must be able to handle 10,000 pounds. This is serious stuff!

The equipment is not all that is required to be certified as safe—climbers must be OSHA certified for communications tower (or equivalent elevated platform) work—easy enough to do if you can find someone to certify you. Further, most communications sites require liability insurance to cover any loss, damage, or injuries that may result from your actions or inactions. Accidents happen, but someone does have to pay for them. Also, your personal medical insurance may not cover you if you are injured in doing this type of work. It also pays to be in reasonably good physical condition (treadmills, stair climbers, and pull-ups are good practice for climbing) and learn to pace yourself for climbing work. Climbing itself is only half the job. You still have work to do when you get up there. Heavy sunscreen and ample hydration are also highly recommended.

If you will be climbing a tower or working on a rooftop, chances are you will not be carrying all of your tools and equipment with you, but will have them hauled up on load lines, preferably through a pulley, with someone on the ground handling the weight. This puts them at risk because they are working below you and with things that are moving above them. A safety helmet is required. Securing loads adequately and *without* any fancy knots or wrappings is a must. When the ground crew is not actively helping to do work, they should step away from the tower, outside of the "drop zone"—an area around and below the tower where things are likely to fall, accounting for wind as well.

Boy Scouts and Mariners Need Not Apply

There is nothing worse than being the climber on the tower and having an antenna hauled up to you that has so many trick knots and loops around it that you cannot safely get the equipment untied and mounted. If you see bits of rope left on antennas and mounting hardware on a tower, that is a clear indication that someone screwed up. Simple loops at the top and bottom of piping and antennas are recommended, as are clips or "beaners" to attach tool bags to hauling ropes.

Ground crew helpers need to think a little differently than when tying a Christmas tree to the roof of the family sedan—you are tying for someone else. The climber and helper(s) should work out these techniques on the ground in advance. Plan the job and what will happen when. Plans are subject to change as the climber advances up a tower, checks the terrain around him, and discovers wind or other issues that have to be worked around. Work together to consider possible alternative plans in advance. Positively, fully, and adequately communicate anything and everything that is or will happen with positive acknowledgment on both ends. It's lonely on the tower, and the climber is almost totally dependent upon helpers to be able to work efficiently and safely. If you do not communicate fully, the wrong things will happen. Consider using high-quality two-way radios (and not the dime-store FRS radio either; the RF signals at most communications sites will render cheap radios unusable) to coordinate efforts beyond shouting distance.

Last but not least—what goes up must come down—and eventually does. The following picture (Figure 16.4) is the aftermath of a fallen tower, one I have climbed and worked on. (No, the damage is not my fault.) Standing about 15 years or more, this tower gave in to 60+ MPH winds during a late winter storm in Northern California. This tower looked and was safe for the most part and was not suspected of suffering damage in the winds it is normally exposed to.

Figure 16.4 High winds broke this communications tower in half. The structure at the right is a temporary tower erected to maintain communications until repairs to or replacement of the existing tower could be done.

Aside from age or rust, the contributing factor to the demise of such towers, normally able to withstand 120 MPH or higher winds for short periods of time, is the amount of stuff mounted above critical structural points. The tower itself presents a wind load factor—that means stress applied laterally to the structure by wind pushing on brackets, pipes, cross-braces, etc. Anything added to a tower presents more wind load. Most cylindrical and exposed dipole antennas add relatively little wind load by themselves, but several of them begin to add up. Panel antennas and dishes add 2 to 10 times the amount of wind load of other conventional antennas. Tower site owners and installers must be very careful to balance wind load factors versus antenna placement. Of course everyone wants their antenna at the top of the tower, but that may not be safe, prudent, or reasonable, due to wind load or physical mounting issues.

Common sense must rule in the absence of anything else. Look around. Listen carefully. Be aware. Pay attention. Think about every action and potential reaction. If something looks, feels, or sounds like it might break loose and fall, it probably will. Stay away from trouble spots and *be safe*!

Materials and Techniques

The number two concern about antennas is doing it right. Why bother if it is not going to work and last longer than a day, a week, or a month. A good friend provided an excellent motto for this and other projects: *The price of quality only hurts once.*

Potato chip can antennas do not survive in the rain and winds.

Anything left outdoors for any period of time is going to be vulnerable to and suffer from the elements—wind, rain, dust, sunlight, salty air, perhaps even snow and ice. Few of us can afford the money to replace and the time to reinstall damaged antennas or feedline.

Face it, an antenna is not like your keyboard, mouse, monitor, computer, or router. Once an antenna is installed, it is often forgotten—as it should be, if you used the right materials and installed it all properly. Select the best material you can find for the job at hand, and for as long as you expect it to last. In some cases, select even better materials to give yourself a margin of safety and longevity. Think about what an antenna system is and what it must endure throughout its expected lifetime.

An antenna system is made up of several pieces of hardware—bits of metal, plastic, cable, nuts, bolts, cable ties, tape, and connectors. Most of these pieces are left outdoors to the whim of the elements. The hardware itself is not considered visually appealing or suitable to most people. Any visual appeal or tolerance diminishes quickly because of the elements, as may the performance, strength, or safety of what you have installed. Corrosion and wind are your antenna system's two worst enemies. Corrosion plus wind makes for an unsafe system.

The Proper Tools and Supplies

The materials you choose and how you install them can minimize the effects of either and make for a longer lasting, safer system. The right tools and supplies make the installation go smoother, and make it more secure and water-resistant, if not waterproof. The following is a list of supplies that should be in your kit of items for antenna installations and repairs:

- Assorted combination wrenches with both open and box ends
- Heavy duty wire cutters
- Utility knife
- 3/16- to 1/4-inch wide black cable ties (not white, clear, or colored)
- 3M Scotch #33 or #88 electrical tape
- 3M Scotch #130 splicing tape
- Spray can of cold-galvanizing paint
- Small- to medium-sized wire brush
- Power drill motor and assorted drill bits

Mast and Antenna Installation Materials

If you have an existing TV or other antenna mast on your roof that you can install your wireless antenna on, check it for damage, rust, and secure fastening. And if guyed, make sure the guy wires and clamps are in good, nonrusted condition. Sometimes replacing what is there is of benefit to everyone. Rusty bolts, pipes, wires, and clamps are unsafe, insecure, and are possible sources of RF noise and interference. If the existing items are in relatively good condition, then wire-brush and overspray all of the joining parts, clamps, and bolts with cold-galvanizing spray paint to protect and make them last longer. Make this a routine.

If you are simply going to install a small omnidirectional antenna for local use, you can probably get by with a set of small clamps to fasten the antenna to a vent pipe or a chimney mount kit (if allowed by local ordinance). If you need to elevate your antenna well above roof level, I do not recommend chimney mount kits and certainly not strapping a mast to a vent pipe. Chimneys are not designed or intended for additional lateral loads, and a mast and antenna can add considerable side load and leverage to them. Vent pipes are not well secured in the walls, and are usually not thick enough to handle any additional loading.

With this in mind, you are limited to using roof-cap mast base plates and guying (tying off) your mast or installing a tripod atop the roof. You can find adequate materials at local electronics outlets and hardware stores, though I prefer to use galvanized steel pipe or heavy duty mast material in place of thin-wall painted steel "TV mast" sections. The latter crimp, crush, bend, and rust quite easily. Yagi and dish type antennas present a higher wind load than omnidirectional antennas, so your choice of materials has to account for this. Antennas for 2.4 and 5 GHz are much smaller than most oddball-looking TV antennas, but we would like the installation to reflect that we know what we are doing—quality, long-lasting workmanship using good materials. With that in mind, the following materials are typical and recommended for good long-term home rooftop installations:

- Thick-wall galvanized steel pipe, 1-1/4 to 1-1/2 inch diameter, water pipe or heavy conduit, but *not* electrical metallic tubing
- Heavy-duty tripod for pitched roof mountings
  - 3-foot model for 5- to 10-foot mast pipes; 5 feet tall for 20 foot pipes
  - 3-foot model fine for 20-foot mast pipes if you add guy wires
- Tilt-over roof anchor plate for mast-only (no tripod) installations; mast to be guyed
- 1/8- to 3/16-inch galvanized steel guy wire (for a 20 foot mast with three guy wires, you need 100 feet)
- Guy wire anchor hooks—to secure the wires from the mast to the roof
- 8 foot long 1/2 inch diameter copper ground rod
- #6 to #10 stranded copper wire (green insulation preferred)
- Assorted grounding clamps to suit the mast size and ground rod
- Galvanized or rust-resistant clamps, nuts, washers, and bolts

- 5/16-inch, 5- to 6-inch long lag bolts to secure base to roof, or 10-inch long carriage bolts for through-roof to backing plates
- 2×4 lumber stock to use as back-plating for fastening the tripod or base plate to the roof
- Roofing caulk to seal holes and apply under mounting plates as they are set

The rooftops of commercial and nonresidential buildings may require significantly different materials. For instance, it is not uncommon to anchor a tripod to a set of 2×6 or 2×8 boards and hold it onto the rooftop with cinder blocks. The parapet or ledge of commercial buildings or the side wall of an elevator penthouse will likely require special brackets and concrete anchors to adequately secure a shorter mast.

Commercial radio towers come in all sizes and shapes—some with 1 inch round legs, some with 1- to 4-inch angled steel pieces, and large hilltop towers designed to carry several microwave dishes have 4- to 10-inch diameter legs at the bottom sections, scaling back to 2- to 4-inch diameter legs at the top sections. You are typically required to use heavy-gauge galvanized steel hardware intended for communications towers. You will not find suitable mast or antenna mounting hard for commercial towers at local hardware stores. For these items, consult a professional communications or tower facility to locate a vendor for commercial brackets and clamps.

In any case, where a metal item extends above the rooftop, there is the potential for lightning discharges, thus the ground rod, #6–#10 wire and clamps. For homes and other low level (one- or two-story) buildings, you should run your own ground wire and drive a ground rod to bleed off any static and avoid lightning strikes. If a ground rod cannot be driven in or the roof is higher than three stories, you will probably be able to find a common safety ground point or cold water pipe to secure the ground wire.

Grounding is important not only for lightning protection, but also to help reduce the overall RF noise level at locations with several radio systems. Grounding also bleeds off any static or induced electrical currents that could cause injury to workers or perching birds.

Good Neighbor Policy and Local Regulations

Part of preparing for the installation of any antenna is figuring out where to put it. Many homeowners association policies and local ordinances prohibit the installation of antennas of any type, anywhere on your home or property. Others limit the height or placement of antennas to minimize their apparent visual impact on the local surroundings and architecture.

Barring local restrictions, it is best to follow a "good neighbor policy" and voluntarily locate your antenna where it will have minimum visual impact on your neighbors. After all, what looks good and works well for you may not appeal to the grouch next door, or those who determine they are suddenly sensitive to or adversely affected by a few microwatts of RF signal. Most homeowners have little choice. The ridge of the roof runs sideways, parallel to the street, yielding maximum visual exposure. In this case, you have to figure out which neighbor will be bothered least looking out the window—the one whose kitchen window would be near the antenna or the other one whose bedroom window (typically with curtains closed) would be closest.

If restrictions prohibit you from installing an antenna so that it can be seen from the street, you will have to determine a mounting position and method behind the peak of the roof line. If you are lucky, you may be able to tuck the antenna in behind the chimney for maximum height and discretion.

Best Practices and Techniques

If you are able to and have decided to use a tripod or tilt-over roof plate to mount the antenna onto, you must find a location to place the mounting surfaces directly above rafters and beam so that lag bolts have something to bite into to be effective. If you cannot find such a place easily, or you prefer a slightly more secure fastening method, you can find a place to set the mounting feet between or on either side of rafters or beams, and use a backing plate inside the attic area to span across sets of rafter. Either method suggests that you survey the inside of your attic (the bottom side of your roof) for obstructions, electrical wiring, or anything else that might interfere with your antenna mounting.

You may want to begin working from within your attic anyway to place and drill at least rafter-locating pilot holes from the inside out for better placement accuracy. Use a 1/8-inch drill bit for the pilot holes—no sense in making your roof look and leak like Swiss cheese. Remember, any hole you drill should be closed up and sealed with roofing caulk as soon as possible to avoid water damage. If it's not obvious yet, this type of work is best done with two people working together, one inside and one outside to coordinate mounting alignment, etc.

After you have selected and prepared your mounting method of choice, set the mount in place. Check the alignment and adjust as necessary. For tripod mounts, you should set the mast in place and check to be sure it is level in all directions before finalizing the mounting location. Drill the holes for the lag or carriage bolts and set the mounting in place. When you are ready to secure the mounting, first lift up the mounting plate or foot and liberally apply roofing caulk to the roof where the mount will set down. This will help seal the hole and the area around it to prevent water leakage. Finally, set the bolts, tighten, and check for security. Even though most of the weight will initially be downward, the mounting should be secure from lateral movement due to wind, and to make sure water cannot seep in between the plate/foot and the roofing material.

With the mounting set in place and secured, you are ready to set the mast and antenna. Unless you are 7 or 8 feet tall and able to reach the top of the mast to install and aim the antenna if necessary, you should mount the antenna onto the mast first before setting the mast into the mounting.

In very rare cases, such as the installation of my Sprint Broadband wireless service "pizza box" antenna, my location and nearby trees required that Sprint's technicians install a 35 foot push-up mast atop my roof. The order of installation was the same for the mounting plate; then the mast was put up and guyed at the first 10 foot level. Once the mast was securely guyed, a ladder was brought up and placed against the mast to allow a technician to climb up, attach the antenna and feedline to the top section of the mast, and then attach guy wires to the three remaining push-up sections. The top/smallest section was raised first and locked into place; then the second section was pushed up and locked, and finally the third. As sections were pushed up, the feedline was secured to the mast with cable ties every 12 to 18 inches. Aiming the antenna in the direction of its main tower was done by turning the raised mast sections carefully. If you can picture this event in your mind, yes, it was as risky as it looks. I am glad I was just watching, and now I know how to take this assembly down when the time comes.

Moisture is an enemy of all things electrical, and especially RF signals. A water-tight seal of all connections is a must for a trouble-free installation. Part of the task of attaching the antenna to the mast is to seal the connection of the feedline cable at the antenna. If you have a Yagi-type antenna with an end-fed connection, you can seal the connectors with tape (Figures 16.5 and 16.6) or use a sealing boot (Figure 16.7).

Figure 16.5
Apply a layer of high-quality electrical tape (3M Scotch #33 or #88), fully covering the connector to the end of the threads and beyond. This layer keeps the connector clean and prevents contamination and 'gunk' from the next layer. Slicing and peeling off the sealing layers to service the connection is also much easier.

If you are using a Yagi antenna that is side-fed parallel to the boom of the antenna, it will not be easy to use tape, so a sealing boot is required. Most sealing boots are made of a heat-shrinkable tubing filled with an electrically safe moisture-proof caulking material. Sealing the boot requires the use of a heat gun or propane torch to shrink the tubing so it adheres firmly around the connector and the caulking oozes out from the edges.

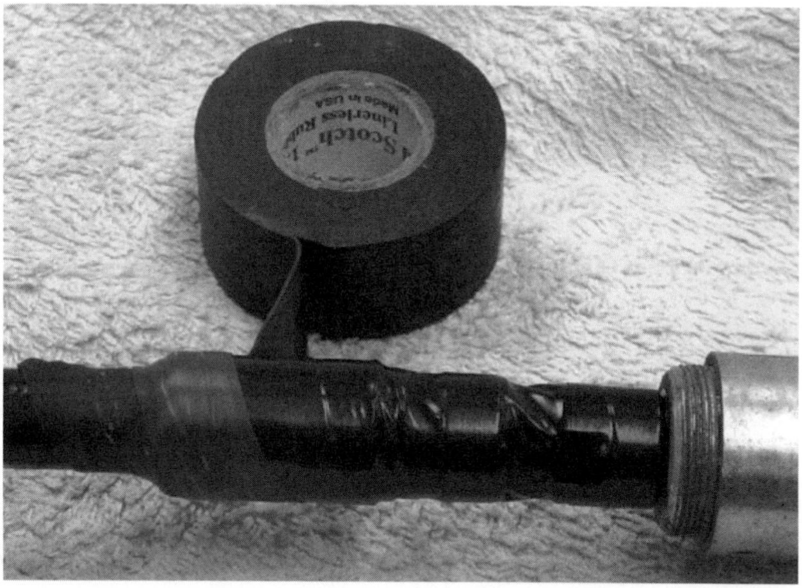

Figure 16.6
Apply a second layer of soft rubber splicing tape (3M Scotch #130) or coax-seal putty over the first layer of electrical tape. Overwrap the splicing tape or putty with a final third layer of Scotch #33 or #88 to protect the splicing tape or sealant.

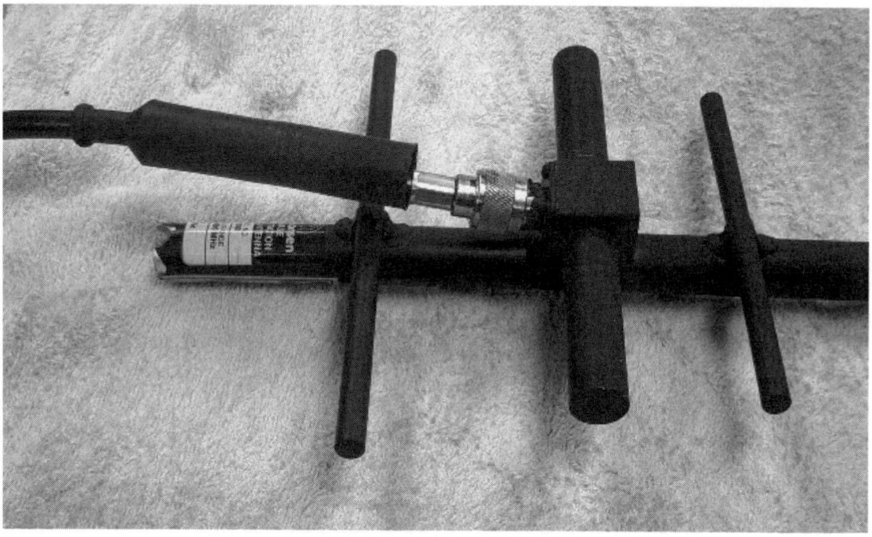

Figure 16.7
A heat-shrinkable sealing boot is recommended in all cases, but especially when it is not possible to properly wrap the connection with tape.

When you have the feedline connection firmly attached to the antenna and sealed tightly, the next step is obviously to attach the antenna to the top of the mast pipe. The instructions for an omnidirectional antenna should indicate the proper mounting dimensions. For a Yagi or dish type antenna, I follow a rule-of-thumb of mounting the antenna so its topmost part is 1 to 2 inches below the top end of the mast. Unless the antenna manufacturer's instructions say otherwise, this rule-of-thumb is intended to offer some protection to the antenna as the mast is positioned—you will hit another object with the tip of the mast before you damage the antenna—or so the theory goes. Another reason for doing this is that the end of the pipe is more likely to rust and weaken, so having the antenna clamped on some distance away from the end gives it a longer-lasting solid mounting position. You are going to this much work to place something out of reach and you probably do not want to have to do this part of the job over. To that end, as you attach the antenna and run the nuts and bolts together, *always* use locking washers at least behind the nuts; otherwise the vibration from the wind will loosen the clamping and the antenna will eventually wobble around.

With the antenna firmly attached to the mast, the next step is to secure the feedline along the mast. Before you do this, make a 6- to 10-inch diameter loop in the feedline about 6 to 12 inches from where it connects to the antenna, and fasten the loop to the mast. This gives you some extra feedline to work with if you have to move the antenna, and serves as a bit of decoupling so RF from the antenna does not radiate along the coax. There are more specific and deliberate methods to do this, but a loop never hurts.

If you have a 20-foot long mast, chances are you will have to provide guy wires to keep it from tilting and snapping over in the wind. I highly recommend this because although your seasonal high winds may be only 10 to 20 MPH, gusts of 40 to 60 are not unusual during storms, and that is enough to stress any material. Attach a guy-ring 1 to 2 feet below the bottom of the antenna and secure the guy wires to it. Guy wires should ideally extend out from the mast at a 45-degree angle for optimum protection, but as little as 20 to 30 degrees may be all you can accomplish, depending on the area of your roof.

You will need to allow 30 to 40 feet of guy wire times three or four guys for the wire to be able to reach and secure to the anchor points at the other end. Locate and set the anchor points in advance of raising the antenna. This will save a lot of time and effort later. The anchor points should be set into rafters or beams, not merely roofing material and backing board, as they will take a lot of lateral stress from winds.

Attach the feedline to the mast every 8 to 16 inches with one to two wraps of electrical tape or plastic cable ties. I indicated black cable ties in the materials list above because the black ones are typically ultraviolet resistant and will not break down and crack off over time. The white nylon and decorator colored ties will disintegrate more quickly in the sunlight. Leave 2 to 4 feet of the mast free so that you can position it into the mounting without the cable being in the way. You are now ready to "raise the mast" and get underway.

A 5- or 10-foot mast pipe is not that heavy and may be raised by one person, but a 20-foot pipe is a lot heavier, and you have the additional weight of the antenna and feedline to consider when you move it around. Get help if you need it! You will need help if you are using a tilt-over base plate and guy wires, as it is impossible (as far as I know) for one person to set and hold the mast upright and move around to set all the guy wires at their anchor points. So, set the mast into the tripod or base plate. If you are using a tripod, secure the mast in place with the anchor bolts provided and you are just about done. For tilt-over methods, ensure the mast is vertical with a level and do a final adjustment and setting of the guy wires.

Complete the fastening of the feedline to the mast, allowing some extra space to loop around the upper brackets of the tripod. And if you have a directional antenna, perform at least a visual alignment of it to the direction of interest. If there is an active access point already operating at the other end, you may be able to use your computer and the NetStumbler program to fine-tune the alignment for highest signal strength. If there is no active signal at the other end, you will probably be doing some trial-and-error realignments at each end to optimize the signal.

Your next step is to attach one end of your ground wire to the mounting base. Whether you then tape or cable-tie the ground wire and signal feedline together or not depends on the destination for each wire—something to consider in setting the ground rod in place. Complete the running of the wires to the edge of your roof, avoiding any places that may scrape the wire and open it up, and avoiding locations where the wire may get stepped on. Before going farther, play out some excess wire to form a downward loop at the edge or below the eaves or awning. This is a drip loop and will keep water from flowing along the remaining length of wires into your house or office. Bring the end of the loop up and fasten the wires under the eaves to complete the loop; then continue to run the wire to its destination. An additional drip loop may be necessary before the wire enters the structure if it is exposed to the elements at that point. Seal any holes with latex caulking and finish the feedline run to your wireless device, and the ground wiring to your ground rod or a cold water pipe.

Clean up your mess, throw away the debris, put your ladder and tools away, and enjoy your new stronger wireless signal!

A Few Final Hints

If you have a choice of materials when buying connectors, choose silver-plated ones. I prefer and would suggest gold-plated, and for some smaller connectors, you may not have a choice. But I don't know of any gold-plated large form connectors like the Type N. Nickel or brass connectors may look nice, but they do not weather well. Silver-plating oxidizes in a good way—silver-oxide is actually a better conductor. When copper, chrome, or brass weather, they tend to lose conductivity, especially when mated with dissimilar metals, and this can adversely affect your signal or create noise spots that allow for interference to your system or those of others.

If you are extra cautious or simply want to add value and longevity to your system, you may consider installing a lightning protector in your feedline at the point it enters or just after it has entered your structure. A lightning protector is a device that detects and reacts quickly to the presence of high voltage on the line and shorts it to ground, hopefully before damaging your equipment. You will need to attach a ground wire to this device for it to be most effective. A variety of protectors are available from most wireless equipment vendors like HyperLink Technologies, but if you are looking for a brand-name recommendation, one manufacturer preferred by most communications sites is Poly-Phaser (http://www.polyphaser.com). Lightning and overvoltage protectors are also available for Ethernet cabling and are highly recommended for power-over-Ethernet applications.

If lightning seems to be the least of your worries, take a look at Figure 16.8. This is a picture of what was a 6-inch diameter, 1/2-inch thick Teflon insulator used as the spacer/dielectric between the center conductor and outer shield pipe of a transmission line from a 1.5

megawatt UHF TV station in the Midwest. This and dozens of similar insulators were destroyed and had to be replaced after a freak severe ice and lightning storm in late spring 1976. The tower and transmitting antenna took a direct lightning hit during the early morning hours while the station was on-the-air. I had the pleasure of working near the tower, with large chunks of ice still falling off of it to repair damaged wiring to other facilities, while a climber began an all day and night journey up the tower to disassemble the 1200 feet of feedline and replace most of the insulators. Significant grounding and the arced-over Teflon is probably what saved the television transmitter and other equipment. This insulator is my reminder and "good luck charm" of what safe practices are all about.

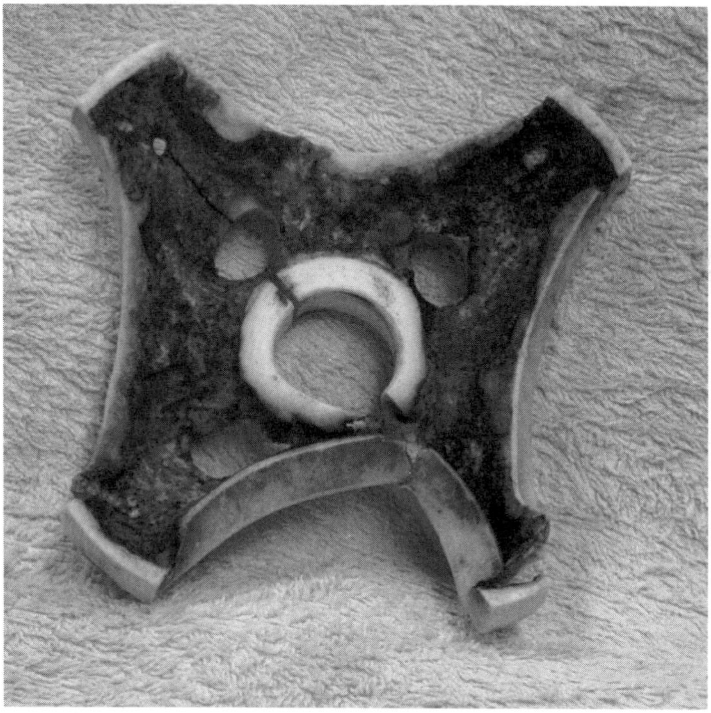

Figure 16.8 Lightning strikes contain enough energy to turn Teflon into charcoal. Teflon emits phosgene gas when burned, which can cause near immediate death if inhaled. (Absolutely do not try to replicate this with a torch!)

I mentioned cold-galvanizing spray paint earlier as one of the tools to keep handy. It is a good idea to spray any and all nuts, bolts, washers, threads, and metal joints, as well as the ends of mast pipes to reduce the chances of corrosion and possible sources of RF noise. Cold-galvanizing spray is a good preventive and remedial maintenance tool for any metal exposed to the elements—except antennas. Use it on new installations. Upon visiting existing installations, wire brush any metal surfaces that do not have spray on them, and especially those that are beginning to corrode; then apply spray. Do your neighbors a favor and clean and spray their hardware too. You will be sparing everyone potential damage and RF noise problems.

While you are at a communications site or on a rooftop full of antennas, be alert to any loose wires, corroded hardware, or just about anything that does not look right. Wires and hardware should be fastened securely and not allowed to flap in the breeze. Anything that is not right is a possible safety hazard and a potential source of interference. Contact the site owner or communications company and report any problems found. You may save someone a lot of money or someone's life.

Summary

The first and last message I have for you in this chapter is *be safe!*

APPENDIX A

Cable Connections

At some point in time, even wireless equipment must be connected to a wired network. To do so, you need either a straight-through or a crossed-over Ethernet cable. The pin connections for each are shown in Figures A.1 and A.2. Straight-through cables are used most frequently, and they interconnect workstations to hub equipment. Crossed-over cables are often used to interconnect two hubs or routers. Note that only pin pairs 1 and 2 and 3 and 6 are necessary.

Pin pairs 4 and 5 (blue/white), and 7 and 8 (brown/white) are available for other uses. These are the pins used, as shown in Table A.1, for power over Ethernet (POE) to supply low-voltage DC from internal power sources to outdoor mounted access points. An excellent how-to article on building your own POE interfaces is available at: http://www.nycwireless.net/poe/.

Because there is resistance and thus voltage drop along the thin wires, POE is probably best implemented with an unregulated power source indoors and a regulator at the endpoint, designed to feed the proper voltage to the access point or whatever is at the other end. If your access point requires 5 volts and that is what you apply indoors, you may find you have only 3 to 4 volts at the access point when it is connected—not enough to operate it properly, if at all. Some manufacturers and many aftermarket equipment suppliers provide specific adapters for specific equipment to be connected via POE.

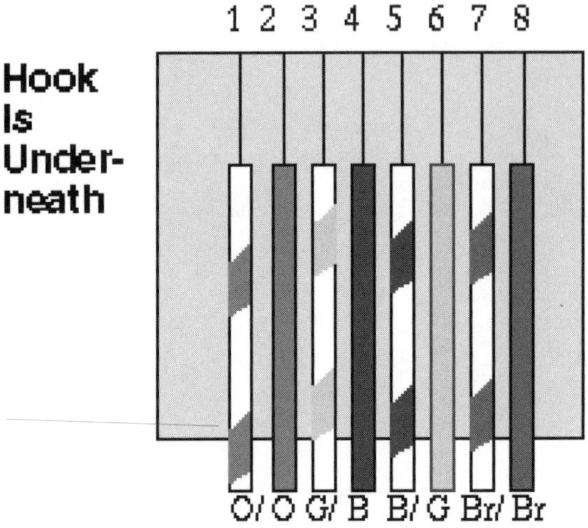

Figure A.1 The proper wire color to pin orientations for building straight-through Ethernet cables.

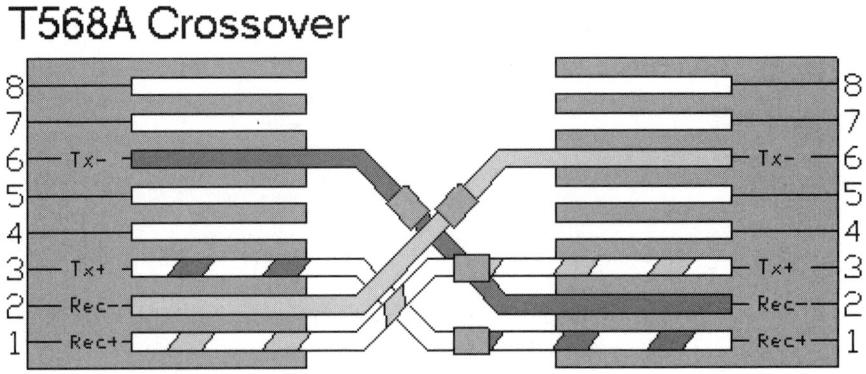

Figure A.2 The proper wire color to pin orientations for building crossed-over Ethernet cables.

TABLE A.1 Power-over-Ethernet DC Supply Wiring

DC Power Lead	Pins/Pair	Pair Color
Positive (+) DC Voltage	Pins 7 and 8	Brown/White
Negative (–) DC Voltage	Pins 4 and 5	Blue/White

APPENDIX B

Assembling RF Connectors

If you're going to "play with RF," sooner or later you're going to have to put a connector onto a piece of coaxial cable. To do that and maintain the correct RF properties of the connector-coax interface, and to ensure that the connector mates properly with the socket, you need to cut and trim the cable accurately and assemble the connector to the cable in a prescribed manner. For those of you who are proficient at wiring trays, cable harnesses, punch-down blocks for network jack panels, and making your own Ethernet cables, assembling RF connectors is not far from what you're familiar with, but the care needed is a notch above basic crimping and punching—and the good news is you can do this even if you're color-blind.

The graphics presented have been provided by Amphenol-Connex (www.amphenolconnex.com) and through the courtesy of Connex Electronics (www.connex-electronics.com), a distributor of their products. Connex is a world leader in connector technology. There are other manufacturers of cables and connectors but the standards and many of the first military-grade connector designations were done by Amphenol and its researchers first. You will likely be able to obtain these or similar connectors through online retailers such as HyperLink Technologies (www.hyperlinktech.com).

All connectors used for 802.11a and 802.11b service should be nickel, silver, or gold with Teflon insulating materials. Avoid low-cost, after-market connectors available with lesser materials or nylon insulators that are mechanically weak, melt at high soldering temperatures, or break down with exposure to the elements. These will degrade the performance of your system—"the price of quality on hurts once."

Certainly, if you are not mechanically inclined or do not want to invest the tools or time to assemble your own, then buying preassembled cables is your best option. Preassembled pigtails are highly recommended for systems that use the smaller MC-Card or MMCX connectors, as these require special tools and care to assemble.

Tools

To work with these and similar connectors you will need a few select tools and the dexterity to use them all properly. In most cases the following tools will be all that you need:

- A sharp razor or Xacto knife
- A pair of 4- to 6-inch needle-nosed pliers
- A pair of 4- to 5-inch fine diagonal or flush-cut wire cutters
- A larger 6- to 8-inch wire cutter for larger cables
- A small-medium 3/8- to 9/16-inch opening adjustable wrench
- A pair of combination or offset pliers to grasp larger connector bodies
- A ruler with fractional and decimal measurements
- A good quality soldering iron (not a gun-type or a tinner's iron)—I recommend a temperature-controlled Weller soldering station with both 700 to 800 degree Fahrenheit 1/16-inch pointed and 3/16-inch flat tips
- A high quality solder—minimally "60/40" but I prefer to use "63/37" mix solder for more reliable soldering, especially for those new to the craft
- A suitable crimping tool for crimp-on connectors

For crimp-type connectors, spending the money on the right crimping tool ($40–80 for generic crimpers, up to $400 or more for manufacturer's specific tools) is one of the best investments you will make towards proper connector work. Sometimes you can work around the absence of a crimping tool by soldering the center pin to the center conductor, or the ferrule to the connector body, making sure that solder flows to the shield as well, but soldering crimp connections is not recommended.

Soldering

Good soldering is a balance between the amount of heat available—a 700 or 800 degree temperature controlled tip is best—the size of the item to be heated, adequate contact to transfer the heat, and the timing of getting the temperature of the item high enough to melt and accept the solder without burning or melting the surrounding insulation. When soldering, make sure the tip of the iron is exceptionally clean, well-tinned, and cleaned again before making contact with the surface to be soldered. Then, when heating the item, make solid contact, get the piece hot as quickly as possible, apply solder, let it flow, and then remove the heat. I suggest practicing the tinning and soldering process with ordinary speaker cable and inexpensive lugs before moving to delicate RF connectors and smaller wires and pins.

If you want to become proficient at soldering, I suggest a careful review of the NASA workmanship standard #8739.3 available at http://workmanship.nasa.gov/ws_8739_3.jsp for tips and graphic examples of efficient reliable soldered connections.

Technique

Connector and cable manufacturers and technicians of various experience levels may provide a variety of methods for trimming cable ends and removing insulation. Your goal here is to make clean, unfrayed, straight cuts without making nicks or gouges in the center conductor of coaxial cables, or leave fragments of shielding foil or braid strands that could weaken or short the connections.

When performing the coax-to-connector assembly inspect the cable closely to ensure that bits of wire do not get in the way to potentially cause short circuits between the outer shielding and the inner signal conductor.

Note: *The two biggest problems with connections are opens and shorts—either can cause your transmitted signal to reflect back to the device, which may destroy the transmitter or an associated receiver circuit.*

For most cables and smaller connectors, start at the end of the cable and expose and clean up the center conductor wire first, working back to remove just insulation and shielding material, then outer insulation to the proper dimensions.

Before making any cuts, study the assembly instructions carefully. Often, it is easier to apply any backing nuts or crimp-on ferrule pieces to the cable first, then cut and apply the other pieces.

Note: *One of the most frequent mistakes made during connector assembly is discovered* after *you've assembled the connector to the cable—solder, crimping and all. You forgot to put the nut or back ferrule piece onto the cable* before *beginning the assembly process. Placing the ferrule onto the cable is actually* part of *the process (see Step 1). The problem isn't too much hassle if you have not assembled a connector to the other end of the cable—you can simply run the ferrule up from the other end of the cable. If the other end of the cable is already terminated, or buried deep in a wall someplace, you will have to un-do or cut off the connector. My high school electronics instructor taught me a simply profound way to prevent this from happening—*"always *put the nut or ferrule onto the cable* first *then you won't forget to do it later." If you remember that part you will save yourself a lot of time, rework, and connector expense.*

For the first cut, expose the center conductor only using a sharp knife blade to slice through the outer insulation, shield, and inside insulation in one smooth careful cut around the circumference of the cable.

To avoid knicks or gouges in the center conductor requires feeling the blade's movement into the material and stopping all pressure when the center conductor is reached. You will similarly feel your way into the shield and insulation sections. Check all dimensions and measure carefully. Expose more center conductor than is required and then cleanly trim it to the proper dimension by itself.

Next, depending on the connector and assembly instructions, slice off outer jacket and shielding to leave the inner dielectric exposed without shielding at the end. For many connectors, it is adequate to leave the shielding extending to the end of the center insulation and remove only the outer insulating jacket.

If soldering any of the pieces is required, you will probably do that step next and let the parts cool so that you can handle them for the final assembly according to the instructions.

For connectors that are to be soldered onto larger cables such as LMR-400 or 9913-type—such as N-type connectors and older PL-259 UHF connectors—the trimming steps and technique is a bit different. For these, a length of outer insulation only is removed from around the shielding material, then the shielding is tinned with solder to form a rigid material to work with. When this has cooled, the shield and inner conductor are trimmed off to expose a suitable length of center conductor material. If the center conductor is made of stranded wire, this is tinned with solder to make it solid. This preparation allows you to place the connector body onto the cable without fear of strands of shield or center wire shorting out to each other. It also makes it easier to complete the soldering of the connector body onto the cable. Again—do not forget to apply any back nuts or crimp ferrules onto the cable before final assembly of the connector body to the cable—50 or 100 feet is a long way to push these items along the cable from the other direction.

If you have a volt-ohm meter to test the connections with, set the meter to measure in the high ohms range—100,000 or 1,000,000 ohms—then test the newly assembled connector to be sure there are no shorts in the wiring. You should have no reading at all.

With connectors at both ends of the cable, test for shorts between center pin and shield then set the meter to measure a low resistance range of 1, 10, or 100. Test for continuity between the shield/connector bodies and then the center pins from end-to-end—there should be a low-resistance indication of less than 10 ohms telling you everything is correct.

It's OK to wiggle the connectors a little as you test them to make sure the connectors are firmly attached and do not have intermittent connections. A connector that is loose or otherwise mechanically unsound will cause you problems now or later.

If you have shorted connections between shield/body and center pin, you must fix these—typically by starting over with new connectors. If you have no connection between both shields/bodies, or no connection between center pins, you will have to start over as well.

The Connectors

Connex has provided us with diagrams for three common connector types—N, SMA, and TNC—in normal and reverse center pin varieties for small diameter and normal size cables. These diagrams should give you a very good representation of the dimensions, cutting techniques, and assembly order for most connectors and cables you will encounter.

Type N Plugs and Jacks

Type N plugs and jacks are identified by their larger exterior body and center pin size. Pin styles may be male or female on either end. Type N connectors are typically used on the larger diameter, low-loss LMR-400 and 9913F7 cables for long cable runs to antenna locations.

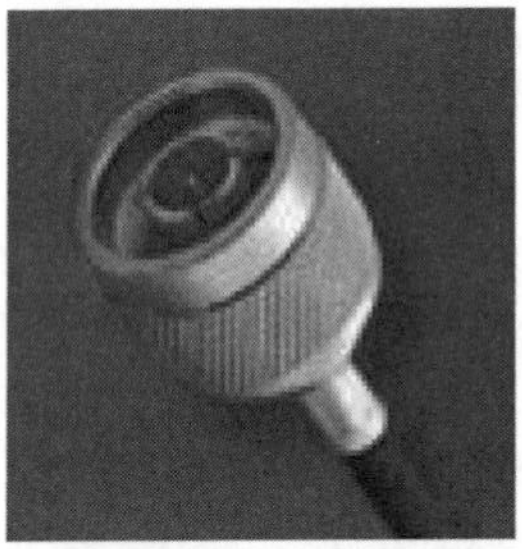

Type N—Crimp-on Plug for RG-58 Size Cables

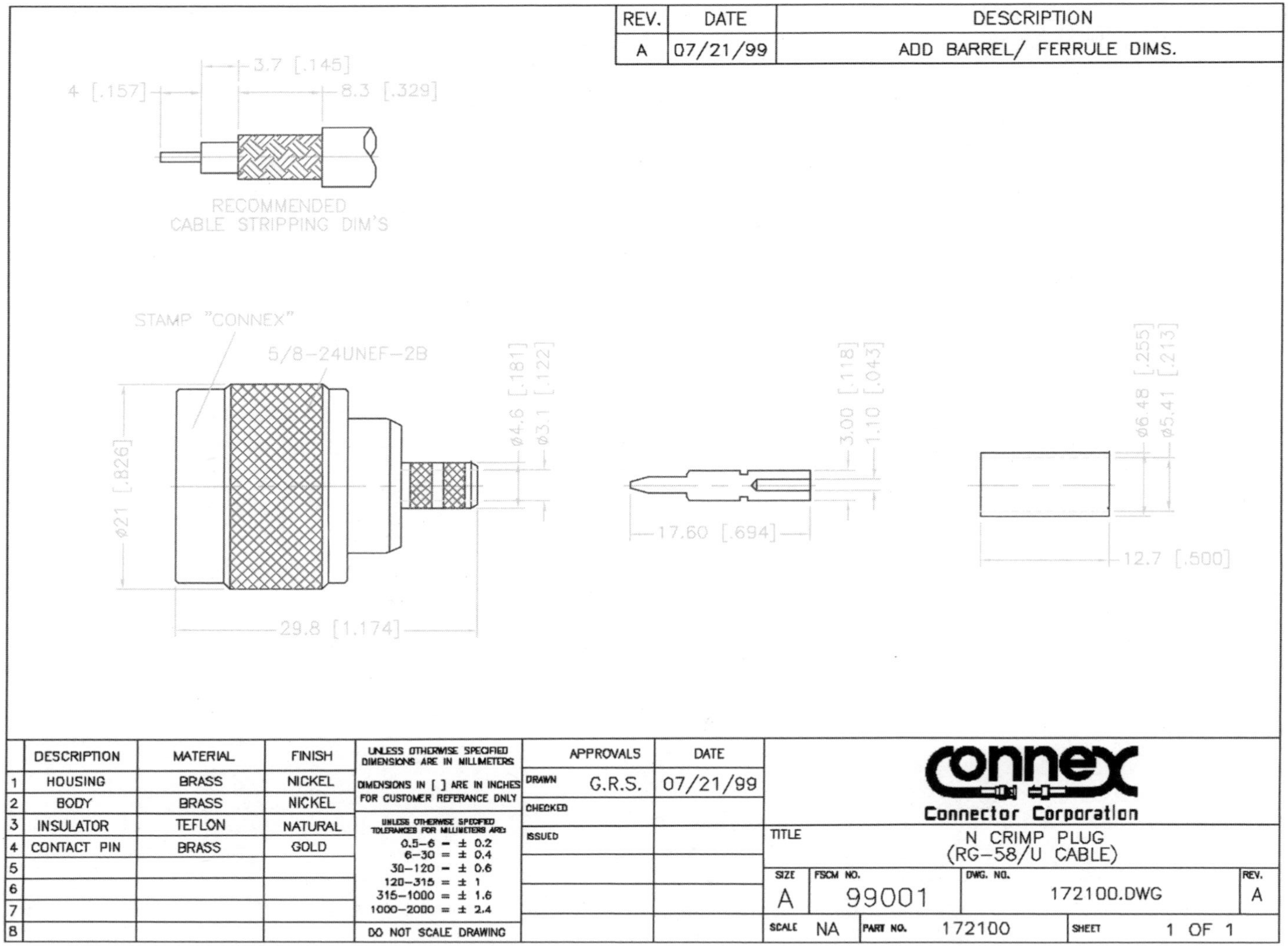

Type N—Crimp-on Jack for RG-58 Size Cables

REV.	DATE	DESCRIPTION
NC	07/21/99	INITIAL RELEASE

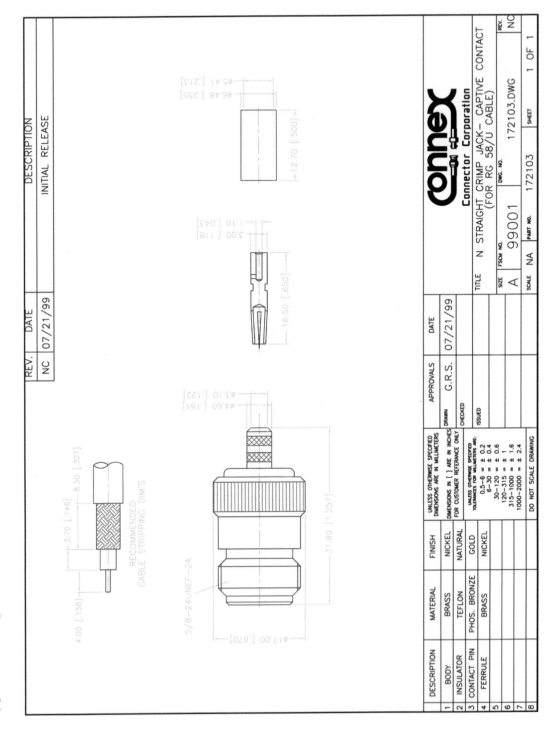

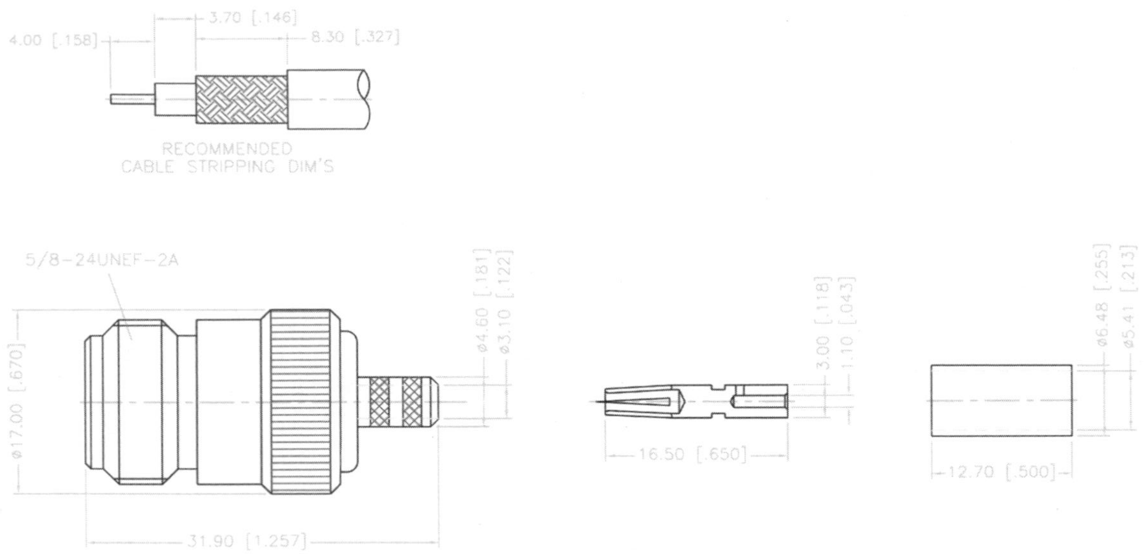

	DESCRIPTION	MATERIAL	FINISH
1	BODY	BRASS	NICKEL
2	INSULATOR	TEFLON	NATURAL
3	CONTACT PIN	PHOS. BRONZE	GOLD
4	FERRULE	BRASS	NICKEL
5			
6			
7			
8			

UNLESS OTHERWISE SPECIFIED DIMENSIONS ARE IN MILLIMETERS

DIMENSIONS IN [ ] ARE IN INCHES FOR CUSTOMER REFERANCE ONLY

UNLESS OTHERWISE SPECIFIED TOLERANCES FOR MILLIMETERS ARE:
0.5–6 = ± 0.2
6–30 = ± 0.4
30–120 = ± 0.6
120–315 = ± 1
315–1000 = ± 1.6
1000–2000 = ± 2.4

DO NOT SCALE DRAWING

	APPROVALS	DATE
DRAWN	G.R.S.	07/21/99
CHECKED		
ISSUED		

Connex Connector Corporation

TITLE N STRAIGHT CRIMP JACK– CAPTIVE CONTACT (FOR RG 58/U CABLE)

SIZE	FSCM NO.	DWG. NO.	REV.
A	99001	172103.DWG	NC

SCALE	PART NO.	SHEET
NA	172103	1 OF 1

Type N—Reverse-Polarity Crimp-on Plug for RG-58 Size Cables

REV.	DATE	DESCRIPTION
A	07/31/98	MODIFY CONTACT PIN INNER DIAMETER

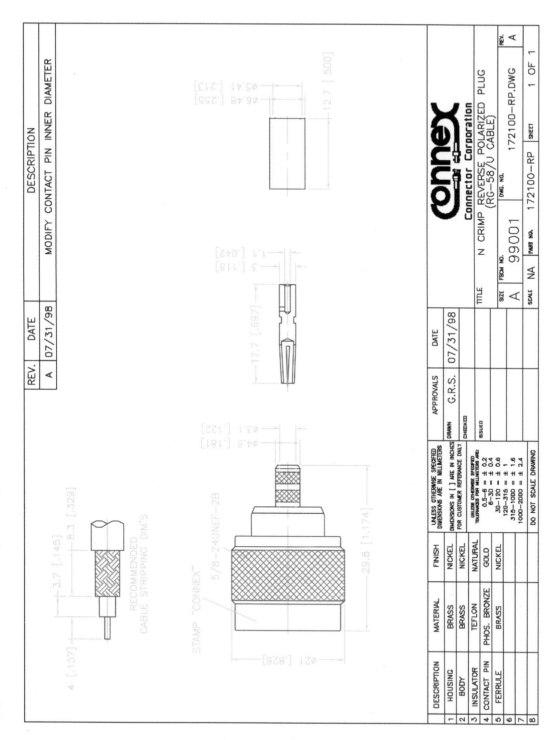

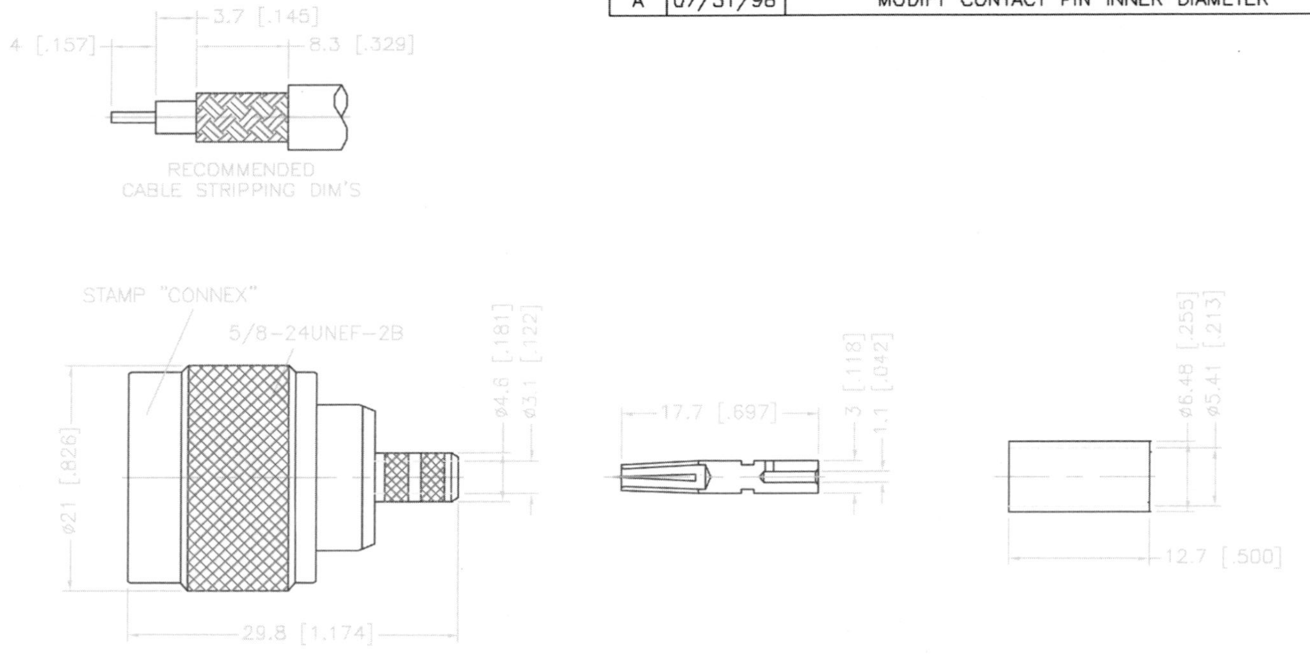

	DESCRIPTION	MATERIAL	FINISH
1	HOUSING	BRASS	NICKEL
2	BODY	BRASS	NICKEL
3	INSULATOR	TEFLON	NATURAL
4	CONTACT PIN	PHOS. BRONZE	GOLD
5	FERRULE	BRASS	NICKEL
6			
7			
8			

UNLESS OTHERWISE SPECIFIED DIMENSIONS ARE IN MILLIMETERS

DIMENSIONS IN [ ] ARE IN INCHES FOR CUSTOMER REFERANCE ONLY

UNLESS OTHERWISE SPECIFIED TOLERANCES FOR MILLIMETERS ARE:
0.5–6 = ± 0.2
6–30 = ± 0.4
30–120 = ± 0.6
120–315 = ± 1
315–1000 = ± 1.6
1000–2000 = ± 2.4

DO NOT SCALE DRAWING

	APPROVALS	DATE
DRAWN	G.R.S.	07/31/98
CHECKED		
ISSUED		

connex
Connector Corporation

TITLE	N CRIMP REVERSE POLARIZED PLUG (RG-58/U CABLE)				
SIZE A	FSCM NO. 99001	DWG. NO. 172100-RP.DWG			REV. A
SCALE NA	PART NO. 172100-RP		SHEET 1 OF 1		

Type N—Reverse-Polarity Crimp-on Jack for RG-58 Size Cables

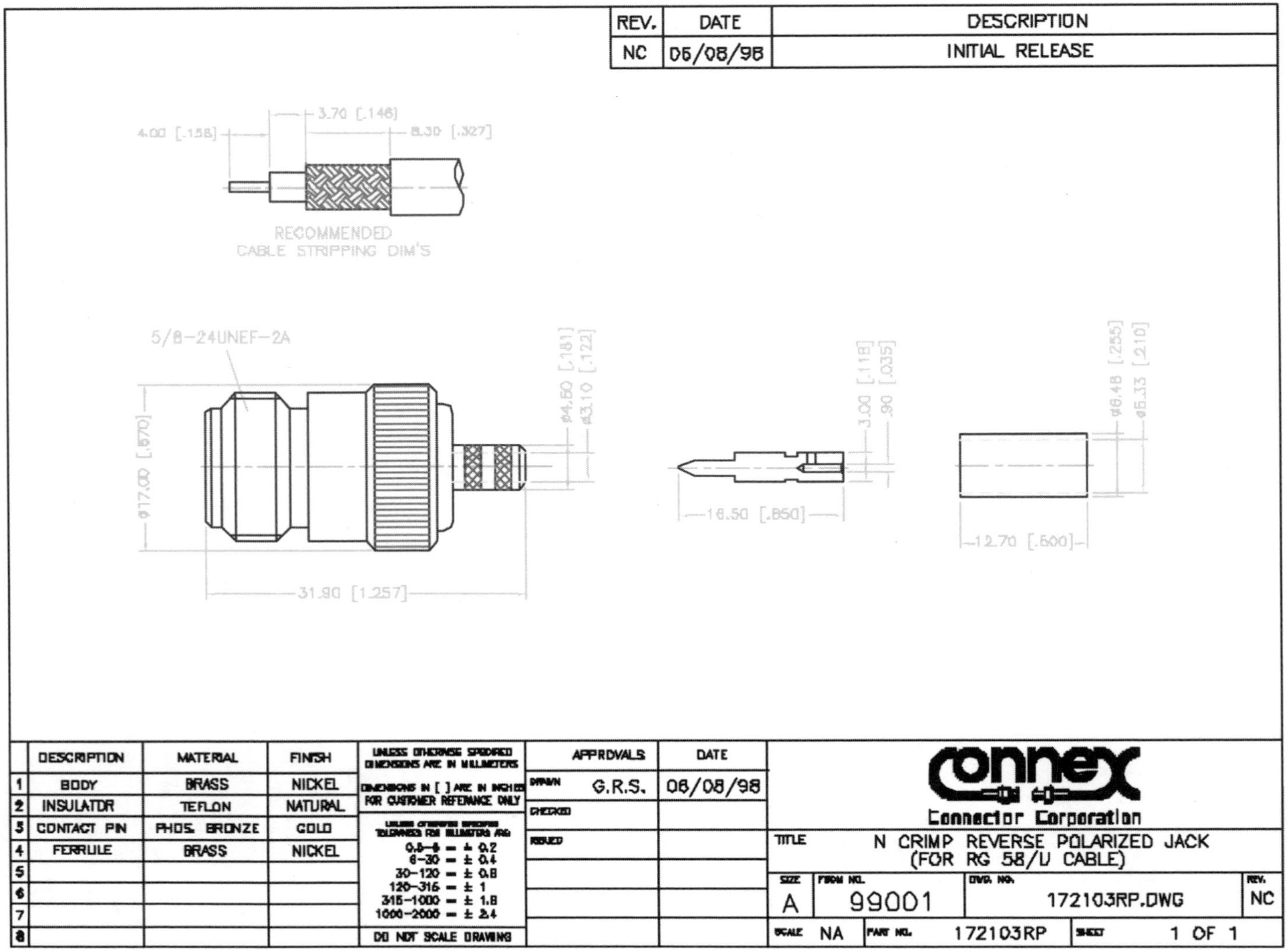

SMA Plugs and Jacks

SMA connectors are identified by their small (1/4-inch) diameter male threading on jacks and 5/16-inch body nut on plugs. Pin styles may be male or female on either end. SMA connectors are typically used on small diameter RG-174/RG-316 cables for pigtails but may be used for medium diameter RG-58/RG-142 cables for moderate length cable runs.

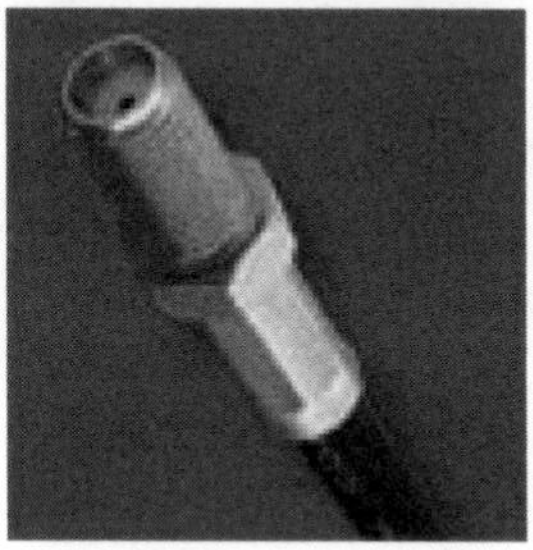

SMA—Reverse-Polarity Crimp-on Plug for RG-174/316 Size Cables

REV.	DATE	DESCRIPTION
NC	10/27/97	INITIAL RELEASE

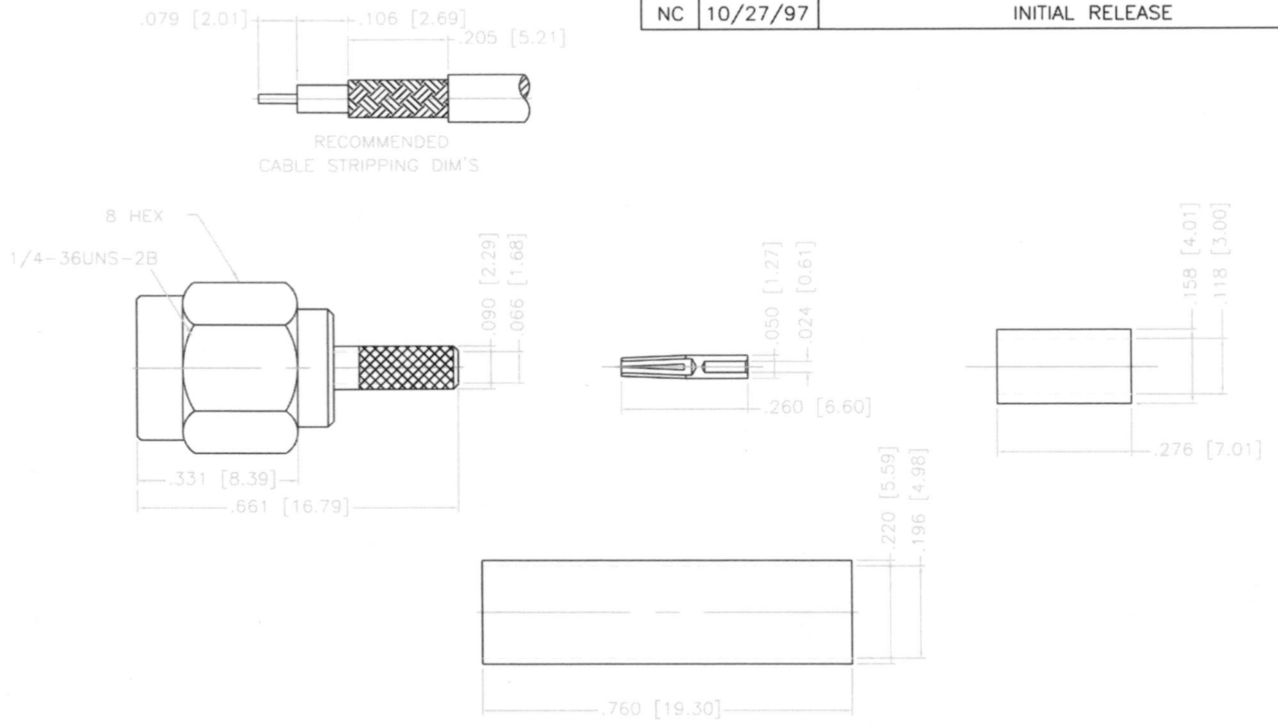

	DESCRIPTION	MATERIAL	FINISH
1	BODY	BRASS	NICKEL
2	INSULATOR	TEFLON	NATURAL
3	CONTACT PIN	BER. COPPER	GOLD
4	FERRULE	COPPER	NICKEL
5	HEAT SHRINK TUBE		
6			
7			
8			

DIMENSIONS ARE IN INCHES. DIMENSIONS IN [ ] ARE IN MILLIMETERS.

UNLESS OTHERWISE SPECIFIED TOLERANCES FOR INCHES ARE:
.XXX = ±
.XX = ±

TOLERANCES FOR MILLIMETERS ARE:
0.5–6 = ± 0.2
6–30 = ± 0.4
30–120 = ± 0.6
120–315 = ± 1
315–1000 = ± 1.6
1000–2000 = ± 2.4

DO NOT SCALE DRAWING

	APPROVALS	DATE
DRAWN	G.R.S.	10/27/97
CHECKED		
ISSUED		

Connex Connector Corporation

TITLE	SMA CRIMP PLUG RG316/U REVERSE POLARIZED		
SIZE: A	FSCM NO.: 99001	DWG. NO.: 132114-RP.DWG	REV.: NC
SCALE: —	PART NO.: 132114-RP	SHEET: 1 OF 1	

SMA—Reverse-Polarity Crimp-on Jack for RG-174/316 Size Cables

REV.	DATE	DESCRIPTION
NC	01/25/00	INITIAL RELEASE

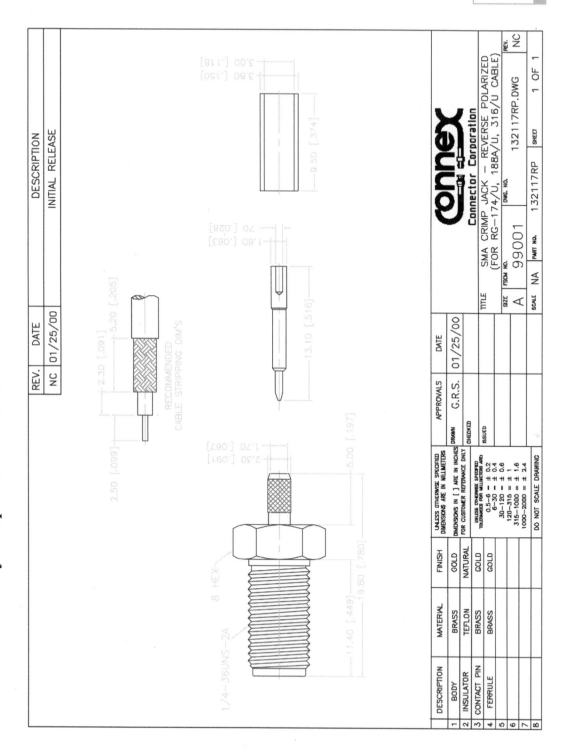

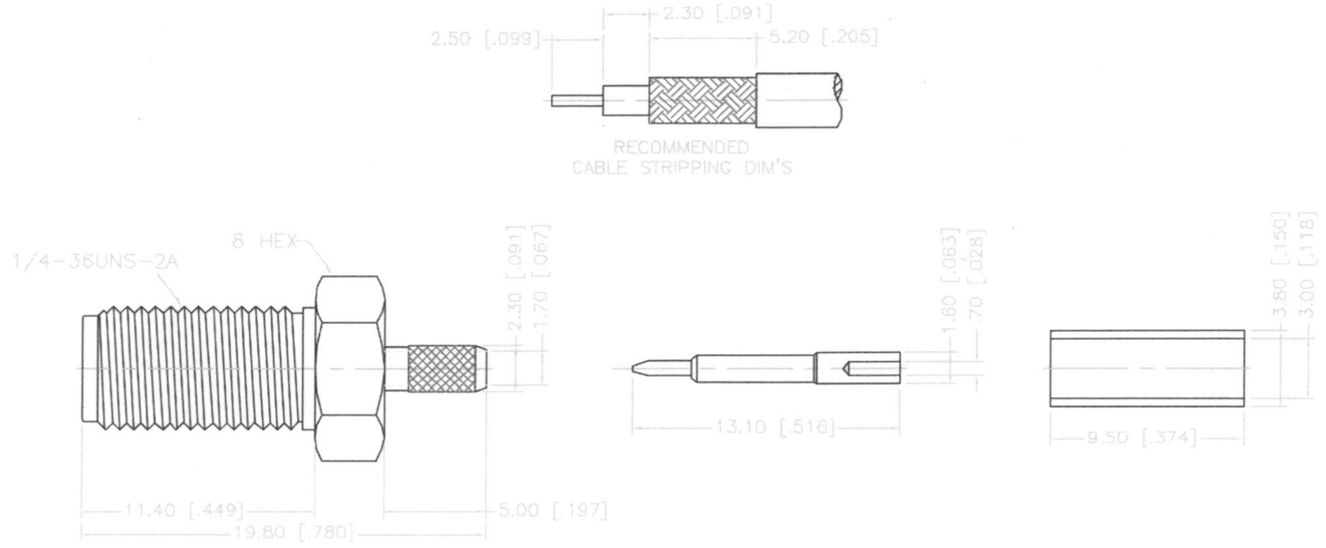

	DESCRIPTION	MATERIAL	FINISH
1	BODY	BRASS	GOLD
2	INSULATOR	TEFLON	NATURAL
3	CONTACT PIN	BRASS	GOLD
4	FERRULE	BRASS	GOLD
5			
6			
7			
8			

UNLESS OTHERWISE SPECIFIED DIMENSIONS ARE IN MILLIMETERS

DIMENSIONS IN [ ] ARE IN INCHES FOR CUSTOMER REFERANCE ONLY

UNLESS OTHERWISE SPECIFIED TOLERANCES FOR MILLIMETERS ARE:
0.5–6 = ± 0.2
6–30 = ± 0.4
30–120 = ± 0.6
120–315 = ± 1
315–1000 = ± 1.6
1000–2000 = ± 2.4

DO NOT SCALE DRAWING

APPROVALS		DATE
DRAWN	G.R.S.	01/25/00
CHECKED		
ISSUED		

connex
Connector Corporation

TITLE	SMA CRIMP JACK — REVERSE POLARIZED (FOR RG-174/U, 188A/U, 316/U CABLE)		
SIZE A	FSCM NO. 99001	DWG. NO. 132117RP.DWG	REV. NC
SCALE NA	PART NO. 132117RP	SHEET 1 OF 1	

TNC Plugs and Jacks

TNC connectors are identified by their medium-size (3/8-inch) diameter male threading on jacks and 7/16- to 1/2-inch knurled body nut on plugs. Pin styles may be male or female on either end. TNC connectors are typically used on medium diameter RG-58/RG-142 cables for moderate length cable runs and access point antennas.

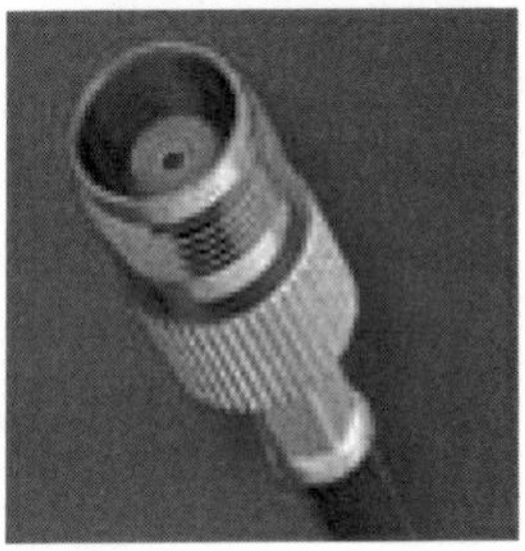

TNC—Crimp-on Plug for RG-58 Size Cables

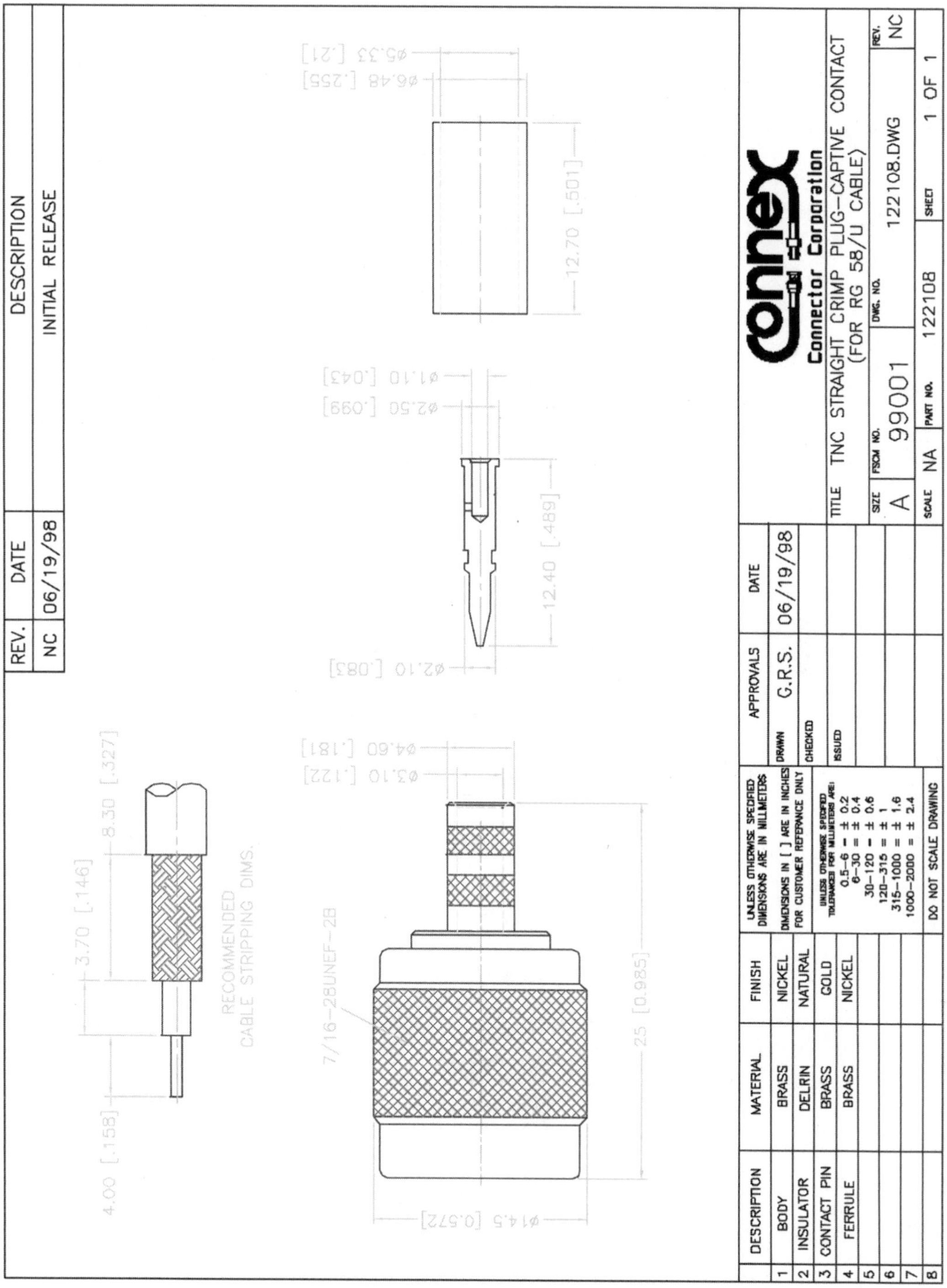

TNC—Reverse-Polarity Crimp-on Plug for RG-58 Size Cables

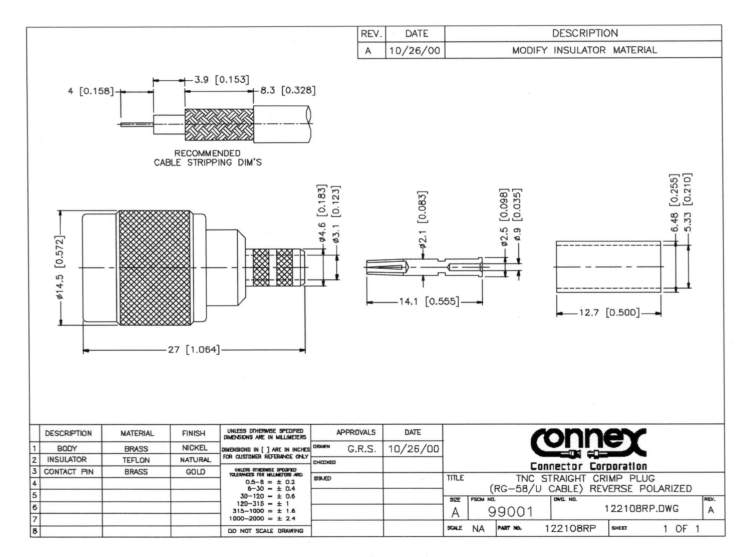

REV.	DATE	DESCRIPTION
A	10/26/00	MODIFY INSULATOR MATERIAL

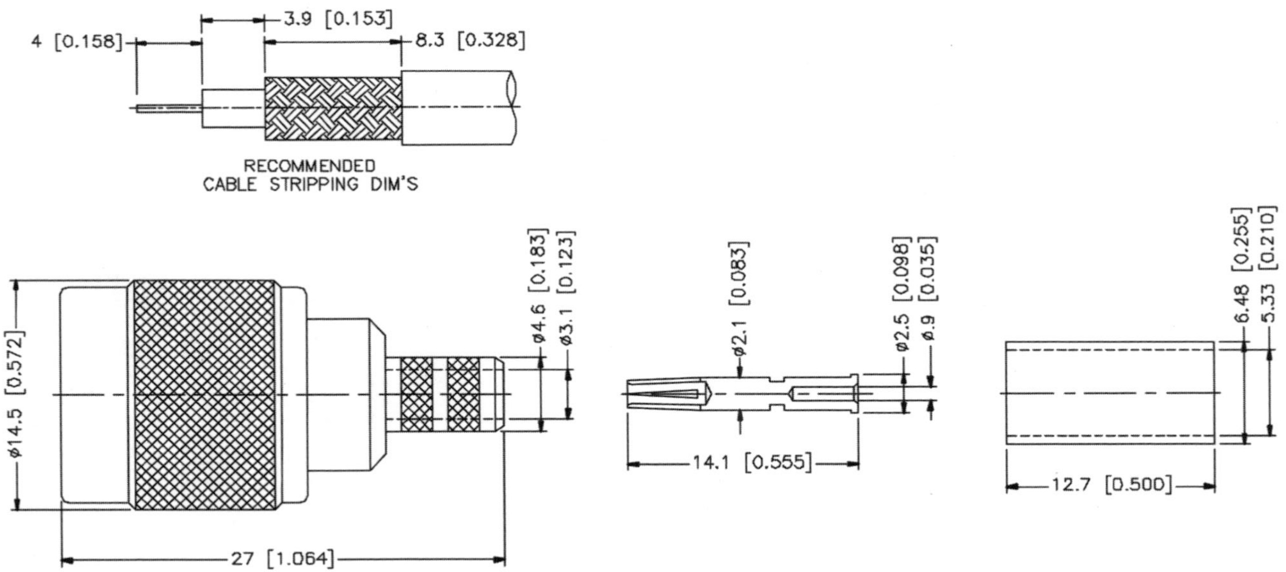

	DESCRIPTION	MATERIAL	FINISH
1	BODY	BRASS	NICKEL
2	INSULATOR	TEFLON	NATURAL
3	CONTACT PIN	BRASS	GOLD
4			
5			
6			
7			
8			

UNLESS OTHERWISE SPECIFIED DIMENSIONS ARE IN MILLIMETERS

DIMENSIONS IN [ ] ARE IN INCHES FOR CUSTOMER REFERANCE ONLY

UNLESS OTHERWISE SPECIFIED TOLERANCES FOR MILLIMETERS ARE:
0.5–8 = ± 0.2
6–30 = ± 0.4
30–120 = ± 0.6
120–315 = ± 1
315–1000 = ± 1.6
1000–2000 = ± 2.4

DO NOT SCALE DRAWING

	APPROVALS	DATE
DRAWN	G.R.S.	10/26/00
CHECKED		
ISSUED		

connex
Connector Corporation

TITLE	TNC STRAIGHT CRIMP PLUG (RG-58/U CABLE) REVERSE POLARIZED				
SIZE A	FSCM NO. 99001	DWG. NO. 122108RP.DWG			REV. A
SCALE NA	PART NO. 122108RP	SHEET 1 OF 1			

TNC—Reverse-Polarity Crimp-on Jack for RG-58 Size Cables

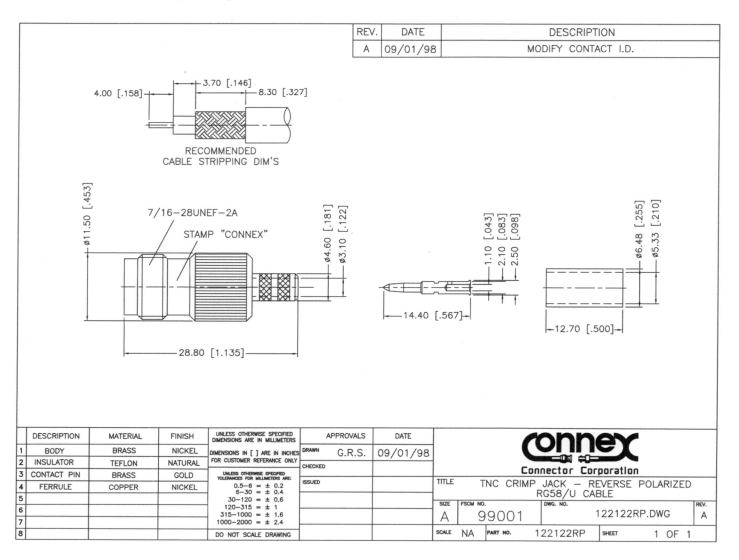

REV.	DATE	DESCRIPTION
A	09/01/98	MODIFY CONTACT I.D.

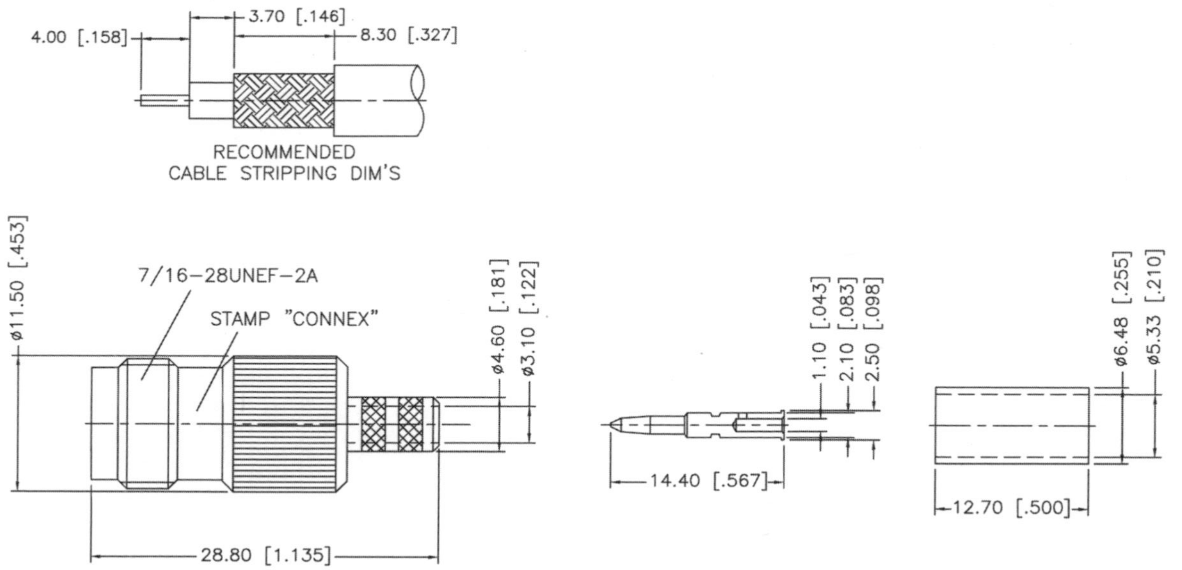

	DESCRIPTION	MATERIAL	FINISH
1	BODY	BRASS	NICKEL
2	INSULATOR	TEFLON	NATURAL
3	CONTACT PIN	BRASS	GOLD
4	FERRULE	COPPER	NICKEL
5			
6			
7			
8			

UNLESS OTHERWISE SPECIFIED DIMENSIONS ARE IN MILLIMETERS

DIMENSIONS IN [ ] ARE IN INCHES FOR CUSTOMER REFERANCE ONLY

UNLESS OTHERWISE SPECIFIED TOLERANCES FOR MILLIMETERS ARE:
0.5–6 = ± 0.2
6–30 = ± 0.4
30–120 = ± 0.6
120–315 = ± 1
315–1000 = ± 1.6
1000–2000 = ± 2.4

DO NOT SCALE DRAWING

	APPROVALS	DATE
DRAWN	G.R.S.	09/01/98
CHECKED		
ISSUED		

connex
Connector Corporation

TITLE	TNC CRIMP JACK – REVERSE POLARIZED RG58/U CABLE		
SIZE A	FSCM NO. 99001	DWG. NO. 122122RP.DWG	REV. A
SCALE NA	PART NO. 122122RP	SHEET 1 OF 1	

TNC—Reverse-Polarity Crimp-on Plug for RG-174/316 Size Cables

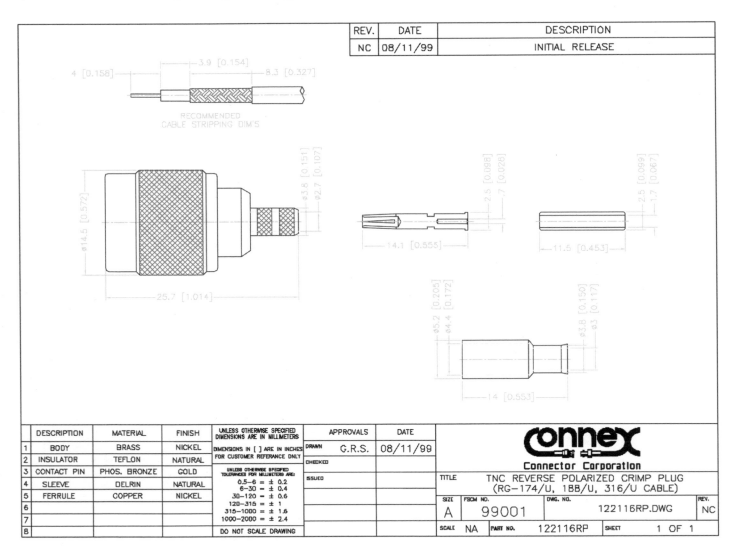

REV.	DATE	DESCRIPTION
NC	08/11/99	INITIAL RELEASE

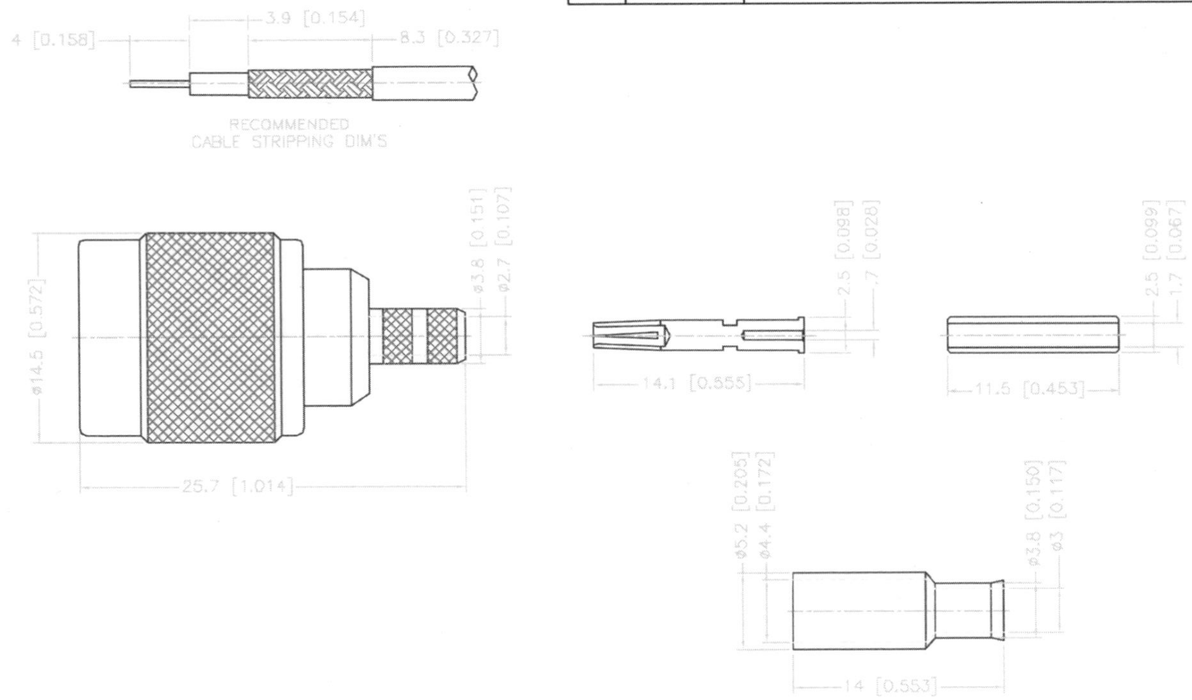

	DESCRIPTION	MATERIAL	FINISH
1	BODY	BRASS	NICKEL
2	INSULATOR	TEFLON	NATURAL
3	CONTACT PIN	PHOS. BRONZE	GOLD
4	SLEEVE	DELRIN	NATURAL
5	FERRULE	COPPER	NICKEL
6			
7			
8			

UNLESS OTHERWISE SPECIFIED DIMENSIONS ARE IN MILLIMETERS
DIMENSIONS IN [ ] ARE IN INCHES FOR CUSTOMER REFERANCE ONLY

UNLESS OTHERWISE SPECIFIED TOLERANCES FOR MILLIMETERS ARE:
0.5–6 = ± 0.2
6–30 = ± 0.4
30–120 = ± 0.6
120–315 = ± 1
315–1000 = ± 1.6
1000–2000 = ± 2.4

DO NOT SCALE DRAWING

	APPROVALS	DATE
DRAWN	G.R.S.	08/11/99
CHECKED		
ISSUED		

connex Connector Corporation

TITLE: TNC REVERSE POLARIZED CRIMP PLUG (RG–174/U, 188/U, 316/U CABLE)

SIZE	FSCM NO.	DWG. NO.	REV.
A	99001	122116RP.DWG	NC

SCALE NA | PART NO. 122116RP | SHEET 1 OF 1

APPENDIX C

On the CD-ROM

Computers chips and most digital devices are nothing without software to make them do something, or to tell us about them. The CD-ROM in the back of the book contains some of the most popular programs for building and peering into wireless networks.

The marketing target for most wireless LAN tools is obviously enterprise deployments for Microsoft Windows clients and servers. I'm including some similar tools for Macintosh, expecting they are as easy for you to use. I've had a lot of fun with and learned a lot from each and every one of them, and I think you will too. Read about them, pop the CD in your drive, install them—preferably extracting and installing them onto your hard drive first as they do create some temporary and log files—and welcome to the world of wireless networking!

Resources for Windows

Aerosol—Aerosol Program

Aerosol is an easy to use Windows-based wireless network detection program for use with WLAN adapters using the PRISM2 chipset such as the ATMEL USB or WaveLAN wireless cards on Windows. Aerosol requires a supported protocol driver to be installed. You can install WinPcap from http://winpcap.polito.it, or the Prism Test Utilities from the main Aerosol page, http://www.stolenshoes.net/sniph/aerosol.html. WinPcap support has been tested with WinPcap_3_0_a4.exe, prior versions are known to have issues. Aerosol like similar WLAN detection products will reconfigure the card to do its job so you cannot wireless network while using it. Please extract the file and place it on your hard drive before running the Aerosol.exe program.

AirMagnet—AirMagnet Demo

AirMagnet is one of a small number of products designed for the guru of corporate WLAN implementations, providing so much information about any and all access points and client adapters within range of its own WLAN adapter you may be overwhelmed. A very useful product for identifying rogue and misbehaving WLAN devices nearby. There are two versions—laptop and Windows CE/PDA—the laptop version is more capable as far as recording and reporting what it finds. The Windows CE version allows you to get much of the same information the laptop version does, with portability and a signal strength indication for direction finding so that you can move about and locate specific WLAN devices. The demo versions of the product on the CD-ROM, for both Windows and Windows CE are canned, with display samples only, as the full product requires a specific Cisco WLAN card to run. These will not show real-time data, but give you a good example of how feature-rich this product is. http://www.airmagnet.com

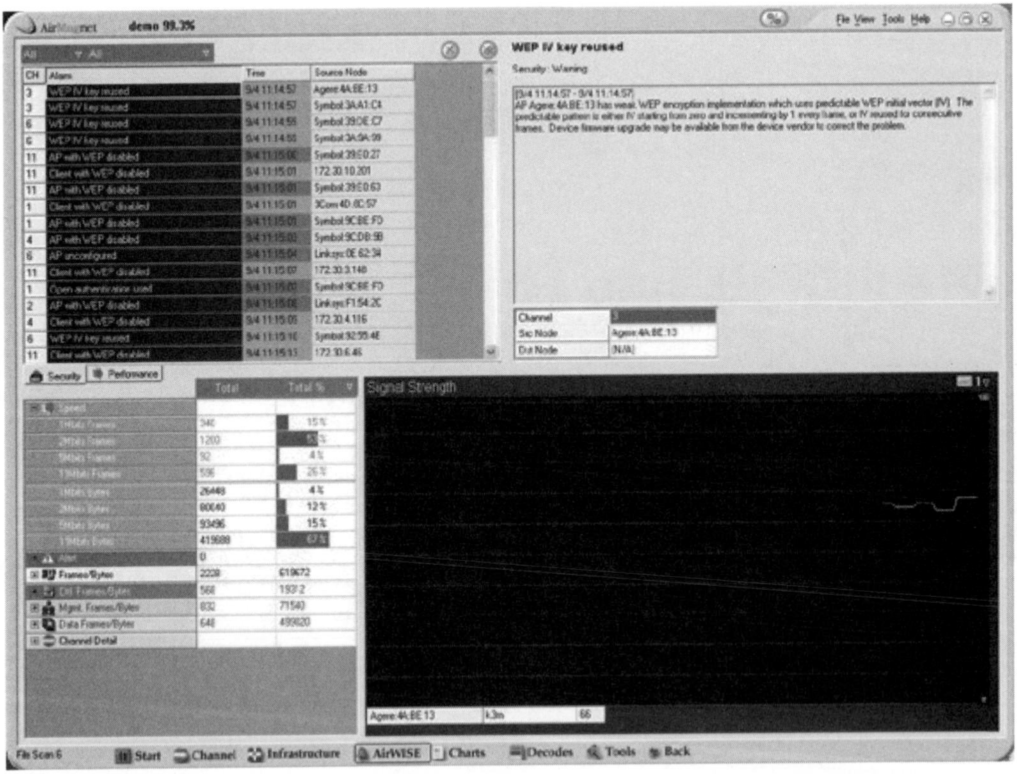

AiroPeek—AiroPeek Packet Sniffer

For the true LAN techie packet sniffing is everything. AiroPeek puts your WLAN adapter into listen-only mode, reveals what it sees, and gives you a lot of filtering to narrow down what you're looking for. Chances are you'll need to update your wireless adapter firmware and drivers to get it to work. If you need to discover an intruder or a new threat to your network, you may have to dig down and look at streams of data packets to determine the cause. http://www.wildpackets.com/products/airopeek

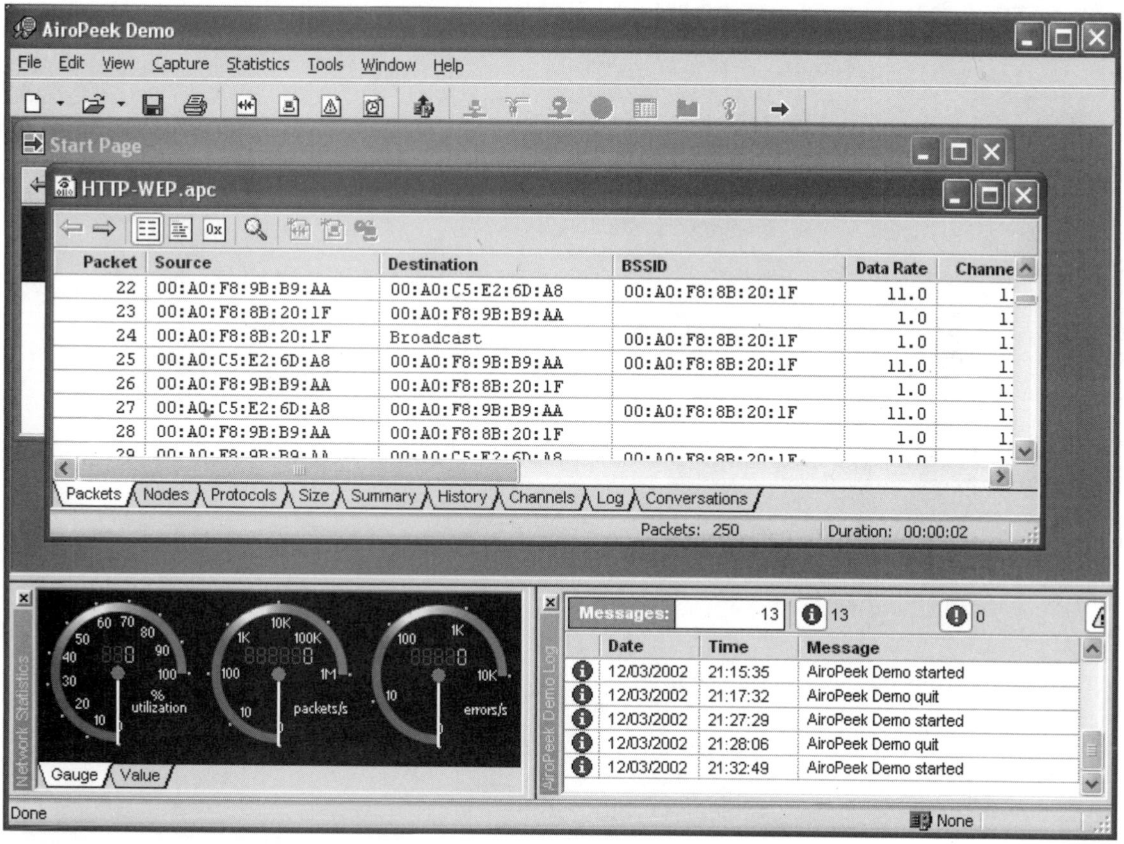

Funk—Funk Software Odyssey Server and Windows Client
Odyssey is an integrated package of the company's Steel-Belted RADIUS remote access authentication software with 802.1x EAP-TLS security for Windows 2000. Odyssey provides a complete access control and security solution for wireless LAN deployments. This product is so easy to use I cannot imagine trying anything else to deploy a secure, controlled WLAN solution. http://www.funk.com/

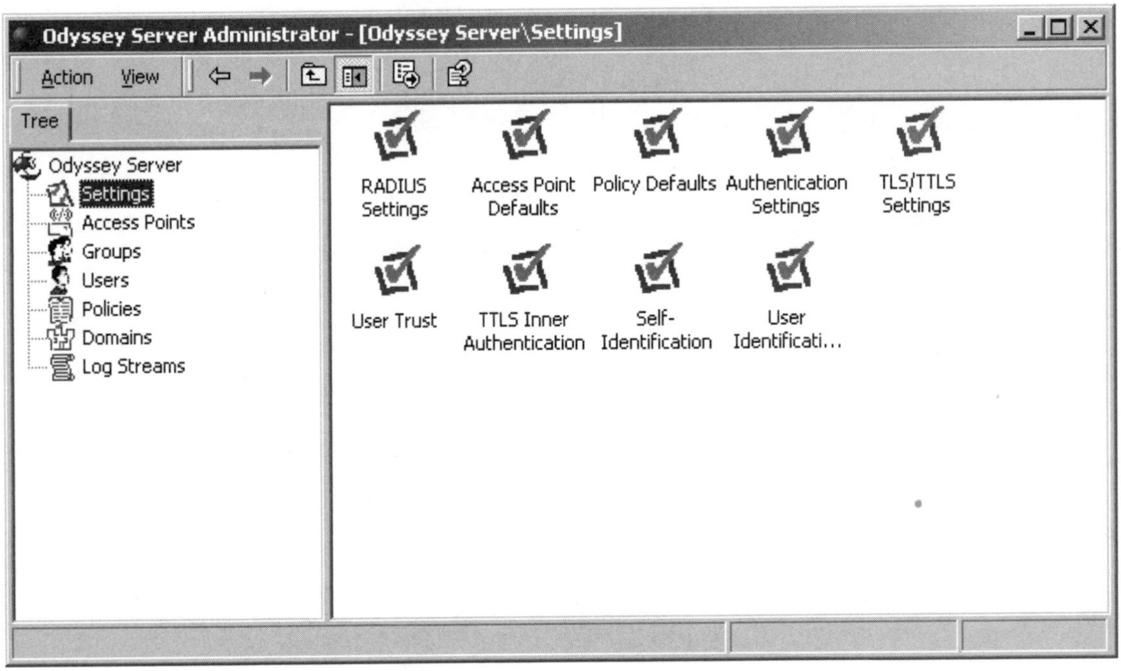

ISS—Internet Security Scanner

Internet Security Systems' Wireless Scanner provides automated detection and security analyses of mobile networks by utilizing 802.11b to determine system vulnerabilities. Fortunately it didn't reveal any security holes in my network other than I broadcast my SSID while I'm testing the WLAN. http://www.iss.net

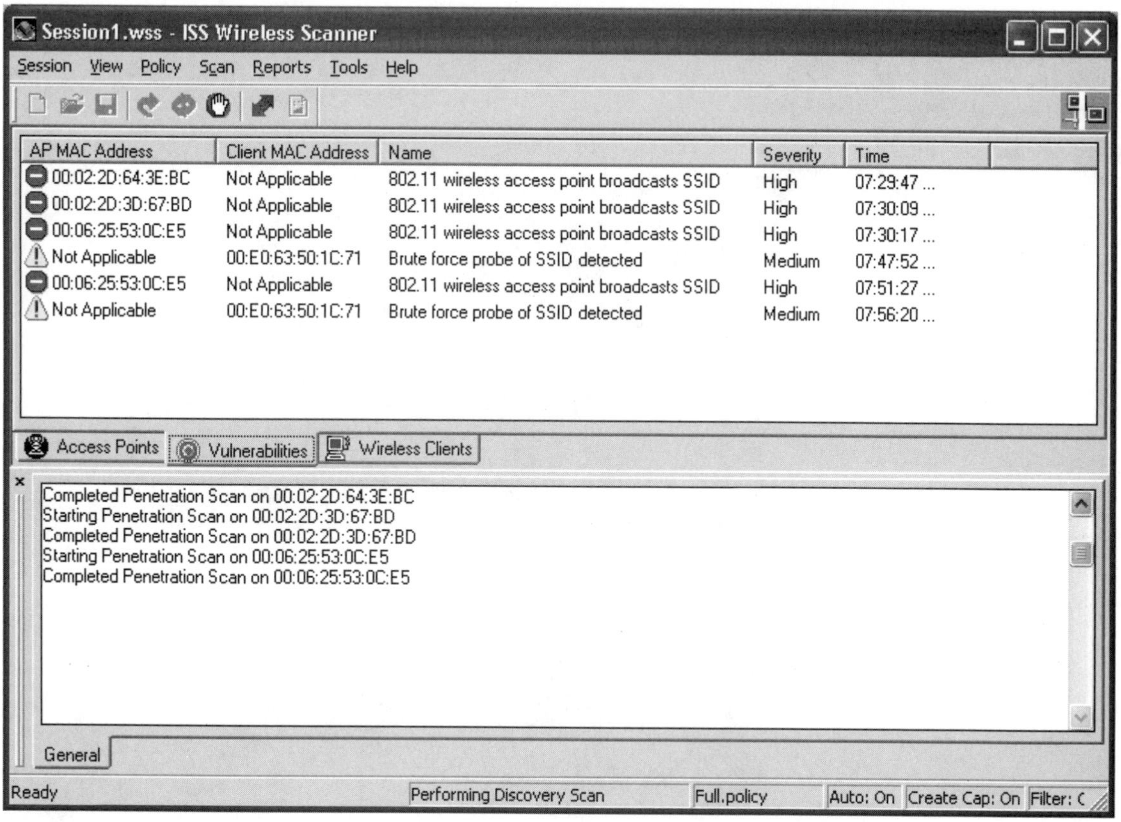

NetStumbler

NetStumbler is a universal tool to use for detecting wireless network activity. It provides significant amounts of data about each wireless access point you can receive. It will reveal the MAC address of active wireless devices, channels used, signal strength, SSIDs or lack thereof, as well as whether encryption is used at a particular access point. http://www.netstumbler.com

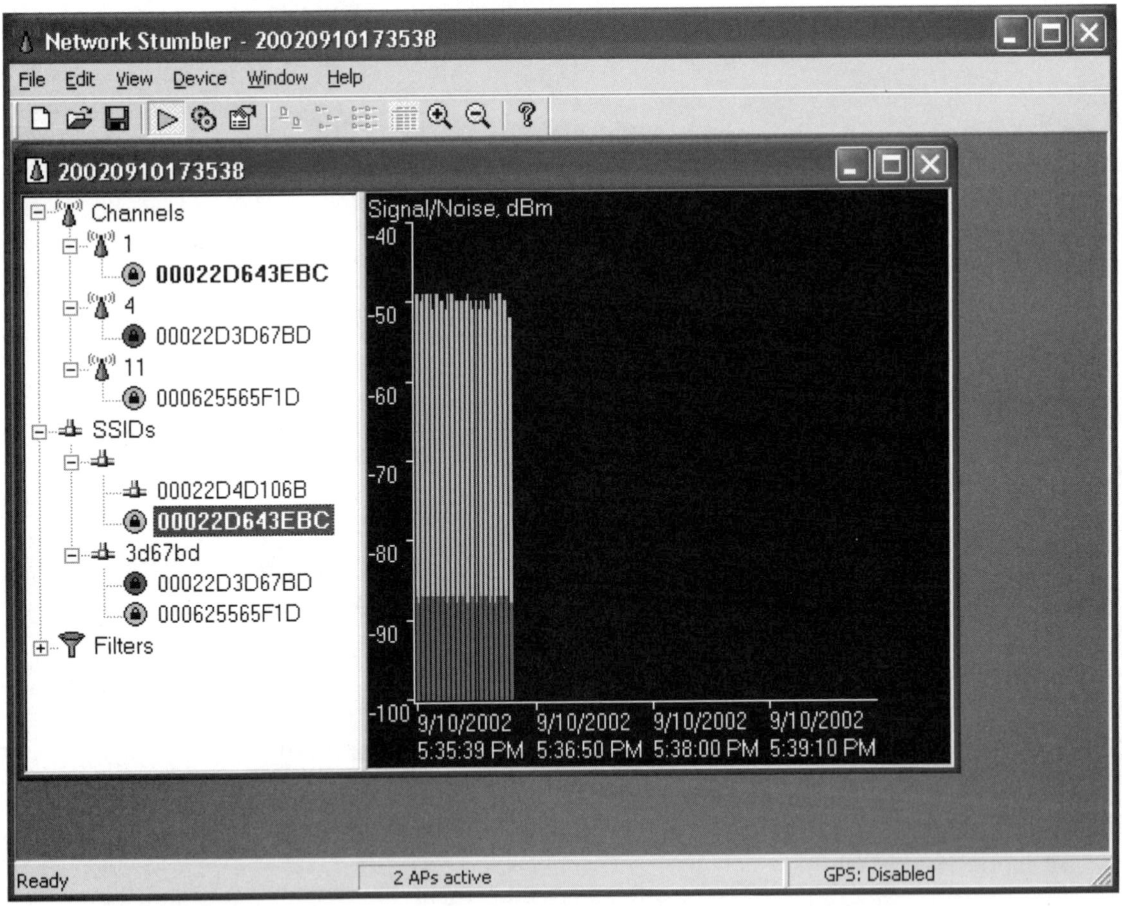

Radio Mobile—Roger Coudé's Radio Mobile

If you are planning numerous or complex wireless networks that have to cover long distances or irregular terrain, you simply cannot do without Radio Mobile. Radio Mobile uses standard geological survey maps containing terrain data to show you the signal strength of a signal throughout a selected area. This is a freeware program providing features similar to very expensive commercial radio site planning and coverage software. You can plot point-to-point paths or point-to-multipoint signal distributions and see the signal strength available. For instructions and links to obtain map files visit the Radio Mobile website: http://www.cplus.org/rmw/english1.html

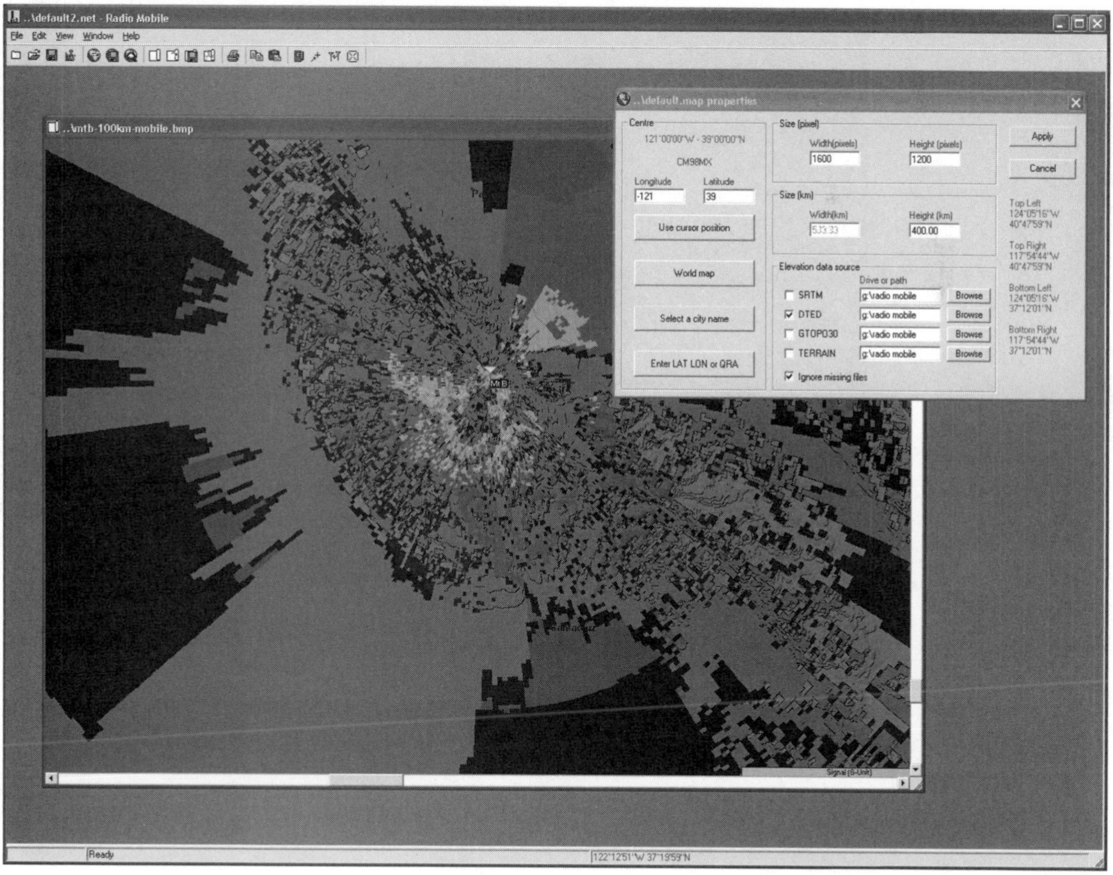

Tardis—Tardis 2000 and K9

I love Tardis. What else can I say. For years this program has been keeping a variety of servers and workstations on-time and in-synch with the National Institutes of Standards and Testing's (NIST) atomic clocks. You can select the time server you prefer to obtain time references from, some public, some private, with many across the world. Tardis acts as both a time-getting client and time server program. For workstations needing to synchronize with a local server, there is the small, less complex K9 program as well. Registering Tardis costs $20 and K9 is a mere $6. http://www.kaska.demon.co.uk/

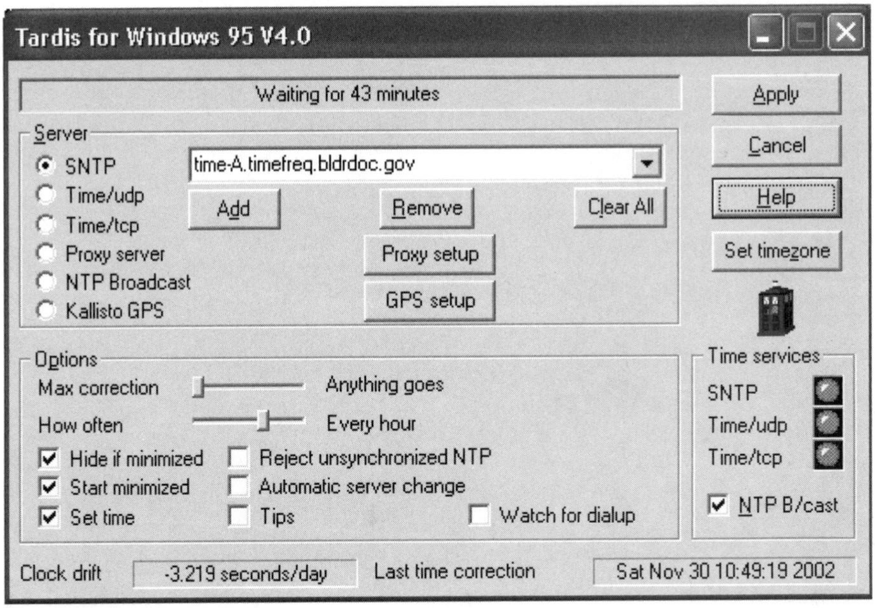

WiMetrics—WiSentry

WiSentry is a LAN-based product that identifies wireless devices connected or trying to gain access to your wired LAN. Set alarms for intrusions and rogue access points, and then sniff them out with the Windows CE/PDA version. Not yet as feature rich as some intrusion detection products, it is not limited to using specific WLAN adapters nor does it require that there be any wireless devices on your LAN, as it will tell you if and when they appear. http://www.wimetrics.com

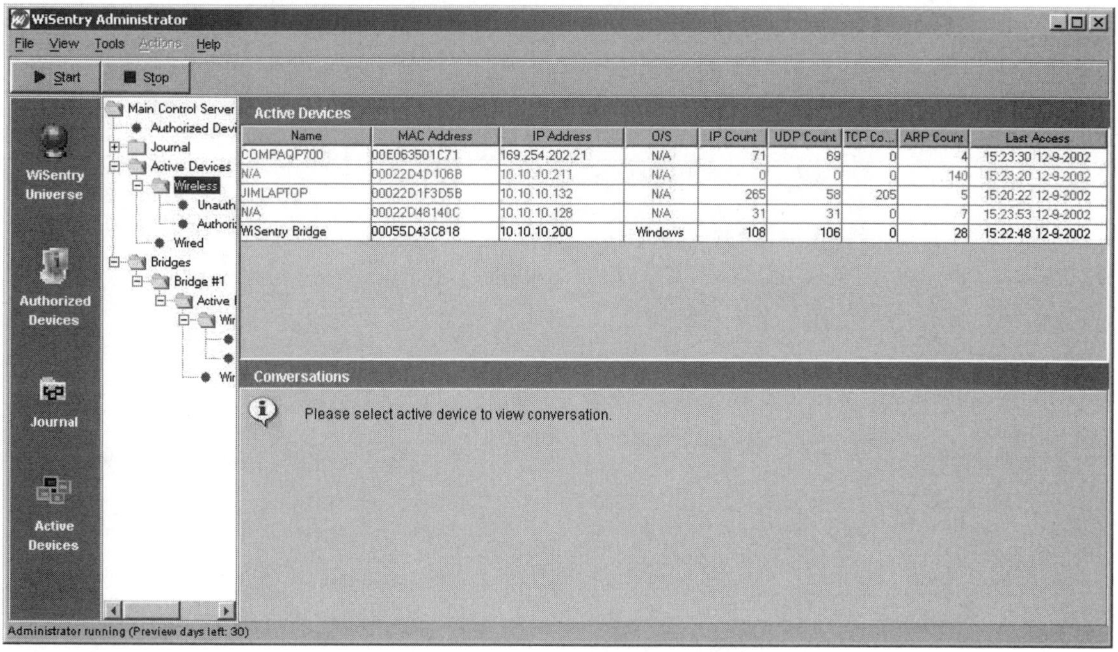

WLANExpert

I really wanted to love WLANExpert until I discovered it does not run on Windows 2000 or XP. If you don't mind running it on Windows 98 or Me you'll be fine, and you may want to so that you can enjoy its features. It works with most Intersil Prism2-based WLAN cards, covering LinkSys and similar products. Two of the best features are built-in antenna testing and reporting on whether your attached antenna is good or bad—most useful for external antenna connections or detecting a broken internal antenna. It has a module that lets you set the transmit power for your LAN card. http://www.allaboutjake.com/network/linksys/wlanexpert.html

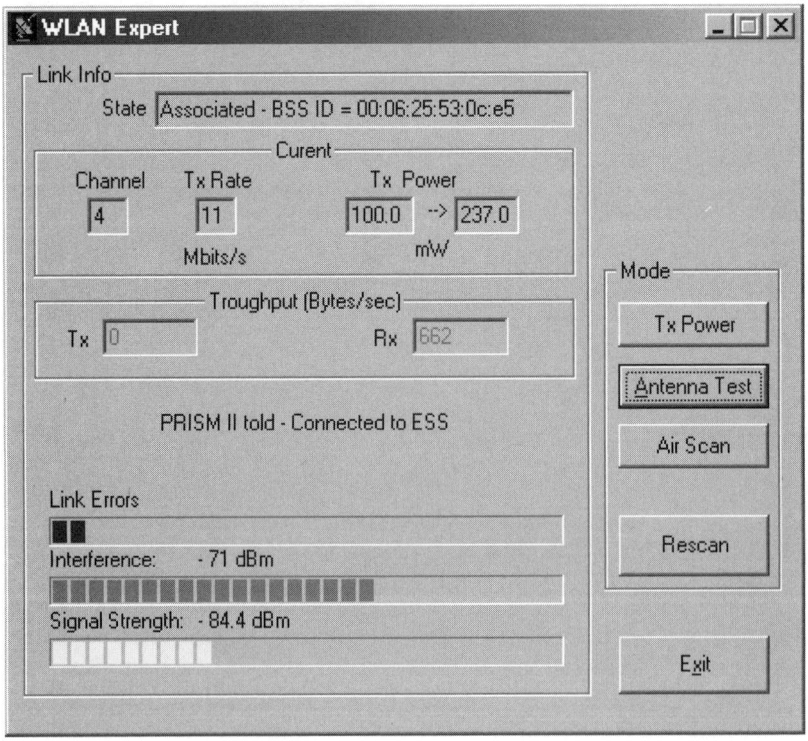

ZEDu—ZoneEdit Dynamic Update
ZoneEdit Dynamic Update is my tool of choice for updating my ZoneEdit DNS services for a couple of the servers I maintain on my residential DSL service. It is simple and effective—which is all you need to do the job. After the 45-day evaluation period you can register the product for $17.95 through the author's Web site. http://glsoft.glewis.com/

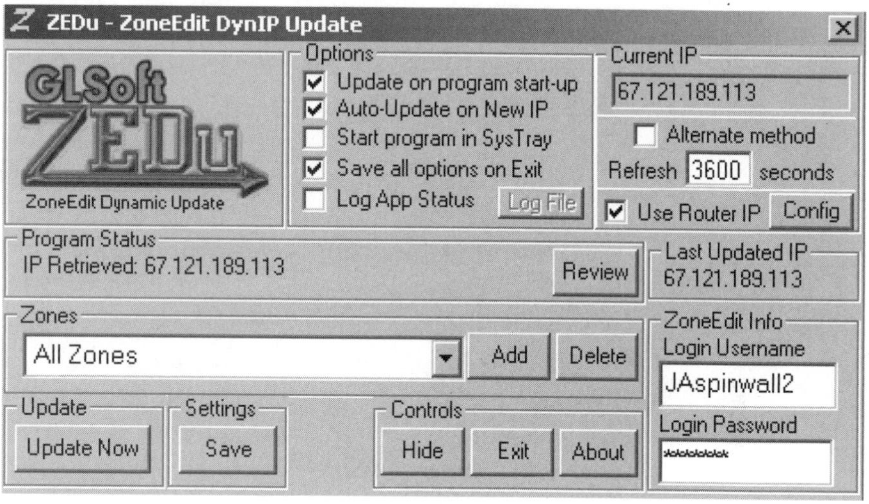

Resources for Macintosh

Macintosh\APScanner—APScanner for Mac
A tool for detecting the presence of nearby wireless LANs. http://www.packetninja.ca/

Macintosh\MACStumbler—MacStumbler
Another tool for detecting the presence of nearby wireless LANs. http://www.macstumbler.com/

Macintosh\ClassicStumbler—ClassicStumbler
http://homepage.mac.com/alk/projects/software/classicstumbler/

Linux Resources

Since the open-source environment is quite dynamic, I chose to not include any of the source or binary files or installers on the CD. You should visit the respective Web site for the program you are interested in to get the latest files for your configuration. After finally getting wireless to work on my Linux system, thanks to files and help from AbsoluteValue Systems and the wlan-ng'contributions, I had the pleasure of trying AirSnort, NoCat, and Sputnik successfully and found them to be all they said they were—effective and useful. If you're a Linux-junkie, dig in!!

AbsoluteValue Systems: http://www.linux-wlan.org/
A must-visit site to obtain source code and relevant information to build into your Linux system for wireless networking.

AirSnort: http://airsnort.shmoo.com/
AirSnort is the most popular tool for grabbing WEP encryption key information from a wireless network. It may be of value as part of a security analysis but it's real purpose is to reveal the keys of other people's wireless LANs.

Kismet Packet Sniffer: http://www.kismetwireless.net/
Kismet sniffs data packets present on a wireless network—valuable stuff if you're into low-level network and data security analysis.

NoCat Authentication: http://nocat.net/
NoCat appears to be the choice of gateway and access control programs for many open/community and closed/commercial wireless network hotspots. It is the foundation for the Sputnik portal program.

SOHOWireless LANRoamer: http://www.lanroamer.net/
LANRoamer is another option for creating a wireless network hotspot similar to the Sputnik project—download the CD-ROM image file, burn a CD, put the CD in a system with a wireless card and access to your network or the Internet. Instant wireless portal site.

Sputnik: www.sputnik.com
Want to provide a community network? Get up and running fast with this CD-ROM–bootable instant portal. The software forces users of a Sputnik-backed access point to log in to the Sputnik.com server. The service is free, and the web site maintains a list of affiliated community hotspots.

SSIDSniff: http://www.bastard.net/~kos/wifi/
SSIDSniff falls into the same category as WAVE Stumbler; it detects and identifies other nearby wireless LANs.

Trustix Firewall: http://www.trustix.com
Finally, a firewall for the rest of us who are not and do not want to be proficient at IPChains and similar scripts to control what goes in and out of our networks. Trustix Firewall is a secure Linux implementation designed to make any x86 system into a firewall appliance, with a graphical interface for configuring it specifically as a firewall to go between your LAN and the Internet or other connections. It also provides IPSec VPN services between two systems that have static IP addresses. While there is no specific wireless component to this product, it treats wireless connections as it would any other Ethernet connection; it's a good tool for any network.

WAVE Stumbler: http://www.cqure.net/tools08.html
WAVE Stumbler allows you to detect and identify other wireless LANs nearby. It is a good tool for doing site surveys, to see who is on which channel, and (perhaps with a directional antenna) find other WLANs.

WEPCrack: http://sourceforge.net/projects/wepcrack
WEPCrack is designed to prove the ease of breaking the WEP key encryption scheme. It does not sniff for packets; instead you must acquire packets using the prismdump program to create a file of captured packets and then feed that file into WEPCrack.

wlan-ng pages: http://prism2.unixguru.raleigh.nc.us/
A must-visit site to get source code and installable wireless networking files for all that is installable for RedHat Linux and common wireless devices.

GLOSSARY

802.11—A family of specifications developed by the Institute of Electrical and Electronics Engineers, Inc. (IEEE) for wireless LAN technology. 802.11 specifies the radio signal interface between a client radio and a base station radio, or between two client radios. There are several specifications in the 802.11 family:

- **802.11**—Wireless LANs providing 1 or 2 Mbps transmission in the 2.4 GHz band, using either frequency hopping spread spectrum (FHSS) or direct sequence spread spectrum (DSSS).
- **802.11a**—A subset of 802.11 that provides an up to 54 Mbps data rate using the 5 GHz band. 802.11a uses an orthogonal frequency division multiplexing encoding scheme, rather than FHSS or DSSS.
- **802.11b**—(Also known as *Wi-Fi or wireless fidelity.*) A subset of 802.11 that provides 11 Mbps data rates with the ability to scale back to 5.5, 2, or 1 Mbps rates, and uses the 2.4 GHz band. 802.11b uses the DSSS modulation scheme. 802.11b allows Ethernet functionality over radio.
- **802.11c**—Relates to 802.11 bridging functions.
- **802.11f**—An interaccess point protocol to help ensure interoperability for roaming access.
- **802.11g**—Provides 20+ Mbps in the 2.4 GHz band.
- **802.11h**—A future standard for wireless spectrum management.
- **802.11i**—A standard for enhancing the security of wireless local area networks (WLANs). Preceded by an interim nonstandard Wi-Fi protected access (WPA) security enhancement.
- **802.1x**—A standard for wireless LAN authentication methods.

Access point—A bridge that provides access for wireless stations to a wired local area network (LAN), and typically onto a wired LAN.

Access point (AP)—A wireless network interface device, acting as or replacing the function of the hub or switch in a wired network, to allow wireless network cards in client systems to connect to a LAN or the Internet.

Access time—The amount of time necessary for data to become available from a disk drive or memory area after a request is issued.

Acknowledge (ACK)—A signal sent by a receiving device confirming that information sent has been received. The opposite of NACK.

Ad-Hoc mode—A group of computers with wireless local area network (WLAN) adapters, connected as an independent peer-to-peer WLAN.

Adapter—A hardware device, usually a set of connectors and a cable, used between two pieces of equipment to convert one type of plug or socket to another, or to convert one type of signal to another. Examples are a 9-to-25 pin serial port adapter cable, a serial-port-to-serial-port null modem, and a PC-printer-interface-to-printer cable.

Adapter card—A plug-in card used to exchange signals between the computer and internal or external equipment such as a parallel printer or serial ports, video adapters or disk controllers.

Add-in card—*See* adapter card.

Address—A location in memory or on a hardware bus of either a specific piece of data or a physical hardware device.

Advanced configuration and power interface (ACPI)—A standard specification and method for the monitoring of system activity and control of system configurations with power applied to or removed from system components, or switched to other components, depending on power states. Accommodates different modes of sleep, suspend, and full-on system readiness of many system components.

Advanced graphics port (AGP)—A high-performance data bus designed specifically to handle digital information from a computer system to a video adapter. AGP is a specific enhancement to the peripheral component interconnect (PCI) bus, allowing the video adapter to directly access main memory.

Advanced power management (APM)—A standard specification and method for the monitoring of system activity and control of power applied to or removed from system components, accommodating different modes of Sleep, Suspend, and Full-On system readiness. Sleep mode allows for maintaining current system activity with reduced power consumption, such as having disk drives and displays powered off, but the central processing unit (CPU) and memory retaining the last activities. Suspend mode allows for maintaining minimal current system activity with no power consumption. APM is expected to be superceded by advanced configuration and power interface (ACPI).

Advanced technology attachments (ATA)—An industry-wide specification for the interfacing of devices, typically hard disk drives, to the PC/AT (advanced technology) standard data bus.

Alt-key codes—A combination of keystrokes using the Alt key, plus one or more letter or number keys, to cause a particular program function or operation. The Alt key acts like a Shift or Ctrl key to change the function or use of a particular key. Alt-key combinations and their uses differ between many programs. One particular and common use for the Alt key is to allow the entry of the decimal value of ASCII characters, especially the upper 128 special characters available with DOS, to draw lines and boxes. These keystrokes require use of any Alt key and the numeric data entry pad (rather than the top-row number keys). One example is pressing and holding the Alt key while entering the number sequence 1, 9, and 7, then releasing the Alt key. This should cause the entry and display of a set of single crossed lines the size of a character.

American National Standards Institute (ANSI)—A governing body managing specifications for the computer industry and other disciplines. In terms of computing, ANSI maintains a set of standards for the coding and displaying of computer information, including certain "escape sequences" for screen color and cursor positioning. A device-driver file, `ANSI.SYS`, can be loaded in your PC's `CONFIG.SYS` file so that your screen can respond properly to color and character changes provided from programs or terminal sessions between computers.

American Standard Code for Information Interchange (ASCII)—ASCII defines the numerical or data representation of characters, numbers, and foreign language characters in computer data storage, text files, and display. There are 128 predefined characters, numbered 0–127, representing the alphabet, numbers, and data-terminal control functions that nearly any computer system will interpret properly. ASCII characters are represented or transferred in decimal or hexadecimal numeric representations, from 0–255 (decimal) or 00–FFh (hex). The upper 128 characters (128–255) vary between computer systems and languages and are known as the symbol set. IBM defined these as Extended ASCII characters, which include a variety of lines and boxes for pseudographical screen displays. ASCII also defines the format of text files. ASCII text files generated on PCs differ slightly from the original ASCII standard, and may appear with extra lines on other computer systems.

Amplifier—An electronic device used to increase the power of a weaker signal. Amplifiers or power amplifiers may be used to create a stronger transmitted signal from a lower power wireless network device.

Antenna—A device used as the terminus for a radio transmitter, converting radio frequency electrical energy into radio waves, and as the Internet to catch transmitted waves, converting them from wave energy to electrical energy, to supply signal to a radio receiver. Antennas come in various shapes, forms, and sizes, with different purposes and effects. An omnidirectional antenna radiates and accepts wave energy from all directions equally. A directional antenna radiates and accepts wave energy from only one or a few directions. Antennas, by special tuning and shaping elements, may have gain or appear to passively amplify a signal's strength. In a bidirectional amplifier, special circuitry allows signal power to be increased in both the transmitting and receiving directions.

Application—A computer program or set of programs designed to perform a specific type or set of tasks to make a computer help you do your work or provide entertainment. Typical applications are games, word processing, database, or spreadsheet programs.

Archive attribute—*See* attributes.

Association—The process of wireless adapters establishing a connection with each other on the same radio channel, but not necessarily being able to communicate via transmission control protocol/Internet protocol (TCP/IP) or other selected network protocol (this requires authentication). Reassociation occurs if the chosen channel gets too noisy or the signal drops out and picks up again.

ATA—AT-attachments—An industrywide specification for the interfacing of devices, typically hard disk drives, to the PC/AT standard data bus.

Attributes—Every DOS file entry, including subdirectories, is accompanied by an attribute byte of information that specifies whether the file is read-only, hidden, system, or archived. Read-only indicates that no program operation should erase or write-over a file with this attribute. Hidden indicates that the file should not be displayed or used in normal DOS `DIR`, `COPY`, or similar operations. The system attribute indicates that a file belongs to the operating system, which typically applies only to the hidden DOS files `IO.SYS` or `IBMBIO.COM` and `MSDOS.SYS` or `IBMDOS.COM` files. The archive attribute indicates that a file has been changed since the last backup, or that it should be backed up during the next backup session. Backup operations clear this attribute.

AUTOEXEC.BAT file—An ASCII text file that may contain one or more lines of DOS commands that you want executed every time you boot-up your PC. Also known as just the autoexec file, this file can be customized using a text editor program, so that you can specify a DOS prompt, set a drive and directory path to be searched when you call up programs, or load terminate-and-stay resident (TSR) programs that you want to have available all of the time.

Backup—The process of copying one, several, or all of the files on one disk to another disk, a set of diskettes, or tape cartridges for archival storage or routine protection against a system failure or loss of files. A backup should be done regularly and often.

Base address—The initial or starting address of a device or memory location.

Base memory—*See* DOS memory.

Base service set (BSS)—Wireless stations communicate directly to each other, peer-to-peer, without an infrastructure or typical network or gateway between them. Also called ad-hoc networks. Any access control or authentication is strictly up to the users of both interconnected systems. Also considered independent base service set (IBSS).

Basic input/output system (BIOS)—The first set of program code to run when a PC system is booted up. The BIOS defines specific addresses and devices and provides software interface services for programs to use the equipment in a PC system. The PC system BIOS resides in a ROM chip on the system board. BIOS also exists on add-in cards to provide additional adapter and interface services between hardware and software.

Batch file—An ASCII text file that may contain one or more lines of DOS commands that you want to execute by calling for one file, the name of the batch file, rather than keying them in individually. Also known as "bat" files, these files can be customized using a text editor program so that you can specify a DOS prompt, set a drive and directory path to be searched when you call up programs, or load and execute specific programs. Batch files are used extensively as shortcuts for routine or repetitive tasks, or those for which you do not want to have to remember each step. These files always have the extension .BAT, as required by DOS.

Battery backup—The facility of retaining power to a system or memory chip from a battery pack when AC power is not available. The battery may be a rechargeable or temporary type.

Bit—A bit is the smallest unit of information or memory possible in a digital or computer system. A bit has only two values—1, or on, and 0, or off. A bit is the unit of measure in a binary (1/0) system. It might be thought of as a binary information term. A bit is one of 8 pieces of information in a byte, one of 16 pieces in a word (16-bit words), or one of 4 pieces in a nibble (half a byte.)

Blue-screen, blue screen of death (BSOD)—The screen appearance commonly associated with the crash or sudden failure of the Windows operating system.

Bluetooth—A short-range radio technology aimed at simplifying communications among various devices. It is most often used for nonnetwork/Internet applications, such as remote controls, wireless headsets, mice and keyboards, and printers.

Boot up—The process of loading and running the hardware initialization program to allow access to hardware resources by applications.

Break—*See* control-break.

Bridge—A network device used to interconnection one or more different networks to act as if they were part of the same network. Bridging different private networks is a typical application. Bridging two Internet connections, or an Internet connection fully onto a LAN together, is considered a no-no. In wireless networking, a bridge may be two wireless networking devices tied back-to-back to interconnect different wireless LANs or act as a repeater for client systems.

BUFFERS—A small area of memory used to temporarily store information being transferred between your computer hardware and a disk drive. This is a settable parameter in the `CONFIG.SYS` file. Common values range from 3–30, as `BUFFERS=x`.

Built-in command—A command or service that loads with and is available as part of the DOS command processor program, `COMMAND.COM`. `DIR`, `COPY`, `DEL`, `TYPE`, and `CLS` are examples of some internal DOS commands. *See also* internal command and your DOS manual.

Bulletin board service (BBS)—Personal or commercial information systems accessible by modems, that were popular before the Internet became a common means of online communications.

Burn-in—The process of running diagnostic or repetitive test software on some or all components of and in a PC system for an extended period of time under controlled conditions. This process helps verify functionality and sort out weak or defective units before they are delivered or used under normal working conditions.

Bus—An internal wiring configuration between the central processing unit (CPU) and various interface circuits carrying address, data, and timing information required by one or more internal, built-in, add-in, or external adapters and devices.

Byte—The common unit of measure of memory, information, file size, or storage capacity. A byte consists of 8 bits of information. There are typically 2 bytes to a word (typically 16 bits) of information. 1,024 bytes is referred to as a kilobyte or K, and contains 8,192 bits of information.

Cache—A reserved storage area used to hold information enroute to other devices, memory, or the CPU. Information that is called for during a disk-read operation can be read into a cache with additional information "stock-piled" ahead of time so that it is available for use faster than having to wait for a disk's mechanical and electronic delays. Caching is becoming common between disks and the computer data bus or CPU, and between the memory and CPU. to speed up a system's operation. Some CPU chips and controller cards include caching as part of their design.

Cellular digital packet data (CDPD)—A technology for transmitting data over cellular phone frequencies. It uses unused cellular channels in the 800- to 900-MHz range. Data transfer rates of 19.2 Kbps are possible.

Central processing unit (CPU)—The main integrated circuit chip, processor circuit, or board in a computer system. For IBM PC-compatible systems, the CPU may be an Intel or comparable 8088, 8086, 80286, 80386 (SX or DX), 80486 (SX or DX), Pentium, NEC V20 or V30, or other manufacturer's chip.

Challenge Handshake Authentication Protocol (CHAP)—A secure authentication method in which a host system sends a one-time data value and ID information, combined with a shared secret both sides know. A hash value is calculated at the client and sent back to the host, and if the correct data matches up, the host allows the client to communicate. CHAP may be used one time or multiple times randomly, to continue assurance of security without intrusion.

Checksum—An error-checking method used in file reading and writing operations to compare data sent with checksum information, sent to verify correct reception of the information.

Cluster—The smallest unit of measure of disk storage space under PC or MS-DOS. A cluster typically consists of four or more sectors of information storage space, and contains 2,048 or more bytes of storage capacity. *See* sector.

CMOS clock—A special clock chip that runs continuously, either from the PC system power supply or a small battery, providing date and time information.

CMOS RAM—A special memory chip used to store system configuration information. Rarely found in PC or XT models and usually found in 286 or higher models.

CMOS setup—The process of selecting and storing configuration (device, memory, date, and time) information about your system for use during boot up. This process may be done through your PC's basic input/output system (BIOS) program or an external (disk-based) utility program.

Coax, coaxial cable, feedline—Cable created with a concentric design, containing one center conductor, to carry signal energy surrounded by a shielding conductor. A practical way to transfer radio energy from transmitter to an antenna or from an antenna to a receiver.

Code division multiple access (CDMA)—A digital cellular phone technology that uses spread-spectrum techniques. Every channel uses the full available spectrum. Individual conversations are encoded.

Command—A word used to represent a program or program function that you want your computer to perform. Commands are issued by you, through the keyboard or mouse, to tell the computer what to do.

Command line—The screen area immediately after a prompt, where you key in commands to the computer or program. This is most commonly the "DOS command line," as indicated by the DOS prompt (C>, **C:\>**, or similar).

Command-line editing—The process of changing displayed commands before entering or starting the commanded activity.

Communications program—An application program that is used to simulate a computer data terminal, when communicating with a computer at another location by modem or data communication line. Such programs often provide color display features, modem command setups, telephone number dialing directories, and script or batch file-like automatic keystroke and file transfer functions.

CONFIG.SYS—An ASCII text file that may contain one or more lines of special DOS commands that you want executed every time you boot up your PC. Also known as the "config" file, this file can be customized using a text editor program, so that you can specify one or more items specific to how your system should operate when it boots up. You may specify device drivers (with DEVICE=), such as memory management programs, disk caching, RAM disks; the number of files and buffers you want DOS to use; the location, name, and any special parameters for your command processor (usually COMMAND.COM), among other parameters. Refer to your DOS manual or device driver software manual for specific information.

Control-Alt-Delete or Ctrl-Alt-Del—The special key sequence used to cause a reboot of a PC system. If there has been no alteration of the special reboot byte code in low memory since the system was turned on, a warm boot or faster reset of the computer will occur. If the reboot code has been changed, the system may restart with a complete Power On Self Test (POST) test, including RAM memory count. Some systems contain special test code that may be activated in place of POST, by setting of the reboot byte and adding a test jumper on the system board. This latter feature is not well documented and may not be available on all systems.

Control-Break—A combination entry of the Control (Ctrl) and Break (also Pause) keys that can interrupt and stop a program's operation and return the computer to the operating system (DOS). This is also a more robust or stronger version of Ctrl-C key sequence to abort a program. Checking for Control-Break is enhanced by setting BREAK ON in CONFIG.SYS or in DOS. Many programs intercept and do not allow Control-Break to pass to DOS because doing so might cause data loss or corrupt a number of open files in use by a program.

Control-C—A keystroke combination of the Control (Ctrl) and C keys that can interrupt and stop the operation of many programs.

Control code—A combination of keystrokes used by many programs, or during on-line sessions, to cause special functions or operations to occur. Commonly used control codes are Ctrl-S to stop a display from scrolling so it can be viewed more easily, and Ctrl-Q to cause the display to continue. These commands are entered by pressing the Ctrl key first, then the accompanying single letter code, much like using the Shift or Alt keys to change the action of a letter or number key.

Controller—*See* adapter.

Conventional memory—Also known as DOS memory, this is the range of your PC's memory from 0–640k, where device drivers, DOS parameters, the DOS command processor (COMMAND.COM), your applications programs and data are stored, when you use your computer. *See* extended, expanded, video, high, and upper memory.

Corner reflector—A special antenna, similar to a dishlike parabolic antenna, that concentrates transmitted and received energy in one direction only. The received and transmitted signals benefit from increased effective power because of the gain in concentrating the signals to the antenna element.

Crash—The unexpected and unwanted interruption of normal computer operations. When a program crashes, all open data files may be corrupted or lost, and it is possible that hardware may get "stuck" in a loop, with the computer appearing dead or "confused." Recovery from a program crash usually requires a reboot or turning off of power for a few seconds, then restarting the system. A disk crash is normally associated with the improper mechanical contact of the read/write heads with the disk platter, although many people consider any disk error or data loss as a crash.

Current directory—This is the subdirectory you or a program has last selected to operate from that is searched first before the DOS PATH is searched when calling a program. *See also* current disk drive and logged drive.

Current disk drive—The drive that you have selected for DOS and programs to use before searching the specified drives and directories in the DOS PATH (if any is specified). This may also be the drive indicated by your DOS prompt (typically C>, or C:\>, or similar) or that you have selected by specifying a drive letter, followed by a colon and the Enter key, as in A Enter. This is also known as the logged drive.

Cursor—A line or block character on your system display screen, usually blinking, that indicates where characters that you type will be positioned or where the current prompting for input is active. When at the DOS command line, the cursor is normally at the end of the DOS prompt string.

Decibel (dB)—A unit of electrical signal measurement, using a logarithmic scale used as a reference to quantify radio and audio signals—either power gain, loss, or signal strength. Decibels are measured with special equipment, such as spectrum analyzers, or may be calculated based on known electrical factors. Typically based on a specific power level (in watts) into a known load impedance (in ohms—600 ohms for audio, 50 ohms for radio). A reference of 0 decibels is typically 1 mW into a 600 ohm load for audio, and 1 mW into a 50 ohm load for radio. Based on a logarithmic scale, a –10 dB signal is 1/10th and a –3dB signal is 1/2 as strong as a 0 dB signal; a 10 dB signal is 10 times, and a 3 dB signal is twice as strong as a 0 dB signal. Most radios require between –60 and –90 dB to receive a signal clearly. The ambient radio frequency noise in a typical clear, clean reception area ranges from –120 to –100 dB, so a receivable signal must be 30 to 40 dB stronger than the noise.

Default—A predetermined or normal value or parameter used by a program or the computer as the selected value, if you do not or cannot change it by a command or responding to a prompt for input.

Defragment—The process of reorganizing disk files so that they occupy contiguous sectors and clusters on a disk. This is done to reduce the access time (movement of the data read/write heads) needed to read a single data file.

Destructive testing—Testing of memory or disk drives that overwrites the original or existing data, without regard for restoring it, upon completion of the test process.

Device—An actual piece of hardware interfaced to the computer to provide input or accept output. Typical devices are printers, modems, mice, keyboards, displays, and disk drives. There are also some special or virtual devices, handled in software, that act like hardware. The most common of these is called NUL, which is essentially nowhere. You can send screen or other output to the NUL device so that it does not appear. The NUL device is commonly used if the actual device to send something to does not exist, but a program requires that output be sent someplace. NUL is a valid "place" to send output to, although the output really does not go anywhere.

Device driver—A special piece of software required by some hardware or software configurations to interface your computer to a hardware device. Common device drivers are `ANSI.SYS`, used for display screen control; `RAMDRIVE.SYS`, which creates and maintains a portion of memory that acts like a disk drive; and `HIMEM.SYS`, a special device driver used to manage a specific area of extended memory called the high memory area (HMA). Device drivers are usually intended to be used in the CONFIG.SYS file, preceded by a DEVICE= statement. With Windows NT, 2000, ME, and XP, device drivers are loaded within the operating system structure, sometimes automatically and dynamically.

Diagnostics—Software programs to test the functions of system components.

Digital rights management (DRM)—A generic term referring to various methods of copy or distribution protection for digital information—MP3 music files, CD- and DVD-based content, and electronically transmitted information. Used to protect copyright and intellectual property distribution to entitle only those authorized to use the information.

Digital subscriber line (DSL [also xDSL, IDSL, ADSL, HDSL])—A technique of providing high-speed digital communications over conventional telephone wires, using signaling above and different from voice-range frequencies. Implemented in various combinations of upward and downward bandwidth, telephone line, and equipment types. Typically lower cost and higher performance than integrated services digital network (ISDN), depending on the implementation. It is possible to carry DSL signaling over some ISDN and Frame Relay circuits for 144 to 192 Kbps transfer rates, or on specially conditioned wire pairs to achieve T-1 (1.54 Mbps) data rates. A symmetric DSL line can operate as fast as a T-1 line, but the data rate is not guaranteed.

Digital versatile disc (DVD)—A CD-ROM storage media capable of handling 4.7 to 17 gigabytes of information. DVD supports rich multimedia information and menu systems to replicate track, scene, and other specific controls to access stored information.

DIN connector—A circular multiwire electronic connector based on international (German) standards. Available in normal and miniature sizes, with 3 to 7 connection pins. The PC uses 5-pin normal and 6 pin mini-DIN connectors for keyboards, and 6-pin mini-DIN connectors for pointing devices.

Direct memory access (DMA)—A method of transferring information between a computer's memory and another device, such as a disk drive, without requiring central processing unit (CPU) intervention.

Direct-sequence spread spectrum (DSSS)—One of two types of spread spectrum radio, the other being frequency hopping spread spectrum. DSSS divides the user data according to a spreading ratio. A redundant bit pattern is included for each bit that is transmitted, to reduce the possibility of interference. If the bit pattern is damaged during transmission, the data can be recovered.

Directory—File space on disks used to store information about files organized and referred to through a directory name. Each disk has at least one directory, called the root directory, which is a specific area reserved for other file and directory entries. A hard disk root directory may contain up to 512 other files or directory references, limited by the amount of disk space reserved for root directory entries. The files and directories referred to by the root directory may be of any size, up to the limit of available disk space. Directories may be thought of as folders or boxes, as they may appear with some graphical user-interfaces, although they are not visually represented that way by DOS. *See* root directory and subdirectories. All directories, except for the root directory, must have a name. The name for a directory follows the one to eight character restrictions that apply to file names for DOS-only systems. Windows 95 and higher systems enjoy both longer file and directory names. *See also* file name. The term directory has been displaced by folder, though the concept and implementation are the same.

Disk—A rotating magnetic medium used for storing computer files. *See also* diskette and hard disk.

Disk-bound servo track—The data used by a disk drive to position and verify the location of the data read/write heads. This data may be mixed with the user's data, or on separate data tracks on the disk medium.

Disk cache—A portion of memory set aside to store information that has been read from a disk drive. The disk cache memory area is reserved and controlled by a disk caching program that you load in `CONFIG.SYS` or `AUTOEXEC.BAT`. The caching program intercepts a program or DOS request for information from a disk drive, reads the requested data, plus extra data areas, so that it is available in memory, which is faster than a disk drive. This is commonly referred to as read-ahead caching. The cache may also be used for holding information to be written to disk, accepting the information faster than the disk can accept it, and then writing the information to disk a short time later.

Disk drive adapter—A built-in or add-in card interface or controller circuit that provides necessary connections between the computer system input/output (I/O) circuits and a disk drive.

Disk label—1. A surface or sticker on the outside jacket of a diskette that is used for recording information about the contents of the disk. This label may contain as much information as you can write or type in the space provided.

2. A specific area on a disk used to record data as the disk's name or volume label. This area is written with the DOS LABEL command, or prompted for input during certain disk format processes. A volume label may be up to 11 characters long. The volume label will appear on-screen during disk directory operations.

References to the disk label may not be clear about which "label" is to be used. You may use the two definitions above to help determine which label is being referred to by the limitations for each, and the reference you are given.

Disk operating system (DOS)—A set of software written for a specific type of computer system, disk, file, and application types to provide control over disk storage services and other input and output functions required by application programs and system maintenance. All computers using disk drives have some form of disk operating system containing applicable programs and services. For IBM PC-compatible computers, the term DOS is commonly accepted to mean the computer software services specific to PC systems.

Diskette—Also called a floppy diskette, this is a disk media contained in a cover jacket that can be removed from a disk drive. The term *floppy* is deemed synonymous or descriptive of the flexible medium that is the magnetically coated disk of thin plastic material.

DOS diskette—A diskette formatted for use with DOS-based PCs and file system.

DOS memory—Temporary memory used for storage of DOS boot and operating system information, programs, and data during the operation of your computer system. DOS memory occupies up to the first 640 K of random access memory (RAM) space provided in your system's hardware. This memory empties out or loses its contents when your computer is shut off.

DOS system diskette—A diskette formatted for use with DOS-based PCs and file system that also contains the two DOS-system hidden files and `COMMAND.COM` to allow booting up your system from a diskette drive.

Download—The process of receiving or transferring information from another computer, usually connected via modem, onto your computer system. Downloading is a common method of obtaining public domain and shareware programs from bulletin board services (BBSs) and on-line services, obtaining software assistance and upgrades from many companies, or retrieving files or electronic mail from others.

Drive—The mechanical and electronic assembly that holds disk storage media and provides the reading and writing functions for data storage and retrieval.

Dual in-line memory module (DIMM)—A high-density memory packaging system consisting of 168-pins, similar to the edge connector used on larger printed circuit cards. DIMM is used in addition to or in place of single in-line memory module (SIMM) memory design.

Dual in-line package (DIP)—A form of integrated circuit housing and connection with two rows of pins on either side of a component body.

Dual in-line package (DIP) switch—A small board-mounted switch assembly resembling a DIP integrated circuit (IC) package in size and form. Used for the selection of system addresses and options.

Dynamic random access memory (DRAM)—Relatively slow (50 to 200 nSec access time) economical memory integrated circuits. These require a periodic refresh cycle to maintain their contents. Typically used for the main memory in the PC system, but occasionally also used for video memory. *See also* random access memory and static random access memory.

Dynamically linked library (DLL)—A file containing executable program functions that are invoked from another program. DLLs may be shared among many applications and are used only when a program requires the functions contained within, reducing program memory and disk space requirements by eliminating duplication of program elements and file size.

EAP-Cisco wireless (LEAP)—An authentication method used primarily for wireless local area network (WLAN) clients connecting to Cisco WLAN access points, such as the Cisco Aironet Series. It provides security during credential exchange, encrypts data transmission using dynamically generated wired equivalent privacy (WEP) keys, and supports mutual authentication and reauthentication.

EAP-MD5—An authentication method that essentially duplicates challenge handshake authentication protocol (CHAP) password protection on a wireless local area network. EAP-MD5 represents a form of extensible authentication protocol (EAP) support among 802.1x devices. EAP-MD5 is supported on Odyssey Client.

EAP-TLS—A follow-on to secure socket layer (SSL). It provides strong security, but relies on client certificates for user authentication.

EAP-TTLS—At this writing EAP-TTLS, an Internet Engineering Task Force (IETF) draft for a data communications security standard. The proposed standard provides strong security, while supporting legacy password protocols, enabling easy deployment.

Edge connector—An electronic connector that is part of the circuit card, made of circuit foil extended to the edge of the board. A circuit card's edge connector mates with the fingers inside a complementary female socket.

Electronic Industries Alliance (EIA)—An organization that provides and manages standards for many types of electronics designs and implementations. The RS-232C standard for serial data terminal and computer interconnection is the most commonly known EIA standard in the PC market.

Enhanced small device interface (ESDI)—A standards definition for the interconnection of older high-speed disk drives. This standard is an alternative to earlier MFM, coincident applications of small computer system interface (SCSI), and recent integrated drive electronics (IDE) drive interfaces.

Enter key—The command or line termination key, also known as Return on your keyboard. There are usually two Enter keys on your keyboard. Under some applications programs, these two keys may have different functions; the numeric keypad Enter key may be used as an "enter data" key, while the alphanumeric keyboard Enter key may be used as a "carriage return."

Environment—An area of memory set up and used by the DOS software to store and retrieve a small amount of information that can be shared or referred to by many programs. Among other information that the DOS environment area can hold are the PATH, current drive, PROMPT, COMSPEC, and any SET variables.

Escape sequence—A set of commands or parameters sent between devices to control operations; print text orientation or fonts, screen colors, and displays; or begin file transfer operations between systems. Many printers accept escape sequences to change typeface or between portrait and landscape modes. Screen displays and the DOS prompt may be controlled by ANSI escape sequences through the device driver ANSI.SYS. These sequences are started with the transmission or issuance of the ASCII ESC character (appearing similar to <-) or the ASCII control code Ctrl-Left Bracket (^[, decimal 27, 1B hex), and follow with lettered or numbered command definitions. A common sequence is ESC-2-j, possibly appearing as ^[2J on your screen, which is the Clear Screen ANSI escape sequence.

Executable file—A program file that may be invoked from the operating system. Dynamically linked libraries (DLLs) and overlay files also contain executable program information, but their functions must be invoked from within another program.

Execute—The action that a computer takes when it is instructed to run a program. A running program is said to "execute" or "be executing" when it is being used.

Expanded memory—This is an additional area of memory created and managed by a device driver program using the Lotus-Intel-Microsoft Expanded Memory Specification, known also as LIMS-EMS. There are three common forms of EMS; that conforming to the LIMS-EMS 3.2 standard for software-only access to this memory, LIMS-EMS 4.0 in software, and LIMS-EMS 4.0 in hardware. With the proper hardware, this memory may exist and be used on all PC systems, from PCs to 486 systems. Expanded memory may be made up of extended memory (memory above 1 MB) on 386 and 486 systems, or it may be simulated in extended memory on 286 systems. LIMS-EMS 3.2, 4.0 (software) and 4.0 (hardware) are commonly used for additional data storage for spreadsheets and databases. Only LIMS-EMS conforming to the 4.0 standard for hardware may be used for multitasking. Expanded memory resides at an upper memory address, occupying one 64 K block between 640 K and 1 MB. The actual amount of memory available depends on your hardware and the amount of memory you can assign to be expanded memory. The 64 K block taken up by expanded memory is only a window or port giving access to the actual amount of EMS available. There may be as little as 64 K or as much as 32 MB of expanded memory.

Expanded memory manager (EMM)—The term often given to the software or that refers to expanded memory chips and cards. *See also* expanded memory.

Expanded memory specification (EMS)—The IBM PC-industry standards for software and memory hardware that makes up expanded memory.

Extended Industry Standard Architecture (EISA)—The definition of a PC internal bus structure that maintains compatibility with IBM's original PC, XT, and AT bus designs (known as the ISA, or industry standard architecture), but offering considerably more features and speed between the computer system and adapter cards, including a definition for 32-bit PC systems that do not follow IBM's MCA (MicroChannel Architecture).

Extended memory—This is memory in the address range above 1 MB, available only on 80286 or higher systems. It is commonly used for random access memory (RAM) disks, disk caching, and some applications programs. Using a special driver called HIMEM.SYS, or similar services provided with memory management software, the first 64 K of extended memory may be assigned as a high memory area, which can be loaded into some programs and DOS.

Extended memory specification (XMS)—A standard that defines access and control over upper, high, and extended memory on 286 and higher computer systems. XMS support is provided by loading the HIMEM.SYS device driver or other memory management software that provides XMS features.

Extended service set (ESS)—Multiple base service set (BSS) devices forming a network.

Extensible authentication protocol (EAP)—An extension of the point-to-point (PPP) protocol that allows different, multiple authentication methods for access control.

External command—A program or service provided as part of DOS that exists as separate programs on disk rather than built into the `COMMAND.COM` program that loads when you boot up your system. These programs have `.COM` or `.EXE` extensions. Some of these are `FORMAT.COM`, `DISKCOPY.COM`, `DEBUG.EXE`, `LABEL.COM`, `MORE.COM`, and `PRINT.COM`.

FDISK—A special part of the hard disk formatting process required to assign and establish usable areas of the disk as either bootable, active, data-only for DOS, or as non-DOS for other operating system use. The `FDISK` process is to be performed between the low-level format and the DOS format of a hard disk prior to its use.

File—An area of disk space containing a program or data as a single unit, referred to by the DOS file directory. Its beginning location is recorded in the file directory, with reference to all space occupied by the file recorded in the DOS file allocation table (FAT). Files are pieces of data or software that you work with on your computer. They may be copied, moved, erased, or modified, all of which is tracked by DOS for the directory and FAT.

File allocation table (FAT)—This is DOS' index to the disk clusters that files or FAT and directories occupy. It provides a table or pointer to the next disk cluster a file occupies. There are two copies of the FAT on a disk, for reliability. When files are erased, copied, moved, reorganized, or defragmented, the FAT is updated to reflect the new position of files or the availability of empty disk space. Files may occupy many different cluster locations on disk, and the FAT is the only reference to where all of the file pieces are.

File attributes—*See* attributes.

File name—The string of characters assigned to a disk file to identify it. A file name must be at least one, and may be up to eight, leading characters as the proper name for DOS-only systems, in which a file name may be followed by a three character extension, separated from the proper name by a period (.). Windows 95, Windows 98, and Windows NT systems may have long file names of up to 256 characters, including multiple period or 'dot' separators. Allowable file name and extension characters are—A-Z, 0-9, !,@,#,$,^,&,_,-,{,},(,).',`,or ~. Also, much of the IBM extended character set may be used. Reserved characters that cannot be used are—%, *, +, =, ;, :,[,], <, >, ?, /, \, |, " and spaces. File names must be unique for each file in a directory, but the same name may exist in separate directories. Filenames are assigned to all programs and data files.

File name extension—A string of one to three characters used after a file name and a separating period (.), with the same character limitations as the file name, for DOS systems. The extension is often used to identify and associate certain types of files to certain applications. DOS uses `BAT`, `EXE`, and `COM` as files it can load and execute, though this does not preclude the use of these extensions for nonexecutable files. The extensions `SYS`, `DRV`, and `DVR` are commonly used for device driver programs that are loaded and used in the `CONFIG.SYS` file prior to loading DOS (as `COMMAND.COM`). Refer to your software documentation for any limitations or preferences it has for file name extensions.

Filespec—Also known as the file specification or file specifier, this is a combination of a drive designation, directory path, and file name used to identify a specific file in its exact location on your system's disk drive. References to filespec may appear in examples or as prompts as—`d:\path\filename.ext`, where `d:` indicates that you are supposed to place you disk drive information here, `\path\` indicates that you should specify the proper directory and subdirectory information here, and `filename.ext` indicates that you should specify the file's exact name and extension. In use, this might actually be `C:\DOS\COM\FORMAT.COM`.

Firewire—Texas Instrument's name-brand for the Institute of Electrical and Electronics Engineers, Inc. (IEEE)-1394 high-speed serial interconnection standard. Firewire connections are typically used between high-end digital video cameras and peripheral storage devices.

Firmware—Software embedded into a device such as a disk drive, video, or network adapter; wireless access point; or PC card, that controls and supports the functions of the device. The PC's basic input/output system (BIOS) and the startup code for most computers is firmware specific to the hosting computer board. Firmware resides in either read-only memory chips or in FLASH ROM rewriteable memory chips. The operating system used in personal digital assistants (PDAs) may also be considered firmware.

First-in, first-out (FIFO) or FIFO buffering—A small capacity data storage element, memory or register that holds data flowing between a source and a destination. The data flow moves in the order in which it is received and cannot be accessed directly or randomly as with normal memory storage. A FIFO is commonly used in serial communication (COM) ports to retain data while applications software and storage devices catch up to and can store the incoming stream of data.

Fixed disk—*See* hard disk.

Flag—A hardware bit or register, or a single data element in memory that is used to contain the status of an operation, much like the flag on a mailbox signals the mail delivery person that you have an item to be picked up.

Floppy disk—A slang term. *See* diskette.

Format—The process of preparing a disk,(floppy or hard) with a specific directory and file structure for use by DOS and applications programs. Formatting may consist of making the disk usable for data storage only, providing reserved space to make the disk bootable later on, or making the disk bootable, including the copying of the DOS hidden files and `COMMAND.COM`. `FORMAT` is the final process of preparing a hard disk, preceded by a low-level format and `FDISK`. All disk media require a format. Random access memory (RAM) or virtual disks do not require formatting. Formatting, unless performed with certain types of software, erases all data from a disk.

Fragmentation threshold—A parameter available in some access point and client wireless devices. If you experience a high packet error rate, a slight increase in this value to the maximum of 2,432 may help. Too low a value may result in very poor performance.

Frame relay—A data communications circuit between two fixed points, a user and a Frame Relay routing service, capable of transfer rates between 64 Kbps up to T-1 rates. May be carried over part of a "Fractional T-1" circuit.

Frequency-hopping spread spectrum (FHSS)—FHSS is one of two types of spread spectrum radio; the other being direct sequence spread spectrum. FHSS is used where the data signal modulates a narrowband carrier that "hops" in a random, but predictable sequence from frequency to frequency. The signal energy is spread in time domain rather than chopping each bit into small pieces across multiple frequencies. FHSS is not as prone to interference because a signal from another system will only affect this signal if both are transmitting at the same frequency at the same time.

Gateway—1. The Internet protocol (IP) address of the router, switch, cable, or digital subscriber line (DSL) modem through which your PCs gain access to the Internet or foreign (nonlocal) networks. 2. Network equipment that either bridges, repeats, or otherwise relays network traffic from one connection to another.

Gigabyte (GB)—A unit of measure referring to 1,024 MB or 1,073,741,824 bytes of information, storage space, or memory. Devices with this capacity are usually large disk drives and tape backup units with 1.2 to well over 12 GB of storage area.

Global system for mobile (GSM) communications—One of the leading digital cellular phone systems, using narrowband time division multiple access (TDMA), which allows eight simultaneous calls on the same radio frequency. It has little to do with wireless networking, but is one of many technologies tossed into the generic wireless arena.

Hard disk—A sealed disk drive unit with platters mounted inside on a fixed spindle assembly. The actual platter is a hard aluminum or glass surface coated with magnetic storage media. This definition also suits removable hard disks in which the hard platters are encased in a sealed casing and mate with a spindle similar to the attachment of a floppy diskette to the drive motor. The platters are sealed to keep foreign particles from interfering with and potentially damaging the platters or the read/write heads that normally maintain a small gap between them during operation.

Hardware interrupt—A signal from a hardware device connected to a PC system that causes the central processing unit (CPU) and computer program to act on an event that requires software manipulation, such as controlling mouse movements, accepting keyboard input, or transferring a data file through a serial input/output (I/O) port.

Head crash—The undesired, uncontrolled mechanical contact of a disk drive's read/write heads with the disk surface. A minor crash may be recoverable with minimal data loss. A severe crash can render a disk or the head assembly completely useless. Minor to severe head crashes may be caused by mechanical shock, excessive vibration, or mishandling of a drive while it is operating. Not all disk errors or loss of data are the result of a physical crash and disk surface damage. Actual head crashes with disk damage are very rare, compared with loss of data due to the weakening of magnetic properties of an area of the disk, and program or operational errors.

Hexadecimal—A base-16 numbering system made up of four digits or bits of information, where the least significant place equals one and the most significant place equals eight. A hexadecimal, or hex, number is represented as the numbers 0–9 and letters A–F, for the numerical range 0–15 as 0–F. A byte of hex information can represent from 0 to 255 different items, as 00 to FF.

Hidden file—*See* attributes.

High memory area (HMA)—A 64 K region of memory above the 1 MB address range created by `HIMEM.SYS` or a similar memory utility. The HMA can be used by one program for program storage, leaving more space available in the DOS or the low memory area from 0 to 640 K.

High performance file system (HPFS)—A secure hard disk file system created for OS/2 and extended into the NT file system for Windows NT.

Host adapter—A built-in or add-in card interface between a device, such as a small computer system interface (SCSI) hard disk or CD-ROM drive, and the input/output (I/O) bus of a computer system. A host adapter typically does not provide control functions, instead acting only as an address and signal conversion and routing circuit.

Hub—A network device used to connect several network client devices onto the same network segment. *See also* switch.

IBM PC compatible—A description of a personal computer (PC) system that provides the minimum functions and features of the original IBM PC system and is capable of running the same software and using the same hardware devices.

IEEE-1394—An Institute of Electrical and Electronics Engineers, Inc. (IEEE)-1394 standard for high-speed serial interconnection between computer peripherals—typically cameras and data storage systems.

Industrial, scientific, and medical (ISM)—ISM applications are the production of physical, biological, or chemical effects such as heating, ionization of gases, mechanical vibrations, hair removal, and acceleration of charged particles. Uses include ultrasonic devices such as jewelry cleaners and ultrasonic humidifiers, microwave ovens, medical devices such as diathermy equipment and magnetic resonance imaging equipment, and industrial uses such as paint dryers. Radio frequency should be contained within the devices, but other users must accept interference from these devices. These devices can affect 802.11a and 802.11b services at 2.4 and 5 GHz.

Industry Standard Architecture (ISA)—The term given to the IBM PC, XT, and AT respective 8- and 16-bit PC bus systems. Non-32-bit, non-IBM MicroChannel Architecture systems are generally ISA systems.

Infrastructure mode—An integrated wireless and wired LAN is called an infrastructure configuration. Infrastructure is applicable to enterprise scale for wireless access to central database, or wireless application for mobile workers.

Input/output (I/O)—The capability or process of software or hardware to accept or transfer data between computer programs or devices.

Insulation displacement connector (IDC)—The type of connector found on flat ribbon cables, used to connect input/output (I/O) cards and disk drives.

Integrated drive electronics (IDE)—A standards definition for the interconnection of high-speed disk drives, in which the controller and drive circuits are together on the disk drive and interconnect to the PC input/output (I/O) system through a special adapter card. This standard is an alternative to earlier MFM, ESDI, and SCSI drive interfaces, and it is also part of the ATA standard.

Integrated services digital network (ISDN)—A technique of providing high-speed digital communications over conventional telephone wires, using signaling above and different from voice-range frequencies. ISDN uses three different signal channels over the same pair of wires, one D-channel for digital signaling such as dialing, and several enhanced, but seldom used telephone calling features, and two B-channels, each capable of handling voice or data communications up to 64 Kbps. ISDN lines may be configured as Point-to-Point (both B-channels would connect to the same destination) or multipoint (allowing each B-channel to connect to different locations), and Data+Data (B-channels can be used for data-only) or Data+Voice, where either B-channel may be used for data or voice transmission. Interconnection to an ISDN line requires a special termination/power unit, known as an NT-1 (network termination 1), which may or may not be built into the ISDN modem or router equipment at the subscriber end. An ISDN modem may be used and controlled quite similarly to a standard analog modem, and may or may not also provide voice-line capabilities for analog devices. An ISDN router must be configured for specific network addresses and traffic control and may or may not provide voice/analog line capabilities.

Interlaced operation—A method of displaying elements on a display screen in alternating rows of pixels (picture elements) or scans across a display screen, as opposed to noninterlaced operation, which scans each row in succession. Interlacing often indicates a flickering or blinking of the illuminated screen.

Interleave—The property, order, or layout of data sectors around disk cylinders to coincide with the speed of drive and controller electronics, so that data can be accessed as quickly as possible. An improper interleave can make a sector arrive too soon or too late at the data heads, and thus be unavailable when the drive and controller are ready for it, slowing disk system performance. An optimal interleave will have the rotation of the disk, placement of a data sector, and electronics coincident, so there is little or no delay in data availability. Interleave is set or determined at the time of a low-level format, which sets the order of the data sectors. Reinterleaving consists of shuffling data sectors to a pattern optimal for best performance.

Internal command—A command that loads with and is available as part of the DOS command processor program, `COMMAND.COM`. `DIR`,

COPY, DEL, TYPE, and CLS are examples of some internal DOS commands. Internal command is the same as Built-in command. Also see your DOS manual.

International Standards Organization (ISO)—A multifaceted, multinational group that establishes cross-border/cross-technology definitions for many industrial and consumer products. Related to the PC industry, it helps define electronic interconnection standards and tolerances.

Internetwork packet exchange (IPX)—1. A networking protocol, IPX is a datagram protocol used for connectionless communications.

2. A device driver-type TSR program that interfaces a network interface card to the operating system. *See also* NETX.

Interrupt—*See* hardware interrupt, interrupt request, and software interrupt.

Interrupt request (IRQ)—This is a set of hardware signals available on the PC add-in card connections that can request prompt attention by the central processing unit (CPU) when data must be transferred to/from add-in devices and the CPU or memory.

Keyboard—A device attached to the computer system that provides for manual input of alpha, numeric, and function key information to control the computer or place data into a file.

Kilobyte (kB)—A unit of measure referring to 1,024 bytes or 8,192 bits of information, storage space, or memory.

Label or volume label—A 1- to 11-character name recorded on a disk to identify it during disk and file operations. The volume label is written to disk with the DOS LABEL or FORMAT programs or with disk utility programs. This may be confused with the paper tag affixed to the outside of a diskette. *See* disk label.

Language—The specifically defined words and functions that form a programming language or method to control a computer system. At the lowest accessible level, programmers can control a central processing unit's (CPU's) operations with assembly language. Applications programs are created initially in different high-level languages, such as BASIC, C, or Pascal, which are converted to assembly language for execution. DOS and applications may control the computer's operations with a batch (BAT) processing language or an application-specific macro language.

Lightweight extensible authentication protocol (LEAP)—An implementation of EAP, providing access control and security.

Liquid crystal display (LCD)—A type of data display that uses microscopic crystals, which are sensitive to electrical energy, to control whether they pass or reflect light. Patterns of crystals may be designed to form characters and figures, as are the small dots of luminescent phosphor in a CRT (display monitor or TV picture tube).

Loading high—An expression for the function of placing a device driver or executable program in a high (XMS, above 1 MB) or upper memory area (between 640 K and 1 MB.) This operation is performed by a `DEVICEHIGH` or `LOADHIGH` (DOS) statement in the `CONFIG.SYS` or `AUTOEXEC.BAT` file. High memory areas are created by special memory manager programs such as `EMM386` (provided with versions of DOS) and Quarterdeck's `QEMM386`.

Local area network (LAN)—An interconnection of systems and appropriate software that allows the sharing of programs, data files, and other resources among several users.

Local bus—A processor to input/output (I/O) device interface alternative to the PC's standard I/O bus connections, providing extremely fast transfer of data and control signals between a device and the central processing unit (CPU). It is commonly used for video cards and disk drive interfaces to enhance system performance. Local Bus is a trademark of the Video Electronics Standards Association. Local Bus has since been displaced by peripheril component interconnect (PCI) and advanced graphics port (AGP).

Logged drive—The disk drive you are currently displaying or using, commonly identified by the DOS prompt (`C>` or `A:\>`). If your prompt does not display the current drive, you may do a `DIR` or `DIR/p` to see the drive information displayed.

Logical devices—A hardware device that is referred to in DOS or applications by a name or abbreviation that represents a hardware address assignment, rather than by its actual physical address. The physical address for a logical device may be different. Logical device assignments are based on rules established by IBM and the read-only memory basic input/output system (ROM BIOS) at bootup.

Logical drive—A portion of a disk drive assigned as a smaller partition of larger physical disk drive. Also a virtual or nondisk drive created and managed through special software. Random access memory (RAM) drives (created with `RAMDRIVE.SYS` or `VDISK.SYS`) or compressed disk/file areas (such as those created by older Stacker, DoubleDisk, or SuperStor disk partitioning and management programs) are also logical drives. A 40 MB disk drive partitioned as drives C and D is said to have two logical drives. That same disk with one drive area referred to as C has only one logical drive, coincident with the entire physical drive area. DOS may use up to 26 logical drives. Logical drives may also appear as drives on a network server or mapped by the DOS `ASSIGN` or `SUBST` programs.

Logical pages—Sections of memory that are accessed by an indirect name or reference, rather than by direct location addressing, under control of a memory manager or multitasking control program.

Loopback plug—A connector specifically wired to return an outgoing signal to an input signal line for the purpose of detecting if the output signal is active or not, as sensed at the input line.

Loss—The reduction of signal intensity as a function of distance from the transmitting station, electrical characteristics of transmission line (transmitter to antenna or antenna to receiver cabling), attenuation of signals due to natural and man-made obstructions, as well as intervening connectors and adapters in antenna cabling systems. Loss is a major factor when cabling to external antennas to client-side adapter cards or access point devices, and in many forms of construction. Because wireless networking uses very, very high frequencies, loss factors are considerable at every step.

Lotus-Intel-Microsoft Standard (LIMS)—*See* Expanded memory.

Lower memory—*See* DOS memory.

Math coprocessor—An integrated circuit designed to accompany a computer's main central processing unit (CPU) and speed floating point and complex math functions that would normally take a long time if done with software and the main CPU. Allows the main CPU to perform other work during these math operations.

Media access control (MAC) address—A hardware address that uniquely identifies each node of a network. In IEEE 802 networks, the data link control (DLC) layer of the open systems interconnect (OSI) reference model is divided into two sublayers—the logical link control (LLC) layer and the media access control (MAC) layer. The MAC layer interfaces directly with the network media. Consequently, each different type of network media requires a different MAC layer.

Megabyte (MB)—A unit of measure referring to 1,024 K or 1,048,576 bytes of information, storage space, or memory. One MB contains 8,388,608 bits of information. One MB is also the memory address limit of a PC- or XT-class computer using an 8088, 8086, V20, or V30 CPU chip. 1 MB is 0.001 GB.

Megahertz (MHz)—A measure of frequency in millions of cycles per second. The speed of a computer system's main central processing unit (CPU) clock is rated in megahertz.

Memory—Computer information storage area made up of chips (integrated circuits) or other components, which may include disk drives. Personal computers use many types of memory, from dynamic random access memory (RAM) chips for temporary DOS, extended, expanded, and video memory, to static RAM chips for central processing unit (CPU) instruction caching, to memory cartridges and disk drives for program and data storage.

Memory disk—*See* RAM disk.

Metropolitan area network (MAN)—A network connection between two locations, typically a T-1 circuit, but may be integrated services digital network (ISDN), Frame Relay, or other (possibly a virtual private network [VPN] over any Internet connection type) used to bridge local area networks in related office facilities. There is typically a shorter distance between locations than a wide area network (WAN), such as within a city or community.

Microchannel—An input/output (I/O) card interconnection design created by IBM for use in the IBM PS/2 series systems.

Microchannel architecture (MCA)—IBM's system board and adapter card standards for the PS/2 (Personal System/2) series of computers. This is a nonindustry standard architecture (ISA) bus system, requiring the use of different adapter cards and special configuration information than is used on early PC, XT, and AT compatible systems.

Microprocessor—A computer central processing unit contained within one integrated circuit chip package.

Milliwatt (mW)—A unit of power measurement equal to one-thousandth of a watt. Most unlicensed and "Part 15" devices (FRS walkie-talkies) have a transmitted power limit of 100 mW. A portable cellular telephone transmitter output is typically 600 mW.

Modem—An interface between a computer bus or serial input/output (I/O) port and wiring, typically a dial-up telephone line, used to transfer information and operate computers distant from each other. Modem stands for modulator/demodulator. It converts computer data into audible tone sounds that can be transferred by telephone lines to other modems that convert the tone sounds back into data for the receiving computer. Early modems transfer data at speeds of 110 to 300 bits per second (11 to 30 characters per second). Recent technology allows modems to transfer data at speeds of 56,700 bits (5,670 characters or bytes) per second and higher, often compressing the information to achieve these speeds and adding error-correction to protect against data loss due to line noise. Modems typically require some form of universal asynchronous receiver/transmitter (UAR/T) as the interface to the computer bus.

Monochrome display adapter (MDA)—The first IBM PC video system, providing text-only on a one-color (green or amber) display. If you have one of these adapters, you own an antique!

Motherboard—The main component or system board of your computer system. It contains the necessary connectors, components, and interface circuits required for communications between the central processing unit (CPU), memory, and input/output (I/O) devices.

Multicolor graphics array (MCGA)—An implementation of CGA built into IBM PS/2 Model 25 and 30 systems using an IBM analog monitor and providing some enhancements for higher resolution display and gray-scale shading for monochrome monitors.

Multipath—Multiple reflections of a radio frequency signal between a receiver and transmitter that can often cause multiple signals to arrive at the receiving station at the same time, occasionally canceling out each other and the main, direct line-of-sight signal. Multipath instances are one of the major causes of failure of wireless networking.

Multipoint microwave distribution system, multichannel multipoint distribution system (MMDS)—A wireless technology used to transmit large amounts of data, video, or other information within 6 MHz wide channels. MMDS has been used for a variety of subscription-based television systems, and more recently, for high-speed Internet access. MMDS systems are closed/private and require special equipment and authorization from the provider to access the system's content.

Multitasking—The process of software control over memory and central processing unit (CPU) tasks allowing the swapping of programs and data between active memory and CPU use to a paused or nonexecuting mode in a reserved memory area, while another program is placed in active memory and execution mode. The switching of tasks may be assigned different time values for how much of the processor time each program gets or requires. The program you see on-screen is said to be operating in the foreground and typically gets the most CPU time, while any programs you may not see are said to be operating in the background, usually getting less CPU time. DESQview and Windows are two examples of multitasking software in common use on PCs.

Musical instrument device interface (MIDI)—An industry standard for hardware and software connections, control, and data transfer between like-equipped musical instruments and computer systems.

Negative acknowledge (NACK)—A signal sent by a receiving device indicating that sent information was not received. The opposite of ACK.

Neighborhood area network (NAN)—Typical ad hoc wireless network installed by a neighbor with an 802.11x access point at a location providing a high-speed Internet connection (cable, digital subscriber line [DSL], T-1 or other wireless service), to provide wireless Internet access within a block or two of home. With greater coverage, a NAN may also be considered a community or campus area network (CAN).

Network—The connection of multiple systems together or to a central distribution point for the purpose of information or resource sharing.

Network interface card (NIC)—Typically an ISA, PCI, or PC card plug-in adapter used to connect a wired network to a computer. Wireless NICs are used to replace the wires.

NETX—A TSR program that interfaces a network interface card driver program to an active network operating system, for access to LAN services.

Nibble—A nibble is one-half of a byte, or 4 bits, of information.

Nicad battery—An energy cell or battery composed of nickel and cadmium chemical compositions, forming a rechargeable, reusable source of power for portable devices.

Noninterlaced operation—A method of displaying elements on a display screen at a fast rate throughout the entire area of the screen, as opposed to interlaced operation, which scans alternate rows of display elements or pixels, the latter often indicating a flickering or blinking of the illuminated screen.

Norton or Norton Utilities—A popular suite of utility programs used for PC disk and file testing and recovery operations, named after their author, Peter Norton. The first set of advanced utilities available for IBM PC-compatible systems.

NT file system (NTFS)—The NT file system for hard disk drives in Windows NT, 2000, and XP workstations and servers provides security and recoverability, using a secure indexed file structure linked to the security access manager of the operating system. It is nonreadable by any version of DOS.

Null modem—A passive, wire-only data connection between two similar ports of computer systems, connecting the output of one computer to the input of another, and vice versa. Data flow control or handshaking signals may also be connected between systems. A null modem is used between two nearby systems, much as you might interconnect two computers at different locations by telephone modem.

Offsets—When addressing data elements or hardware devices, often the locations that data are stored or moved through is in a fixed grouping, beginning at a known or base address, or segment of the memory range. The offset is that distance, location, or number of bits or bytes that the desired information is from the base or segment location. Accessing areas of memory is done with an offset address, based on the first location in a segment of memory. For example, an address of 0:0040h represents the first segment, and an offset of 40 bytes. An address of A:0040h would be the 40th (in hex) byte location (offset) in the tenth (Ah) segment.

Omnidirectional antenna—An antenna that receives and transmits in all directions equally. Some omnidirectional antennas are constructed to concentrate the transmitted and received signals into a narrow horizontal pattern to create passive amplification or gain for the signals.

Online—A term referring to actively using a computer or data from another system through a modem or network connection.

Online services—These are typically commercial operations, much like a bulletin board service (BBS) that charge for the time and services used while connected. Most online services use large computers designed to handle multiple users and types of operations. These services provide electronic mail, computer and software support conferences, online game playing, and file libraries for uploading and downloading public domain and shareware programs. Often, familiar communities or groups of users form in the conferences, making an online service a favorite or familiar places for people to gather. Access to these systems is typically by modem, to either a local data network access number or through a WATS or direct-toll line. America Online, Prodigy, and CompuServe are among the remaining online services available in the United States and much of the world at large. Online services have given way to the World Wide Web and portal sites such as Yahoo! and MSN.

Operating system—*See* disk operating system.

Operational support systems (OSS)—A term originally coined by telephone companies to describe the systems used to provision, manage and bill for telephone-related services. Today such systems include customer relationship management and workforce administration. In relation to wireless networking, these systems tie-together customer orders, installations, customer support and service maintenance record-keeping.

OS/2—A 32-bit operating system, multitasking control, and graphical user interface developed by Microsoft, currently sold and supported by IBM. OS/2 allows the simultaneous operation of many DOS, Windows, and OS/2-specific application programs.

Orthogonal frequency division multiplexing (OFDM)—A modulation technique for transmitting large amounts of data over radio, and the technique used for 802.11a. OFDM splits the radio signal into multiple smaller subsignals that are transmitted at the same time over different frequencies.

Overlays—A portion of a complete executable program, existing separately from the main control program, that is loaded into memory-only when it is required by the main program, thus reducing overall program memory requirements for most operations. Occasionally, overlays may be built into the main program file, but they are also not loaded into memory until needed. Overlays per se have been made obsolete by Windows and dynamically linked libraries (DLLs).

Page frame—The location in DOS/PC system memory (between 640 K and 1 MB), where the pages or groups of expanded memory are accessed.

Panel antenna—An antenna whose radiating elements are flat, that concentrates transmitted and received energy in a 180 degree pattern around the face of the antenna. The received and transmitted signals may benefit from increased effective power because of signal gain obtained by concentrating the signal to one plane, rather than spread throughout 360 degrees.

Parallel input/output (I/O)—A method of transferring data between devices or portions of a computer, where eight or more bits of information are sent in one cycle or operation. Parallel transfers require eight or more wires to move the information. At speeds from 12,000 to 92,000 bytes per second or faster, this method is faster than the serial transfer of data, where one bit of information follows another. Commonly used for the printer port on PCs.

Parallel port—A computer's parallel input/output (I/O) (LPT) connection, built into the system board or provided by an add-in card.

Parameter—Information provided when calling or within a program specifying how or when it is to run with which files, disks, paths, or similar attributes.

Parity—A method of calculating the pattern of data transferred, as a verification that the data has been transferred or stored correctly. Parity is used in all PC memory structures, as the 9th, 17th, or 33rd bit in 8-, 16-, or 32-bit memory storage operations. If there is an error in memory, it will usually show up as a parity error, halting the computer so that processing does not proceed with bad data. Parity is also used in some serial data connections as an eighth or ninth bit, to ensure that each character of data is received correctly.

Partition—A section of a hard disk drive typically defined as a logical drive, which may occupy some or all of the hard-disk capacity. A partition is created by the DOS `FDISK` or other disk utility software.

Password authentication protocol (PAP)—A standard method of authenticating a user by name and password. A host system requests of a client the log-in information, and the name and password are transmitted back for evaluation by the host. PAP information is transmitted in plain text, unencrypted, and is not secure.

Path—A DOS parameter stored as part of the DOS environment space, indicating the order and locations DOS is to use when you request a program to run. A path is also used to specify the disk and directory information for a program or data file. *See also* filespec.

PC compatible—*See* IBM PC compatible and AT compatible.

Pentium—A 64-bit Intel microprocessor capable of operating at 60–266+MHz, containing a 16 K instruction cache, floating point processor, and several internal features for extremely fast program operations.

Pentium II—A 64-bit Intel microprocessor capable of operating at 200–450+MHz, containing a 16 K instruction cache, floating point processor, and several internal features for extremely fast program operations. Packaged in what is known as Intel's Slot 1 module, containing the central processing unit (CPU) and local chipset components.

Pentium III—A 64-bit Intel microprocessor capable of operating at 450–800+MHz. Packaged in what is known as Intel's Slot 1 module, containing the central processing unit (CPU) and local chipset components.

Pentium IV—An Intel microprocessor operating at speeds between 1.8 and 3 GHz with 512KB Level 2 on-chip processor cache.

Peripheral—A hardware device internal or external to a computer that is not necessarily required for basic computer functions. Printers, modems, document scanners, and pointing devices are peripherals to a computer.

Peripheral component interconnect (PCI)—An Intel-developed standard interface between the central processing unit (CPU) and input/output (I/O) devices, providing enhanced system performance. PCI is typically used for video and disk drive interconnections to the CPU.

Personal computer (PC)—The first model designation for IBM's family of personal computers. This model provided 64 to 256 KB of RAM on the system board, a cassette tape adapter as an alternative to diskette storage, and five add-in card slots. The term generally refers to all IBM PC-compatible models, and has gained popular use as a generic term referring to all forms, makes, and models for personal computers.

Personal Computer Memory Card Industry Association (PCMCIA)—An input/output (I/O) interconnect definition used for memory cards, disk drives, modems, network, and other connections to portable computers. The term has been displaced by the use of PC card instead.

Personal digital assistant (PDA)—Typically a hand-held device used as an electronic address book, calendar, and notepad. Commonly using the Palm OS, Windows CE, or similar dedicated operating system.

Personal system/2 (PS/2)—A series of IBM personal computer systems using new designs, bus, and adapter technologies. Early models did not support the many existing PC-compatible cards and display peripherals, although IBM has provided later models that maintain its earlier industry standard architecture (ISA) expansion capabilities.

Physical drive—The actual disk drive hardware unit, as a specific drive designation (A:, B:, or C:, etc.), or containing multiple logical drives, as with a single hard drive partitioned to have logical drives C:, D:, and so on. Most systems or controllers provide for two to four physical floppy diskette drives and up to two physical hard disk drives, which may have several logical drive partitions.

Pixel—Abbreviation for picture element. A single dot or display item controlled by your video adapter and display monitor. Depending on the resolution of your monitor, your display may have the ability to display 320 × 200, 640 × 480, 800 × 600, or more picture elements across and down your monitor's face. The more elements that can be displayed, the sharper the image appears.

Plug-and-play—A standard for PC basic input/output system (BIOS) peripheral and input/output (I/O) device identification and operating system configuration, established to reduce the manual configuration technicalities for adding or changing PC peripheral devices. plug-and-play routines in the system BIOS work with and around older, legacy, or otherwise fixed or manually configured I/O devices, and reports device configuration information to the operating system. (The operating system does not itself control or affect plug-and-play or I/O device configurations.)

PnP—*See* plug-and-play.

Point-to-point protocol (PPP)—A method of connecting a computer, typically by serial port connection or modem, to a network. The method used to create a dial-up transmission control protocol/Internet protocol (CP/IP) connection between your computer and your Internet service provider.

Pointing device—A hardware input device, a mouse, trackball, cursor tablet, or keystrokes used to direct a pointer, cross-hair, or cursor position indicator around the area of a display screen, to locate or position graphic or character elements, or select position-activated choices (buttons, scroll bar controls, menu selections, etc.) displayed by a computer program.

Port address—The physical address within the computer's memory range that a hardware device is set to decode and allow access to its services through.

Power on self test (POST)—A series of hardware tests run on your PC when power is turned on to the system. POST surveys install memory and equipment, storing and using this information for bootup and subsequent use by DOS and applications programs. POST will provide either speaker beep messages, video display messages, or both if it encounters errors in the system during testing and bootup.

Power over Ethernet (POE)—A wiring method to add DC power supply to standard Ethernet cabling to power an Ethernet device, typically a wireless access point or amplifier, without having to add separate power cabling to the interconnection.

Professional graphics adapter (PGA), professional graphics controller (PGC), professional color graphics system—This was an interim IBM high-resolution color graphics system in limited distribution between EGA and VGA.

Program, programming—A set of instructions provided to a computer specifying the operations the computer is to perform. Programs are created or written in any of several languages that appear at different levels of complexity to the programmer, or in terms of the computer itself. Computer processors have internal programming, known as microcode, that dictates what the computer will do when certain instructions are received. The computer must be addressed at the lowest level of language, known as machine code, or one that is instruction-specific to the processor chip being used. Programming is very rarely done at machine-code levels, except in development work.

The lowest programming level that is commonly used is assembly language, a slightly more advanced and easier-to-read level of machine code, also known as a second-generation language. Most programs are written in what are called third-generation languages such as BASIC, Pascal, C, or FORTRAN, more readable as a text file. Batch files, macros, scripts, and database programs are a form of third-generation programming language specific to the application or operating with which system they are used. All programs are either interpreted by an intermediate application or compiled with a special program to convert the desired tasks into machine code.

Prompt—A visual indication that a program or the computer is ready for input or commands. The native DOS prompt for input is shown as the a disk drive letter and "right arrow," or "caret," character (C>). The DOS prompt may be changed with the DOS `PROMPT` internal command, to indicate the current drive and directory, include a user name, the date or time, or more creatively, flags or colored patterns.

Public domain—Items, usually software applications in this context, provided and distributed to the public without expectation or requirement of payment for goods or services, although copyrights and trademarks may be applied. Public domain software may be considered as shareware, but shareware is not always in the public domain for any and all to use as freely as they wish.

Radiation pattern—The effective fingerprint or profile of the theoretical or practical path radio signals project from an antenna. The pattern is shaped by calculated mechanical and structural elements and construction of an antenna to project a signal in an omnidirectional or unidirectional pattern.

RAM disk or RAM drive—A portion of memory assigned by a device driver or program to function like a disk drive on a temporary basis. Any data stored in a random access memory (RAM) drive exists there as long as your computer is not rebooted or turned off.

Random access memory (RAM)—A storage area that information can be sent to and taken from by addressing specific locations in any order at any time. The memory in your PC and even the disk drives are a form of random access memory, although the memory is most commonly referred to as the RAM. RAM memory chips come in two forms, the more common dynamic RAM (DRAM), which must be refreshed often to retain the information stored in it, and static RAM, which can retain information without refreshing, saving power and time. RAM memory chips are referred to by their storage capacity and maximum speed of operation in the part numbers assigned to them. Chips with 16 K and 64 K capacity were common in early PCs; 256 K and 1 MB chips in the early 1990s; but 8, 16, 32, and 64 MB RAM components are now more common.

Read only—An attribute assigned to a disk file to prevent DOS or programs from erasing or writing over a file's disk space. *See* attributes.

Read-only memory (ROM)—This is a type of memory chip that is preprogrammed with instructions or information specific to the computer type or device in which it is used. All PCs have a ROM-based basic input/output system (BIOS) that holds the initial bootup instructions that are used when your computer is first turned on or when a warm-boot is issued. Some video and disk adapters contain a form of ROM-based program that replaces or assists the PC BIOS or DOS in using a particular adapter.

Read-only memory basic input/output system (ROM BIOS)—The ROM chip-based start-up or controlling program for a computer system or peripheral device. *See also* BIOS and ROM.

Received signal strength indicator (RSSI)—A feature of many wireless integrated circuits to provide a means of measuring the relative strength of the signals you are receiving.

Refresh—An internal function of the system board and central processing unit (CPU) Memory refresh timing circuits to recharge the contents of dynamic random access memory (RAM) so that contents are retained during operation. The standard PC RAM refresh interval is 15 microseconds. *See also* DRAM, RAM, SRAM, and wait states.

Request to send threshold (RTS)—A configurable parameter available in some access point and client wireless devices. This parameter controls what size data packet the low-level radio frequency protocol issues to a RTS packet. Default is 2432. Setting this parameter to a lower value causes RTS packets to be sent more often, consuming more of the available bandwidth, reducing the apparent throughput. The more often RTS packets are sent, the quicker the system can recover from interference or collisions.

Return—*See* Enter key.

Roaming—Roaming is the ability of a portable computer user to communicate continuously, while moving freely between more than one access point.

Root directory—The first directory area on any disk media. The DOS command processor and any `CONFIG.SYS` or `AUTOEXEC.BAT` file must typically reside in the root directory of a bootable disk. The root directory has space for a fixed number of entries, which may be files or subdirectories. A hard disk root directory may contain up to 512 files or subdirectory entries, the size of which is limited only by the capacity of the disk drive. Subdirectories may have nearly unlimited numbers of entries.

Router—A network interface device used to connect and control the path data can take between one or more devices, over one or more connection paths. Typically used at the subscriber end of a T-1, digital subscriber line (DSL), cable, or other high-speed connection. As an example, an office may have a DSL connection to the Internet and a private/dedicated T-1 to a remote office, and a connection to the office local area network (LAN)—the router decides or is told to transfer Internet traffic (browsing, etc.) to the DSL circuit only and office LAN traffic to the other office over the T-1 only, but preventing Internet traffic from appearing on the private T-1, while keeping private T-1 traffic off the Internet. Effectively, the two office LANs become virtually bridged, while the Internet traffic is routed onto the LAN only.

Segments—A method of grouping memory locations, usually in 64 K increments or blocks, to make addressing easier to display and understand. Segment 0 is the first 64 K of random access memory (RAM) in a PC. Accessing areas of memory within that segment is done with an offset address, based on the first location in the segment. An address of 0:0040h would be the 40th (in hex) byte location in the first 64 K of memory. An address of A:0040h would be the 40th (in hex) byte location in the tenth (Ah) 64 K of memory.

Serial input/output (I/O)—A method of transferring data between two devices one bit at a time, usually within a predetermined frame of bits that makes up a character, plus transfer control information (start and stop or beginning and end of character information). Modems and many printers use serial data transfer. One-way serial transfer can be done on as few as two wires, with two-way transfers requiring as few as three wires. Transfer speeds of 110,000 to 115,000 bits (11,000 to 11,500 characters) per second are possible through a PC serial port.

Serial port—A computer's serial input/output (I/O) (COM) connection, built into the system board or provided by an add-in card.

Service set identifier (SSID)—A unique identifier sent at the front end of data sent over a wireless LAN (WLAN). The SSID differentiates one WLAN from another. An SSID is also called the network name because it identifies a wireless network.

Shadow random access memory (RAM)—A special memory configuration that remaps some or all of the information stored in basic input/output system (BIOS) and adapter read-only memory (ROM) chips to faster dedicated RAM chips. This feature is controllable on many PC systems that have it, allowing you to use memory management software to provide this and other features.

Shareware—Computer applications written by noncommercial programmers, offered to users with a try-before-you-buy understanding, usually with a requirement for a registration fee or payment for the service or value provided by the application. This is very much like a cooperative or user-supported development and use environment, as opposed to buying a finished and packaged product off the shelf with little or no opportunity to test and evaluate if the application suits your needs. Shareware is not public domain software. Payment is expected or required to maintain proper, legal use of the application.

Single inline memory module (SIMM)—A dense memory packaging technique with small memory chips mounted on a small circuit board that clips into a special socket.

Single inline package (SIP)—Typically a dense memory module with memory chips mounted on a small circuit board with small pins in a single row that plugs into a special socket.

Small computer system interface (SCSI)—An interface specification for interconnecting peripheral devices to a computer bus. SCSI allows for attaching multiple high-speed devices, such as disk and tape drives, through a single cable.

Small office/home office (SOHO)—A marketing term for a class of customers with offices at home for self-employment or telecommuting or small businesses with typically fewer than ten employees.

Software interrupt—A (nonhardware) signal or command from a currently executing program that causes the central processing unit (CPU) and computer program to act on an event that requires special attention, such as the completion of a routine operation or the execution of a new function.

Many software interrupt services are predefined and available through the system basic input/output system (BIOS) and DOS, while others may be made available by device driver software or running programs. Most disk accesses, keyboard operations, and timing services are provided to applications through software interrupt services.

Spectrum analyzer—A piece of expensive, specialized test equipment used to view and measure a variety of signals within a narrow or broad spectrum. Typically used by engineers to help design and tune radio equipment, or survey sites for radio interference.

ST506/412—The original device interface specification for small hard drives, designed by Seagate, and first commonly used in the IBM PC/XT.

Start bit—The first data bit in a serial data stream, indicating the beginning of a data element. In the old days, when mechanical tele-printers were used, the Start bit signaled the motor and mechanical elements of the printer to start running.

Static random access memory (SRAM)—Fast access (less than 50 nanoseconds), somewhat expensive, memory integrated circuits that do not require a refresh cycle to maintain their contents. Typically used in video and cache applications. *See also* DRAM and RAM.

Stop bit—The last data bit or bits in a serial data stream, indicating the end of a data element. Like the Start bit, the Stop bit signaled the time when the tele-printer mechanics should stop running.

Subdirectory—A directory contained within the root directory or in other subdirectories, used to organize programs and files by application or data type, system user, or other criteria. A subdirectory is analogous to a file folder in a filing cabinet or an index tab in a book. The concept is the same, but the term subdirectory has been displaced by folder.

Surface scan—The process of reading and verifying the data stored on a disk to determine its accuracy and reliability, usually as part of a utility or diagnostic program's operation to test or recover data.

Switch—A network device, much like a hub, that interconnects several network devices onto the same network segment, but that can automatically keep interdevice traffic within its own network segment to reduce overall LAN traffic. In effect, switches can act like routers without complex router setup and instructions.

Sysop—The system operator of a bulletin board service (BBS), on-line service forum, or network system.

System attribute or system file—*See* attributes.

T or X fastener and tool—A four-point special fastener and tool for same that differs from a normal slotted/flat edge, cross-head, or hexagonal fastener.

T-1—A 1.5 megabits-per-second high-speed four-wire data or voice circuit used to convey multiple channels of data or voice traffic between two points. A T-1 line is generally expensive and requires special 'modem' equipment to support interconnection to computers, routers, or telephone systems. In voice service, a T-1 line carries 23 64 kilobyte-per-second channels of discrete call information and voice traffic.

Terminate-and-stay-resident program (TSR)—Also known as a memory-resident program. A program that remains in memory to provide services automatically or on request through a special key sequence (also known as hot keys). Device drivers (`MOUSE`, `ANSI`, `SETVER`) and disk caches, RAM disks, and print spoolers are forms of automatic TSR programs. SideKick, Lightning, and assorted screen-capture programs are examples of hot-key-controlled TSR programs. Under Windows, TSRs or resident programs run at the same time and within the same operating system control as other programs.

Time division multiple access (TDMA)—A technology for delivering digital wireless service, typically related to digital cellular telephone services. TDMA divides a radio frequency signal into time slots, and then allocates slots to multiple calls. With TDMA, a single frequency can support multiple, simultaneous data channels. (Author's note: depending on how the cellular system operator set the system and how crowded it is, TDMA systems often sound robotic/synthesized or distorted, compared to CDMA cellular systems.)

Tiny file transfer protocol (TFTP)—A nonsecure data communication protocol, similar to the Internet's FTP, that is used to transfer firmware or operating parameters to dedicated devices, such as routers and firewalls.

Transmission control protocol/Internet protocol (TCP/IP)—The combination of the standard inter-networking addressing communications protocol—IP and transmission reliability protocol (TCP) that make up the majority of Internet traffic we transmit and receive. TCP provides error recovery and re-transmission services to nearly guarantee data delivery over a network. *See also* UDP.

Transport layer security (TLS)—Provides privacy and data security between client and server (or client and access point).

Twisted-pair cable—A pair of wires bundled together by twisting or wrapping them around each other in a regular pattern. Twisting the wires reduces the influx of other signals into the wires, preventing interference, as opposed to coaxial (concentric orientation) or parallel wire cabling.

Universal asynchronous receiver/transmitter (UAR/T)—This is a special integrated circuit or function used to convert parallel computer bus information into serial transfer information and vice versa. A UAR/T also provides proper system-to-system on-line status, modem ring, and data carrier detect signals, as well as start/stop transfer features. The most recent version of this chip, called the 16550A, is crucial to high-speed (greater than 2400 bits per second) data transfers under multitasking environments.

Universal serial bus (USB)—A high-speed two-wire interconnection between a host system and up to 256 discrete separate devices. USB allows for connection, disconnection, and reconnection of peripheral devices from a computer, supported by Plug and Play or a similar auto-configuring device support system. USB is commonly used to connect printers, cameras, scanners, and some network devices to PCs and Apple Macintosh computers. It is supported in Windows 98, 98SE, Me, 2000, and XP, as well as Mac OS9.x, OS X and later versions of Linux.

UNIX—A high performance multitasking operating system designed by AT&T/Bell Laboratories in the late 1960s. Today UNIX has several offshoots and derivatives, including LINUX, Sun OS/Solaris, FreeBSD, and others. UNIX is the operating system of choice for many 'enterprise' business applications, and most of the servers and Internet services we enjoy today.

Unlicensed national information infrastructure (U-NII)—(also referred to as UNI)—The general reference for the bands of frequencies available for unlicensed device operation occupying 300 MHz of spectrum, divided into three 100 MHz sections. The "low" band runs from 5.15 GHz to 5.25 GHz, the "middle" band runs from 5.25 GHz to 5.35 GHz, and the "high" band runs from 5.725 GHz to 5.825 GHz.

Upload—The process of sending or transferring information from your computer to another, usually connected by modem or a network. Uploading is done to bulletin board system (BBS) and on-line services when you have a program or other file to contribute to the system or to accompany electronic mail you send to others.

Upper memory and upper memory blocks—Memory space between 640 K and 1 MB that may be controlled and made available by a special device or UMB (EMM386.SYS, QEMM386, 386Max, etc.) for the purpose of storing and running TSR programs and leaving more DOS random access memory (RAM) (from 0 to 640 K) available for other programs and data. Some of this area is occupied by basic input/output system (BIOS), video, and disk adapters.

User datagram protocol (UDP)—Used primarily for broadcasting or streaming multi-media content over the Internet. Unlike TCP (TCP/IP), UDP does not contain robust error correction or data re-transmission services.

Utilities—Software programs that perform or assist with routine functions, such as file backups, disk defragmentation, disk file testing, file and directory sorting, etc. *See also* diagnostics.

Variable—Information provided when calling or within a program specifying how or when it is to run with which files, disks, paths, or similar attributes. A variable may be allowed for in a batch file, using %1 through %9 designations to substitute or include values keyed-in at the command line when the Batch file is called.

Video adapter card—The interface card between the computer's input/output (I/O) system and the video display device.

Video graphics array (VGA)—A high-resolution text and graphics system supporting color and previous IBM video standards using an analog-interfaced video monitor.

Video memory—Memory contained on the video adapter, dedicated to storing information to be processed by the adapter for placement on the display screen. The amount and exact location of video memory depends on the type and features of your video adapter. This memory and the video adapter functions are located in upper memory between 640 K and 832 K.

Virtual disk—*See* RAM disk.

Virtual local area network (VLAN)—An interconnection between devices (client PCs, servers, printers, etc.) as if they were part of another LAN some distance away. Typical of a wireless LAN (WLAN) connection used for making a virtual private network (VPN) connection to a LAN, with the ability to roam between different WLAN connections and still be part of the LAN.

Virtual memory—Disk space allocated and managed by an operating system that is used to augment the available random access memory (RAM) memory, and is designed to contain inactive program code and data when switching between multiple computer tasks.

Virtual private network (VPN)—An encrypted connection from a client workstation, a server, or local area network (LAN), to another server or LAN at a different location over the public Internet. Typically used for telecommuting or roaming workers accessing a corporate network, but also useful in securing wireless LAN connections. May also be used to create a VLAN.

Volume label—*See* disk label and label.

Wait states—A predetermined amount of time between the addressing of a portion of a memory location and when data may be reliably read from or written to that location. This function is controlled by the basic input/output system (BIOS), and it is either permanently set or changed in CMOS setup. Setting this parameter too low may cause excessive delays or unreliable operation. Setting this parameter too high may slow down your system. *See also* DRAM, RAM, refresh, and SRAM.

War-chalking, war-walking, war-driving—The activities of surveying an area looking for wireless network hot spots and access points, then marking the direction and type of services available in chalk on sidewalks or walls. Derived from the war-dialing of the "War Games" movie fame.

Wide area network (WAN)—A network connection between two locations, typically a T-1 circuit but may be integrated services digital network (ISDN), Frame Relay, or other (possibly a virtual private network [VPN] over any Internet connection type) used to bridge local area networks in related office facilities.

Wi-fi protected access (WPA)—A pre-802.11i wireless LAN security method.

Windows—A Microsoft multitasking and graphical user interface that allows multiple programs to operate on the same PC system and share the same resources.

Windows NT—A Microsoft 32-bit multitasking operating system and graphical user interface.

Wired equivalent privacy (WEP)—A scheme used to make the data traveling on wireless networks unreadable by those not authorized to use your network. "Wired equivalence" indicates the scheme proposes to make wireless signals as relatively secure from intrusion as using a wired system. Both the 64- and 128-bit WEP encryption schemes can be deciphered by commonly available software tools, so WEP is not to be trusted for secure, valuable, or private data. Consider a virtual private network (VPN) solution to help secure your wireless network.

Wireless local area network (WLAN)—Interconnections between client computers, servers, and other devices over radio waves versus Ethernet cabling connections.

Workstation—A user's computer system attached to a network. Workstations do not necessarily contain diskette or hard disk drives, instead using built-in programs to boot up and attach to a network server, from which all programs and data files are obtained.

World wide web (WWW)—A term used to describe multiple internetworked computer systems providing text and graphical content through the hypertext transfer protocol (HTTP), usually over Internet protocol (IP) networks.

Write protected—The status of a diskette with a write-protection tab or slot. All 3.5 inch diskettes use a sliding window cover over a small hole in the near left corner of the casing (shutter door facing away from you). If the hole is uncovered, the disk is write protected.

XT—The second model of IBM PC series provided with "extended technology," allowing the addition of hard disks and eight add-in card slots. The original XT models had between 64 K and 256 K of random access memory (RAM) on board, a single floppy drive, and a 10 MB hard disk.

Yagi—A form of directional antenna, also referred to as a beam antenna, that concentrates transmitted and received energy in one direction only. The received and transmitted signals benefit from increased effective power because of the gain in concentrating the signals to the antenna element.

INDEX

Note: Boldface numbers indicate illustrations.